INTERNATIONAL CENTRE FOR MECHANICAL SCIENCES

COURSES AND LECTURES - No. 323

STABILITY PROBLEMS OF STEEL STRUCTURES

EDITED BY

M. IVANYI
TECHNICAL UNIVERSITY OF BUDAPEST

M. SKALOUD
CZECHOSLOVAK ACADEMY OF SCIENCES

Springer-Verlag   Wien GmbH

Le spese di stampa di questo volume sono in parte coperte da
contributi del Consiglio Nazionale delle Ricerche.

This volume contains 337 illustrations.

In order to make this volume available as economically and as
rapidly as possible the authors' typescripts have been
reproduced in their original forms. This method unfortunately
has its typographical limitations but it is hoped that they in no
way distract the reader.

ISBN 978-3-211-82398-9   ISBN 978-3-7091-4332-2 (eBook)
DOI 10.1007/978-3-7091-4332-2

PREFACE

The need to save steel has led to the development of new kinds of steel structures. New types of frames and bar systems, thin-walled plated structures and shells play a very important part among them. However, their successful practical application requires adequate scientific data for reliable design. It is therefore not surprising that a great number of research teams in various parts of the world (the authors playing a very important role in these activities) have directed, for a good many years, their attention to investigations into the stability problems of the above structures, since it is very frequently these phenomena which in the case of currently used thin-walled constructional steelwork govern design.

For this reason, an International Advanced School on "Stability Problems of Steel Structures" was held at the International Centre for Mechanical Sciences in Udine with the view to give complete information about all aspects of the design of up-to-date steel (i) frames and bar systems, (ii) plated structures, and (iii) shells from the point of view of buckling. In following this aim, the course presented not only complete scientific background, but also established recommendations, procedures and formulae for practical design, profiting also from the fact that over the last years the invited lecturers took part in the preparation of several national and international Design Codes for structural steelwork.

Thus the course, challenging current concepts and encouraging progress in the field, aimed at establishing newly developed advanced guidelines to design, thereby making constructional steelwork more economic and, consequently, more competitive with respect to structures made of other materials.

The objective of this monograph is to summarize the most important parts of the lectures presented during the aforementioned International Advanced School and thus substantially enlarge the very positive results and impact of the School.

The monograph will therefore be of interest for a wide range of readers dealing with research, teaching, and design of all kinds of steel structures.

M. Iványi
M. Skaloud

CONTENTS

PART 1

PLATED STRUCTURES

BEHAVIOUR OF PLATE COMPONENTS

R. Maquoi
University of Liege, Liege, Belgium

FOREWARD TO PART I

Because modern steel construction makes a wider use of thin-walled structures, consideration of plate buckling has become of paramount importance. That is the subject with which Part I is mainly concerned.

First the behaviour of unstiffened plate components is reviewed in terms of linear theory of plate buckling and of postbuckling behaviour (chapter 1). The relation between local plate buckling and the classification of cross-sections is then analysed with a view to specify the appropriate methods of analysis and of cross-section verification (chapter 2). The tension field models for webs in shear are examined in order to select the one which will be used as design model for practice purposes (chapter 3).

Whereas the three first chapters are basically devoted to the background and the physical understanding of plate behaviour, the chapters 4 to 6 are more especially concerned with design rules for plate girders. Unstiffened and transversely stiffened plate girders are first examined (chapter 4). How to entend basic design concepts to longitudinally stiffened plate girders is the aim of chapter 5. Last, some guidelines regarding the design of web stiffeners are given (chapter 6).

The contents of chapters 1 to 6 is mainly inspired from the work carried out within Task Working Group 8/3 "Plate Buckling" of the European Convention for Constructional Steelwork (in short E.C.C.S.) and reflected by ECCS Publications n° 44 and 60. Some amendments and/or improvements have however been brought in order to reach a compliance with specifications of Eurocode 3 for steel buildings.

The basic references are thus :

[1] Eurocode 3 - Design of Steel Structures - Part 1 - Design Rules and Rules for Buildings. Commission of the European Communities, Brussels, Nov. 1990.

[2] E.C.C.S. - Technical Committee 8 "Structural Stability": Behaviour and Design of Steel Plated Structures. Edited by P. DUBAS and E. GEHRI, E.C.C.S. Publication n° 44, E.T.H. Zürich, 1986.

[3] E.C.C.S. - European Recommendations for the Design of Longitudinally Stiffened Webs and of Stiffened Compression Flanges. E.C.C.S. Publication n° 60, Brussels, 1990.

The reader interested in a more detailed bibliography is begged to refer to [1] where a very large number of references are listed.

Attention has been focused on bare steel plate girders only. Though the scientific background is expected to be extendable to composite plate girders, there are still problems to be solved in this respect due to the composite action of the upper flange and to the different behaviour when hogging or sagging bending moment. Some information on how to approach the design of composite girders may be found in

[4] Eurocode 4 - Design of Composite Steel and Concrete Structures - Part 1 - Design Rules and Rules for Buildings. Commission of the European Communities, Brussels, Oct. 1990.

CHAPTER 1
BEHAVIOUR OF UNSTIFFENED COMPONENTS

1. INTRODUCTION.

Plate buckling is the instability phenomenon that is specific to thin plates subject to in-plane loads. It can occur when the plate is totally or partially subject to compressive stresses.

For instance, let us consider (fig. 1) a simply supported rectangular plate of dimensions a x b and aspect ratio α = a/b close to unity, subject to uniform uniaxial compression. The plate is assumed to have a constant thickness, be perfectly flat, made of an isotropic elastic material without residual stresses and be loaded exactly in its middle plane. Under small values of compressive stresses σ_x, the plate remains flat and any imposed transverse deflection w disappears as soon as its cause is removed. At a magnitude $\sigma_x = \sigma_{x,cr}$ of the compression load, the plate keeps the deformed shape produced by the imposed perturbation when the cause of this latter is removed. This magnitude is termed *critical plate buckling stress*; it characterizes a possible state of *neutral equilibrium*. Such a plate buckling proceeds by *bifurcation of*

equilibrium: the plate remains flat for compression lower than critical
plate buckling load while it is prone to exhibit a deflected equilibrium
state as soon as this load is reached.

Fibres parallel to compression shorten because of elastic strain and bow
effect. The latter is the cause of fibre lengthening in the direction
perpendicular to compression; this membrane effect tends to stabilize
the plate and may result in a possible increase in strength capacity.
As this favourable contribution develops once the plate has buckled, it
is termed *postbuckling strength*. The better the membrane action is
anchored - because of edge restraints -, the larger the postbuckling
strength. The ideal plate response is represented by a load -
shortening diagram (curve 1 in figure 2). The response of an *imperfect
plate* is somewhat different; indeed the unavoidable *out-of-flatness*
prevents the plate from remaining in its original configuration as soon
as the load is applied (curve 2 in figure 2). The effect of usual
magnitudes of out-of-flatness decreases rapidly when the load grows up
so that curve 2 approaches curve 1 asymptotically. There is a limit to
the plate response because of *material yielding*, which results in a
decrease of the plate stiffness and in a corresponding increase in the
rate of deflection; response curves progressively flatten till they
reach a maximum load N_u, termed *collapse (or ultimate) load*, where the
plate stiffness vanishes (fig. 3). Last *residual stresses*, due to the
fabrication process, cause prematurous yielding and accelerate the drop
in plate stiffness. In an actual plate, plate buckling develops by
divergence of the equilibrium.

Above behaviour is not restricted to uniaxial compression; it is
observable in any plate subject, at least partially, to direct
compressive stresses or to load including principal compressive stresses
(pure shear for instance).

The critical plate buckling stress of an unstiffened plate may be
basically found by solving the well-known differential equilibrium
equation :

$$\nabla^2\nabla^2 w = \frac{\partial^4 w}{\partial x^4} + 2\frac{\partial^4 w}{\partial x^2 \partial y^2} + \frac{\partial^4 w}{\partial y^4}$$

$$= \frac{t}{D} \left[\sigma_x \frac{\partial^2 (w+w_0)}{\partial x^2} + 2\tau_{xy} \frac{\partial^2 (w+w_0)}{\partial x \partial y} + \sigma_y \frac{\partial^2 (w+w_0)}{\partial y^2} \right] \tag{1}$$

where $D = Et^3/12(1 - \nu^2)$ is the flexural stiffness of the plate, t the thickness, E the YOUNG modulus and ν the POISSON ratio; w(x,y) is the transverse deflection of the plate. Basic loading cases are normally investigated by considering separately each load component in the second member of equation (1).

The postbuckling stability can only be analysed by implementing above equation in view to account for the initial out-of-flatness and the membrane action ; as the latter produces stretching of the middle plane, both in-plane and out-of-plane behaviours are interacting. The solution of the problem requires to integrate a set of two coupled equations:

$$\nabla^2 \nabla^2 w = \frac{t}{D} \left[\frac{\partial^2 \phi}{\partial x^2} \frac{\partial^2 (w+w_0)}{\partial y^2} - 2 \frac{\partial^2 \phi}{\partial x \partial y} \frac{\partial^2 (w+w_0)}{\partial x \partial y} + \frac{\partial^2 \phi}{\partial y^2} \frac{\partial^2 (w+w_0)}{\partial x^2} \right] \tag{2}$$

$$\nabla^2 \nabla^2 \phi = E \left\{ \left[\frac{\partial^2 (w+w_0)}{\partial x \partial y} \right]^2 + \frac{\partial^2 w_0}{\partial x^2} \frac{\partial^2 w_0}{\partial y^2} - \left(\frac{\partial^2 w_0}{\partial x \partial y} \right)^2 - \frac{\partial^2 (w+w_0)}{\partial x^2} \cdot \frac{\partial^2 (w+w_0)}{\partial y^2} \right\} \tag{3}$$

where $\phi(x,y)$ is the AIRY stress function and $w_0(x,y)$ the initial out-of-flatness. Already integration is not easy. Consideration of the non-linear material behaviour - due to material yielding and residual stresses - should complicate very much the mathematical treatment. Though that is presently possible by using computers, the computations remain lengthy and costly; such a process cannot be contemplated for the daily practice. Nowadays, the trends is to use physical models which are based on some idealizations and simplifications and give the ultimate carrying capacity of plates subject to different kinds of loadings. Such models should be calibrated against test results. Because the critical plate buckling stress is a characteristic of a perfect plate, which is itself a limit case for an imperfect plate, reference is made to it in the design process using ultimate models. Therefore, it is useful to tell somewhat more about the linear theory of plate buckling.

2. LINEAR THEORY OF PLATE BUCKLING.

It is not the place here to review the different methods enabling to derive expressions for plate buckling stresses. In the literature, the critical buckling stress σ_{cr}, or τ_{cr}, of a plate is usually given with reference to the buckling stress of a transverse pin-ended strip of unit width and length b, (fig. 1), termed *EULER reference stress* σ_E, according as:

$$\sigma_{cr} = k_\sigma \, \sigma_E \qquad \qquad (4)$$
$$\tau_{cr} = k_\tau \, \sigma_E \qquad \qquad (5)$$

where :

$$\sigma_E = \pi^2 D/b^2 t \qquad \qquad (6.a)$$
$$= [\pi^2 E/12 \, (1 - \nu^2)] \, (t/b)^2 \qquad \qquad (6.b)$$

writes for steel ($\nu = 0.3$) :

$$\sigma_E \simeq 0.9 \, E \, (t/b)^2 \qquad \qquad (6.c)$$

The dimensionless coefficients k_σ and k_τ, termed *buckling coefficients*, depend on plate aspect ratio, stress distribution, boundary conditions and - for stiffened plates - on relative stiffness properties of the ribs.

2.1. Uniform uniaxial compression.

For a *simply supported long plate subject to uniform uniaxial compression*, the buckling coefficient is found as :

$$k_\sigma = (m/\alpha + n^2\alpha/m)^2 \qquad \qquad (7)$$

where m is the number of half buckling waves in the direction of the compression, while n is the number of such waves in the perpendicular direction. Because only the smallest value of k_σ merits consideration, one has of course to select n = 1; the plate buckles with one single half wave in the transverse direction. The relevant curve is made of successive festoons, the lower envelope of which is solely significative (fig. 4). It appears thus that the buckling mode changes when the aspect ratio increases. The minimum of each festoon is reached when m = α, in which case $k_{\sigma,min}$ = 4; to adopt k_σ = 4 for $\alpha \geq 1$ is a useful and only slightly safe approximate.

Boundary conditions better than simple supports should result in a higher buckling resistance; that is reflected by the comparative buckling coefficient curves A, B and C corresponding to clamped and/or simply supported edges (fig. 7). Other interesting cases are plates having a longitudinal free edge and the other longitudinal edge simply supported or clamped (curves D and E in figure 7). Most often the edges are neither simply supported nor fully clamped, but elastically restrained. It is usually difficult to assess accurately the degree of restraint, so that the conservative assumption of simple supports is usually made.

2.2. Linear direct stress distribution.

When two opposite edges of the rectangular plate are subject to a *linear direct stress distribution* such that a width portion is subject to compressive direct stresses, the buckling coefficient depends in addition on the stress ratio :

$$\psi = \sigma_2/\sigma_1 \tag{8}$$

where σ_2 and σ_1 are the minimum and maximum direct stresses respectively; thus one has $\psi = 1$ for uniform compression and $\psi = -1$ for pure bending. The chart drawn for pure bending ($\psi = -1$) demonstrates again that the buckling mode changes when α grows up and that clamped longitudinal edges provide with a gain of 70 % in the buckling coefficient, compared to simply supported edges (fig. 6). For sake of simplicity, k_σ is taken as $k_{\sigma,min}$ as soon as α exceeds 2/3 when simply supported edges (with $k_\sigma = k_{\sigma,min} \approx 40$). Such charts can be drawn for any value of ψ; only the chart of $k_{\sigma,min}$ versus the aspect ratio α (fig. 7) is given here for $\psi = 1$.

For practice purposes, the buckling coefficient for *pure bending* writes:
- for simply supported edges :
$$\alpha \leq 2/3 : k_\sigma = 15,87 + 1,87/\alpha^2 + 8,6\,\alpha^2 \tag{9.a}$$
$$\alpha > 2/3 : k_\sigma = 23,9 \tag{9.b}$$
- for longitudinal clamped edges :
$$\alpha \leq 0.475 : k_\sigma = 21,3 + 2/\alpha^2 + 42\,\alpha^2 \tag{10.a}$$
$$\alpha > 0.475 : k_\sigma = 39.6 \tag{10.b}$$

When plate elements are subject to direct stress distributions other than uniform compression (ψ = 1) or pure bending (ψ = -1), the buckling coefficient k_σ for *long plates* is approximately given by the expressions listed in Table 1.

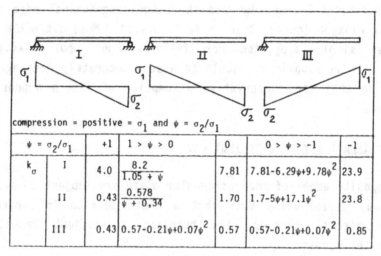

$\psi = \sigma_2/\sigma_1$		+1	$1 > \psi > 0$	0	$0 > \psi > -1$	-1
k_σ	I	4.0	$\dfrac{8.2}{1.05 + \psi}$	7.81	$7.81-6.29\psi+9.78\psi^2$	23.9
	II	0.43	$\dfrac{0.578}{\psi + 0.34}$	1.70	$1.7-5\psi+17.1\psi^2$	23.8
	III	0.43	$0.57-0.21\psi+0.07\psi^2$	0.57	$0.57-0.21\psi+0.07\psi^2$	0.85

Table 1

2.3. Shear.

A rectangular plate experiencing *pure shear* is prone to buckle because pure shear results in equal but opposite principal stresses sloped at ± 45° on the directions of pure shear. Principal compressive stresses may cause plate buckling while principal tensile stresses may help to delay it. The buckling coefficient curve for pure shear is monotoneously decreasing (fig. 8) and is given as follows :
- for all simply supported edges :

k_τ = 4 + 5,34/α^2 ($\alpha \le 1$)   (11.a)

     = 5,34 + 4/α^2 ($\alpha \ge 1$)   (11.b)
- for all clamped edges :

k_τ = 5,6 + 8,98/α^2 ($\alpha \le 1$)   (12.a)

     = 8,98 + 5.6/α^2 ($\alpha \ge 1$)   (12.b)

2.4. Combined shear and direct stresses.

When a rectangular plate is subject to combined shear and direct stresses, the interaction between these load components influences the buckling resistance. Therefore it is distinguished between :

a) critical stresses $\sigma_{cr}^{\bullet} = k_{\sigma} \sigma_E$ and $\tau_{cr}^{\bullet} = k_{\tau} \sigma_E$, which should cause
   plate buckling, when acting separately ;

b) critical stresses σ_{cr} and τ_{cr}, which should produce buckling, when
   acting coincidently.

The latter are drawn from interaction relations, which are most often
only approximates; for instance, in the stress range $-1 < \psi < 1$:

$$0.25 \, (1+\psi) \, (\sigma_{cr}/\sigma_{cr}^{\bullet}) + \sqrt{[0.25 \, (3-\psi) \, \sigma_{cr}/\sigma_{cr}^{\bullet}]^2 + (\tau_{cr}/\tau_{cr}^{\bullet})^2} = 1 \qquad (13)$$

or, alternatively :

$$\sigma_{c,cr}/\sigma_{c,cr}^{\bullet} + (\sigma_{b,cr}/\sigma_{b,cr}^{\bullet})^2 + (\tau_{cr}/\tau_{cr}^{\bullet})^2 = 1 \qquad (14)$$

where σ_c and σ_b are the respective components of axial force and pure
bending of the direct stress distribution.

In normalized coordinates ($\sigma_{cr}/\sigma_{cr}^{\bullet}$, $\tau_{cr}/\tau_{cr}^{\bullet}$), the interaction curve (13)
is an ellipse which is centered on the vertical axis and symmetrical
with respect to this axis, and intersects the axes at the unit values.
Both curves (13) and (14) degenerate into a parabola with vertical axis
for combined shear and uniaxial uniform compression and into a circle
for combined shear and pure bending (fig. 9). Actually, the interaction
curves depend on the aspect ratio (fig. 10); above approximates are
found quite satisfactory for practice purposes.

Of course a second condition is needed to enable the determination of
σ_{cr} and τ_{cr} ; therefore it is assumed that the loading is proportional,
so that :

$$\sigma_{cr}/\tau_{cr} = \sigma/\tau = \bar{s} \qquad (15)$$

The relevant ratio $\bar{s} = \sigma/\tau$ is presumably known for a reference state -
service conditions for instance -.

3. POSTBUCKLING BEHAVIOUR.

It must be stressed that *the critical load of a plate is not a realistic
measure of the carrying capacity* because the plate is far from perfect.

Indeed the behaviour is affected on the one hand, by unavoidable initial geometric and material imperfections, and, on the other hand, by the effects of plasticity and post-buckling stability.

It is explained above that the *postbuckling stability* can only be investigated when allowance is made for out-of-plane deflections and membrane action. The larger the out-of-plane deflection, the more dominant the membrane action. As membrane stresses are mostly tensile stresses, the plate remains stable in the postbuckling regime and exhibits a postbuckling strength reserve. This strength reserve is however only felt in slender plates ; for plates of low and intermediate stockiness, some other parameters - imperfections and plasticity - may reduce the plate strength below the critical load. The imperfections are mainly the initial lack of flatness and the initial residual stresses, due to the manufacturing process (rolling and/or welding).

The *geometric imperfection* consists mainly in out-of-flatness of the plate; it produces a growth of out-of-plane deformations from the onset of loading, with the result of a loss in plate stiffness ; however, at high level of strains, the behaviour is nearly not affected by the level of imperfection usually met in practice (fig. 11).

The *onset of plasticity* is influenced not only by in-plane stresses but also by plate bending stress components. The larger the out-of-flat-ness, the earlier the yielding and thus the more effect plasticity has on the collapse mode. The consequence of plasticity is a loss in plate stiffness too and a reduction in collapse load compared to a similar perfect elastic plate (fig. 12).

The *initial residual stresses* precipitate first yielding and thus contribute a further reduction in stiffness ; it affects also the collapse strength to some degree.

3.1. Uniaxial compression.

Plates subject to uniaxial compression have been the most extensively studied for what regards the elasto-plastic behaviour. Mostly numerical approaches have been used in this respect, the results of which have been compared with a lot of experimental results.

All these investigations were aimed at studying the effects of the main parameters on the stress-strain response of the plate, whose normalized slenderness $\bar{\lambda}_p$ is given as :

$$\bar{\lambda}_p = \sqrt{f_y/\sigma_{cr}}$$

$$= (1.05/\sqrt{k_\sigma})\ (b/t)\ \sqrt{f_y/E} \tag{16}$$

with k = 4 for uniform compression of a long plate ($\alpha > 1$). Compared to the elastic critical buckling load, the loss of strength due to both geometrical imperfections and residual stresses is the greatest in the vicinity of $\bar{\lambda}_p \simeq 1$, where the interaction between yielding and plate buckling is the largest (fig. 13); that is the range of intermediate slenderness. On the other hand, the greatest loss in pre-collapse rigidity due to residual stresses is observed for plates of moderate and low slenderness (b/t < 60). Last, initial geometrical imperfections reduce generally the compressive strength and can produce a change in the failure shape from a gradual process to a more sudden buckling event with a significant subsequent loss in load carrying capacity, especially in the range of intermediate slenderness. It has been found that the minimum strength in pure uniaxial compression is obtained for an *aspect ratio* close to unity. Long plates show a higher initial stiffness, a higher strength but a steeper unloading after the maximum load is reached (fig. 14).

Regarding *in-plane restraint* along the unloaded edges, these latter are considered successively either fully restrained - edge forces and non deformable boundaries -, or unrestrained edges - free to move in -, or constrained edges - free from edge forces but remaining straight through symmetry-. No significant difference is observed between constrained and unrestrained edges when the plate is stocky ; on the contrary, for slender plates, constrained strength approaches restrained strength, the latter being the higher in any case (fig. 15).

Large *initial out-of-plane deflections* affect the strength for any slenderness; the larger the out-of-flatness, the smaller the ultimate load. The residual carrying capacity after unloading is nearly independent of the imperfection level in the range of intermediate and large plate slenderness ratios (fig. 16).

The effect of *residual stresses* is more marked for stocky plates, because of the larger interaction with overall yielding ; it tends to be masked by geometrical imperfections in slender plates. Thus the effect shall be the greatest in the medium range of plate slenderness.

3.2. Shear.

The behaviour of a plate subject to shear is not the same as in pure compression. Prior to buckling, the stress is mainly a combination of principal (diagonal) tensile and compressive stresses of equal magnitude. Once critical shear buckling has occured, the behaviour is stable again in the elastic regime because the loading is resisted by an increase in the diagonal tensile force, provided the panel be surrounded by stiffening members, onto which the membrane tensile stresses can anchor. (fig. 17).

Initial out-of-flatness again causes growth of the out-of-plane deformations from the onset of loading. Its effect is however less marked because of the presence of the tensile element of load resistance. For slender plates deformations grow progressively and the ratio of diagonal tensile to compressive stresses gradually increases. Because tensile component is imperfection sensitive, it may dominate the behaviour to some extent and reduce the sensitivity of shear strength to imperfection magnitude.

The effect of plasticity and residual stresses on collapse is essentially similar to that observed in pure compression. Yield shear load is an upper bound for stocky panels; for slender panels, yield at the outer fibres results from the growth of deformation, increases subsequent deformations and can lead to lower collapse loads.

It is while stressing that yield limits the degree of tensile resistance that can be built up and therefore influences directly the collapse load contributed by tension field action.

3.3. Combined shear and direct stresses.

As the buckling modes of the two load components are dissimilar, the imperfection sensitivity of the plate is between those produced by the basic stresses separately. That explains that a restrained panel subject to certain proportions of *compression and shear* can exhibit a higher strength in compression than without shear loading (fig. 18).

Few numerical studies are available in this field. Nevertheless, they have produced interaction curves of ultimate stress derived from sets of panel stress-strain responses. These curves are established for restrained and unrestrained panels, the direct stresses being uniform tensile or compressive in one direction only. The small imperfection sensitivity in the compression zone is quite observable, while in the tension zone, the different curves tend quickly to merge because buckling becomes irrelevant.

Similarly interaction curves have been calculated for combined in-plane bending and shear, the moment being expressed in terms of fully plastic moment M_u or yield moment M_y (fig. 19). These curves show a significant dependance of strength on imperfection level for stocky plates, especially when unrestrained.

3.4. Biaxial compression.

The collapse behaviour of *plates in biaxial compression* is affected by factors similar to those discussed in section 3.1. From a design point of view the factor of most relevance is the collapse interaction between the two applied stresses in perpendicular directions given the amplitude of residual stresses in both directions and of the initial out-of-plane deflection. Because the stress response is different in both directions - ϵ corresponding to the higher level of strain applied -, it is possible to plot interaction curves of peak σ_x against coincident σ_y and of peak σ_y against coincident σ_x.

3.5. Design formulation.

From the results of the aforementioned studies, it is possible to suggest an interaction formula of the similar format as for elastic critical buckling stresses.   For instance for uniaxial uniform compression combined with bending and shear, this formula writes :

$$\sigma_c/S_c \, f_y \; + \; (\sigma_b/S_b f_y)^2 \; + \; (\tau\sqrt{3}/S_s f_y)^2 \; = \; 1 \tag{17}$$

where σ_c in the uniform compression stress component, σ_b the maximum value of the bending stress component and τ is the coincident shear stress. S_c, S_b and S_s are numerical multipliers of the yield stress determined to provide the best fit to the analytical interaction curves (fig. 20).

A possible alternative approach for the interaction of bending and shear stress is the tension field mechanism approach - it is discussed in a next chapter - which is normally considered as a design approach for plate girder webs.

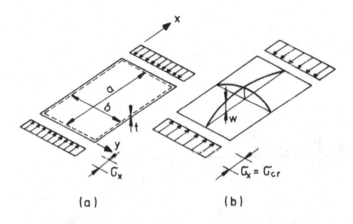

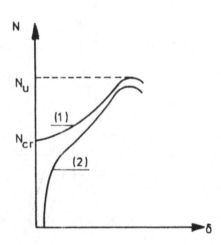

Fig. 1 - Buckling of an ideal rectangular plate

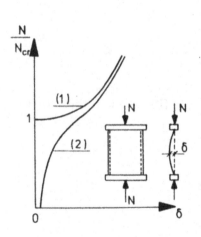

Fig. 2 - Normalized load - shortening curve for an ideal
compression plate

Fig. 3 - Influence of plasticity on
the load - shortening curve

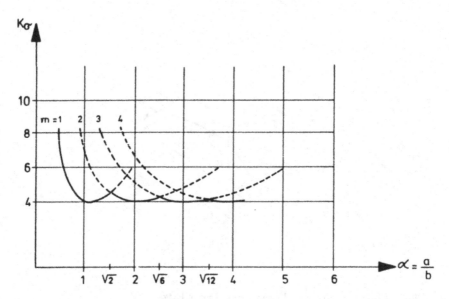

Fig.4 - Buckling chart for a uniformly compressed rectangular plate with
simple supports.

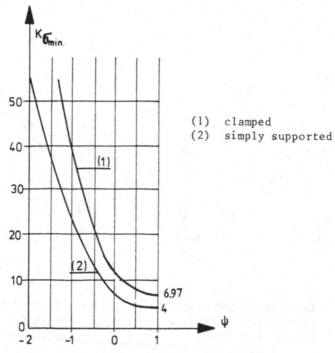

Fig.5 - Buckling chart for a uniformly compressed rectangular plate :
influence of the boundary conditions.

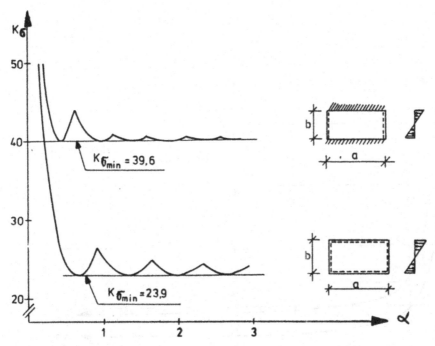

Figure 6 - Buckling chart for a rectangular plate subject to pure
         bending:influence of boundary conditions

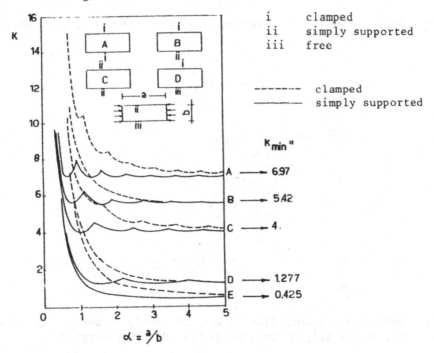

Figure 7 - Plot of minimum values of the buckling coefficient versus
         direct stress ratio

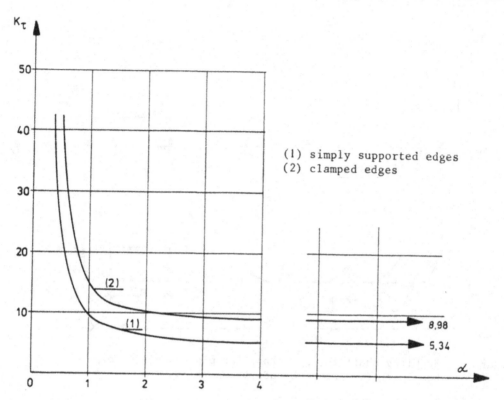

Fig. 8 - Buckling chart for a rectangular plate subject to pure shear.

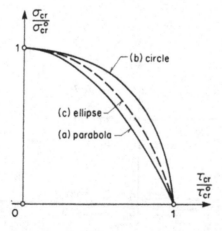

Fig. 9 - Approximate interaction curves for simply supported rectangular plates subject to combined direct and shear stresses.

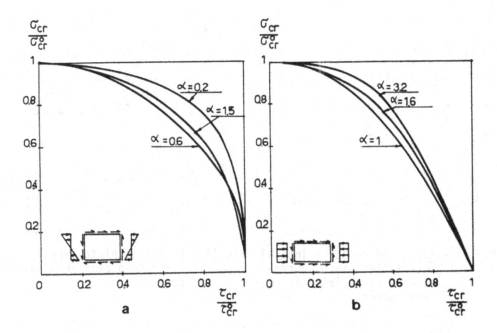

Fig. 10 - Actual interaction curves for simply supported rectangular plates : a) bending and shear ; b) compression and shear

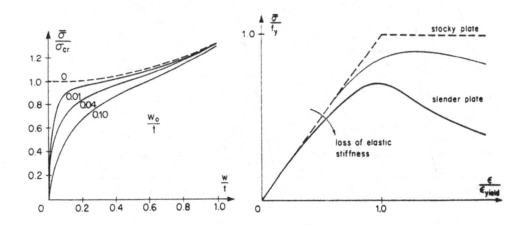

Fig.11 - Effect of relative initial out-of-flatness on load-displacement behaviour.

Fig. 12 - Normalized stress-strain curves.

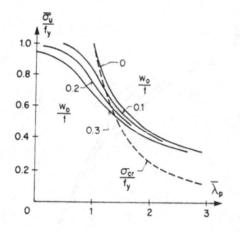

Fig.13 - Effect of relative initial out-of-flatness on normalized load - plate slenderness curves.

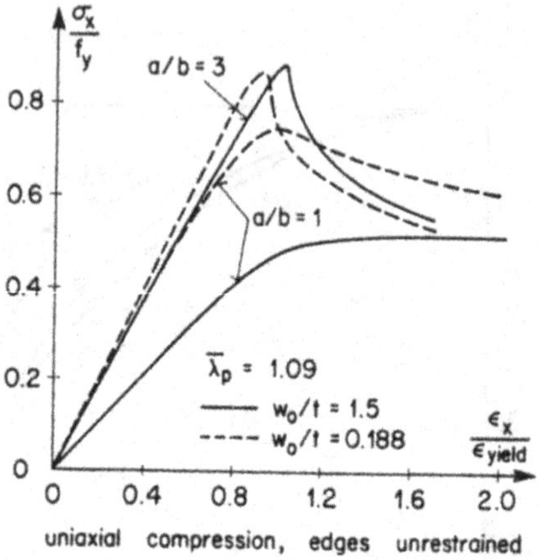

Fig.14 - Effect of plate aspect ratio on plates in uniaxial compression.

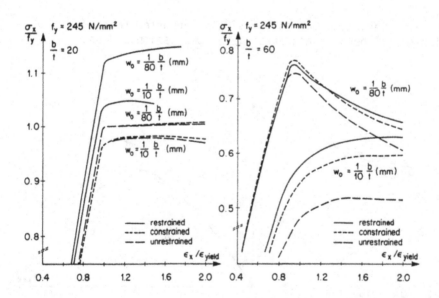

Fig.15 - Effect of in-plane restraint on plates in uniaxial compression.

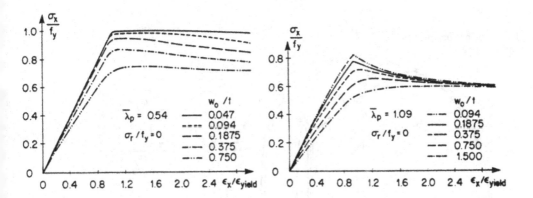

Fig.16 - Effect of out-of-flatness on square constrained plates in
uniaxial compression.

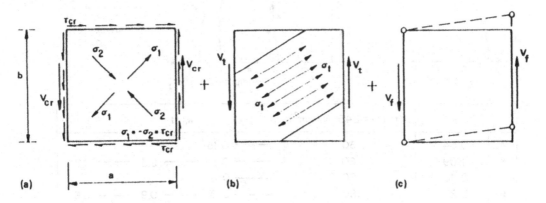

Fig. 17 - Ultimate shear.

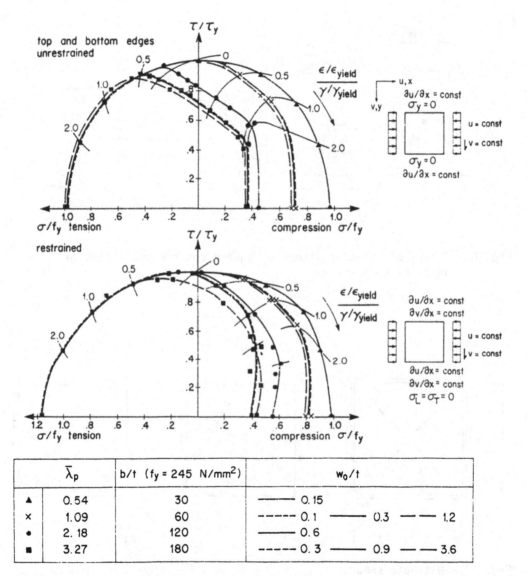

Fig. 18 - Interaction curves for plates under shear and uniform uniaxial direct displacement.

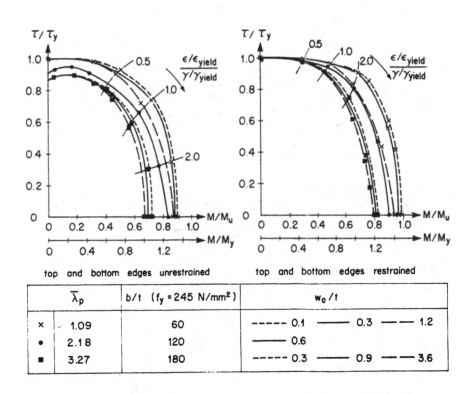

Fig. 19 - Interaction curves for plates under shear and in-plane bending displacement.

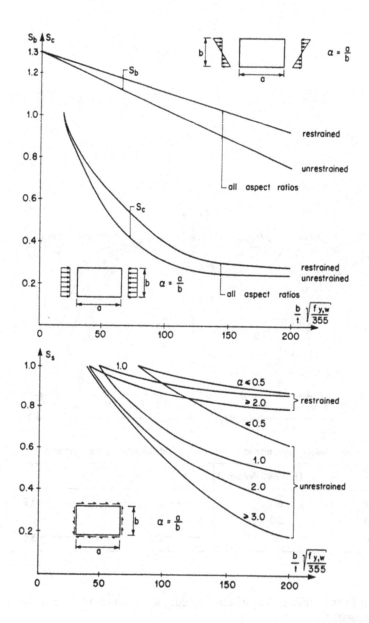

Fig.20 - S Coefficients

CHAPTER 2
CLASSIFICATION OF CROSS-SECTIONS

1. INTRODUCTION.

When designing a structure and its components, the designer is faced to the question of which *model* he is allowed to use. The concept of *model* is related to, on the one hand, i) the *analysis of the structure*, which is aimed at the determination of the stress resultants (also termed internal forces or member forces) under the action of design loads, and, on the other hand, ii) the *verification of the cross-sections*, which consists is checking whether the cross-sections are able to resist these design internal forces. Thus a model implies the use of a *method of global analysis* combined to a *method of cross-section verification*.

Insofar the ultimate limit state design of structures is concerned, there are several possibilities to combine methods of analysis and methods of cross-section verification, according as these methods are referring to elastic or plastic approach. The possible combinations are listed in Table 1.

Model	Method of analysis	Method of cross-section verification
I	plastic	plastic
II	elastic	plastic
IIIa	a. elastic	a. elastic (full section)
IIIb	b. elastic	b. elastic (effective section)

Table 1 - Definition of models.

Model I is related to plastic design of structures. Full plasticity may be developed within cross-sections (i.e. the stress distribution corresponds to a fully rectangular block) so that plastic hinges can form. These have suitable moment rotation characteristics to allow for a sufficient rotation capacity and the formation of a plastic mechanism, as the result of moment redistribution in the structure.

For a structure composed of sections which can achieve their plastic resistance but have not a sufficient rotation capacity to allow for a plastic mechanism in the structure, the ultimate limit state must refer to the onset of the first plastic hinge. Thus, in *Model II*, the internal forces are determined based on an elastic analysis but are compared to the plastic capacities of the corresponding cross-sections. For statically determinate systems, the onset of the first plastic hinge leads to a plastic mechanism; both methods I and II should thus give the same result. For statically indeterminate structures, the moment redistribution is not possible according to Model II, in contrast to what is allowed for with Model I.

When the cross-sections of a structure cannot exhibit their plastic capacity, both analysis and verification of cross-sections must be conducted elastically. The ultimate limit state according to *Model III* is achieved when yielding occurs at the most stressed fibre. Sometimes yielding in the extreme fibre cannot even be attained because of prematurous plate buckling of one component of the cross-section due to an excessive slenderness of this component plate. In such cases, above ultimate limit state should apply to effective cross-sections.

Of course it is not possible to have a model where a plastic method of analysis is combined to an elastic cross-section verification. Indeed, the moment redistribution which is required by the plastic analysis cannot take place without some cross-sections being fully yielded.

2. APPLICABILITY OF THE MODELS TO MEMBERS SUBJECT TO BENDING.

In the previous section, the models are defined in terms of structural design criteria. Actually these are governed by conditions related to stability problems. Indeed plastic redistribution between cross-sections and/or within cross-sections can take place provided that no prematurous local buckling occurs, which is liable to cause a drop-off in carrying capacity.

According as the model considered, it must be warranted that no local instability can occur before either the elastic bending capacity (Model III), or the plastic bending capacity (Model II) of the cross-section, or the formation of a complete plastic mechanism (Model I) is achieved.
Such a mechanism shall form provided that any plastic hinge at first developed be able not only to rotate but, in addition, to *rotate sufficiently*; that presupposes that the cross-sections where plastic hinges form may exhibit the rotation capacity required by the formation of a plastic mechanism.

To ensure a sufficient rotation capacity, the extreme fibres must be able to exhibit very large strains without any drop-off in capacity occurs. In tension, usual steel grades exhibit a ductility which is largely sufficient to allow for the desired amount of tensile strains; in addition, no drop-off is to be feared before the ultimate tensile strength be reached. When compressive stresses, it is not so much a question of material ductility properly but well of ability to sustain these stresses without instability occurs. Therefore, the more simple manner to warrant a sufficient rotation capacity is to limit the width-to-thickness ratio b/t of the component plates which are subject to compressive stresses due to bending moment and/or axial load. Indeed, the larger the b/t ratio, the sooner plate buckling shall occur and consequently, the smaller the available rotation capacity. The range of the b/t ratios to which a specified model is applicable is illustrated in figure 1. As it can be seen, four classes are identified when reference is made to bending :

a) Class 1 cross-sections : *Plastic cross-sections*.
   Such sections can develop a plastic hinge with the rotation capacity required for plastic analysis.

b) Class 2 cross-sections : *Compact cross-sections*.
   Such sections can develop their plastic moment resistance but have a limited rotation capacity.

c) Class 3 cross-sections : *Semi-compact cross-sections*.
   The calculated stress in the extreme compression fibre of the steel member can reach the material yield strength but local buckling is liable to prevent development of the plastic moment resistance.

d) Class 4 cross-sections : *Slender cross-sections*.
   Here it is necessary to make explicit allowance for the effects of local buckling when determining the moment resistance.

The response of the different classes of cross-sections, when subject to bending, is usefully represented by dimensionless moment-rotation curves. Differences in behaviour are well reflected by the plots of figure 2.

3. CLASSIFICATION CRITERIA.

Where to classify a specified cross-section depends on the proportions of each of its compression elements. Compression elements include every element of a cross-section which is either totally or partially in compression, due to axial force and/or bending moment under the load combination considered; the class to which a specified cross-section belongs shall thus depend, amongst others, on the type of loading experienced by this section. In general, the various compression elements (web, flange,...) in a cross section can belong to different classes. The cross-section is normally classified by quoting the least favourable class of its compression elements.

The most important limiting b/t ratios applicable to the compression elements of a cross-section, which enable the appropriate classification to be made, are listed in tables 2 to 5. These limit values, which are justified in the next section, are related to the material yield strength. It is especially important, peculiarly when use is made of plastic design, that the sections selected for the various members should not violate the proportions given herein, and that, in all cases, the section selected should be appropriate for the assumed mode of behaviour. Attention is drawn especially on the length b of the plate elements of the cross-section to be considered when referring to the limit values of the b/t ratios.

When any of the compression elements of a cross-section fails to satisfy the limits given for Class 3 cross-sections, that element shall be treated as a Class 4 element and the cross-section shall be treated as a Class 4 cross-section. The cross-section properties of Class 4 sections shall be based on the effective widths of the compression elements; a detailed study of this question is presented in a next lecture.

Some additional comments can be brought :

a) The resistance of a cross-section with a Class 2 compression flange but a Class 3 web may alternatively be determined by treating the web as an effective Class 2 web with a reduced effective area ;

b) When yielding first occurs on the tension side of the neutral axis, the plastic reserves of the tension zone may be utilised when determining the resistance of a Class 3 cross-section.

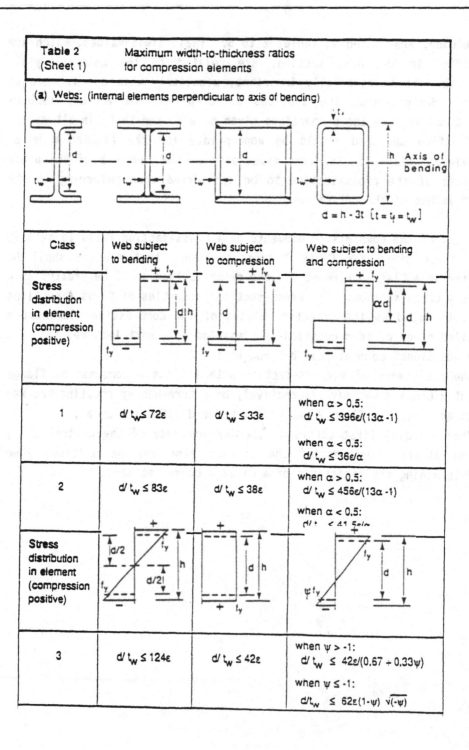

Table 2
(Sheet 1)

Maximum width-to-thickness ratios
for compression elements

(a) Webs: (internal elements perpendicular to axis of bending)

$d = h - 3t$ $[t = t_f = t_w]$

Class	Web subject to bending	Web subject to compression	Web subject to bending and compression
Stress distribution in element (compression positive)			
1	$d/t_w \leq 72\varepsilon$	$d/t_w \leq 33\varepsilon$	when $\alpha > 0.5$: $d/t_w \leq 396\varepsilon/(13\alpha - 1)$ when $\alpha < 0.5$: $d/t_w \leq 36\varepsilon/\alpha$
2	$d/t_w \leq 83\varepsilon$	$d/t_w \leq 38\varepsilon$	when $\alpha > 0.5$: $d/t_w \leq 456\varepsilon/(13\alpha - 1)$ when $\alpha < 0.5$: $d/t \leq 41.5\varepsilon/\alpha$
Stress distribution in element (compression positive)			
3	$d/t_w \leq 124\varepsilon$	$d/t_w \leq 42\varepsilon$	when $\psi > -1$: $d/t_w \leq 42\varepsilon/(0.67 + 0.33\psi)$ when $\psi \leq -1$: $d/t_w \leq 62\varepsilon(1-\psi)\sqrt{(-\psi)}$

Table 3 (Sheet 2)	Maximum width-to-thickness ratios for compression elements

(b) Internal flange elements: (internal elements parallel to axis of bending)

Class	Type	Section in bending	Section in compression
	Stress distribution in element and across section (compression positive)		
1	Rolled Hollow Section Other	$(b-3t_f)/t_f \leq 33\varepsilon$ $b/t_f \leq 33\varepsilon$	$(b-3t_f)/t_f \leq 42\varepsilon$ $b/t_f \leq 42\varepsilon$
2	Rolled Hollow Section Other	$(b-3t_f)/t_f \leq 38\varepsilon$ $b/t_f \leq 38\varepsilon$	$(b-3t_f)/t_f \leq 42\varepsilon$ $b/t_f \leq 42\varepsilon$
	Stress distribution in element and across section (compression positive)		
3	Rolled Hollow Section Other	$(b-3t_f)/t_f \leq 42\varepsilon$ $b/t_f \leq 42\varepsilon$	$(b-3t_f)/t_f \leq 42\varepsilon$ $b/t_f \leq 42\varepsilon$

$\varepsilon = \sqrt{235/f_y}$	f_y	235	275	355
	ε	1	0.92	0.81

Table 4 (Sheet 3)	Maximum width-to-thickness ratios for compression elements

(c)   Outstand flanges:

Rolled sections   Welded sections

Class	Type of section	Flange subject to compression	Flange subject to compression and bending	
			Tip in compression	Tip in tension
Stress distribution in element (compression positive)				
1	Rolled	$c/t_f \leq 10\varepsilon$	$c/t_f \leq \dfrac{10\varepsilon}{\alpha}$	$c/t_f \leq \dfrac{10\varepsilon}{\alpha\sqrt{\alpha}}$
	Welded	$c/t_f \leq 9\varepsilon$	$c/t_f \leq \dfrac{9\varepsilon}{\alpha}$	$c/t_f \leq \dfrac{9\varepsilon}{\alpha\sqrt{\alpha}}$
2	Rolled	$c/t_f \leq 11\varepsilon$	$c/t_f \leq \dfrac{11\varepsilon}{\alpha}$	$c/t_f \leq \dfrac{11\varepsilon}{\alpha\sqrt{\alpha}}$
	Welded	$c/t_f \leq 10\varepsilon$	$c/t_f \leq \dfrac{10\varepsilon}{\alpha}$	$c/t_f \leq \dfrac{10\varepsilon}{\alpha\sqrt{\alpha}}$
Stress distribution in element (compression positive)				
3	Rolled	$c/t_f \leq 15\varepsilon$	$c/t_f \leq 23\varepsilon\sqrt{k_\sigma}$	
	Welded	$c/t_f \leq 14\varepsilon$	$c/t_f \leq 21\varepsilon\sqrt{k_\sigma}$	
			For k_σ see table 5.3.3	

Table 5 (Sheet 4)	Maximum width-to-thickness ratios for compression elements

(d) Angles:

Refer also to (c) "Outstand flanges" (see Sheet 3).

(Does not apply to angles in continuous contact with other components)

Class	Section in compression
Stress distribution across section (compression positive)	
3	$\dfrac{h}{t} \leq 15\varepsilon$: $\dfrac{b+h}{2t} \leq 11.5\varepsilon$

(e) Tubular sections:

Class	Section in bending and/or compression
1	$d/t \leq 50\varepsilon^2$
2	$d/t \leq 70\varepsilon^2$
3	$d/t \leq 90\varepsilon^2$

4. COMPUTATION OF THE b/t LIMIT VALUES.

The link between the b/t ratios and the classes of cross-sections has
been justified qualitatively above and given quantitatively in Tables 2
to 5 without justification. Present section is aimed at giving the
appropriate background.

Let us first refer to *Class 3 sections*. The elastic critical buckling
stress of a compression element is given as :

$$\sigma_{cr} = k_\sigma \, [\pi^2 E/12(1 - \nu^2)](t/b)^2 \tag{1}$$

where k_σ is the buckling coefficient, E the YOUNG modulus and ν the
POISSON ratio. It is thus proportional to $(t/b)^2$, so that the
width-to-thickness ratio b/t plays thus a similar role as the
slenderness ratio (L/i) for column buckling. In accordance with the
definition of Class 3 sections, the proportions of the compression
element, quoted by the b/t ratio, must be such that σ_{cr} would exceed the
material yield strength f_y so that yielding occurs before the
compression element buckles. The ideal elastic-plastic behaviour of a
perfect plate element subject to uniform compression may be represented
by a normalized load - slenderness diagram, where the normalized
ultimate load :

$$\overline{N}_p = \sigma_u/f_y \tag{2}$$

and the normalized plate slenderness :

$$\overline{\lambda}_p = \sqrt{f_y/\sigma_{cr}} \tag{3}$$

are plotted in ordinates and in abscissae respectively (fig. 3). For
$\overline{\lambda}_p < 1$, $\overline{N}_p = 1$ which means that the compression element can develop its
squash load $\sigma_u = f_y$; for $\overline{\lambda}_p > 1$, $\overline{N}_p$ decreases when the plate
slenderness increases, σ_u being equal to σ_{cr}. Account being taken of
the above expression of σ_{cr}, the normalized plate slenderness writes,
with $\nu = 0.3$:

$$\overline{\lambda}_p = 1,05 \; (b/t) \sqrt{f_y/E \, k_\sigma} \tag{4}$$

This expression is general; indeed loading, boundary conditions and aspect ratio are influencing the value of the buckling coefficient k_σ. The actual behaviour is somewhat different from the ideal elastic-plastic behaviour because of : i) initial geometrical and material imperfections, ii) strain-hardening of the material, and iii) the postbuckling behaviour. Initial imperfections result in prematurous plate buckling, which occurs for $\bar{\lambda}_p < 1$. The corresponding limit plate slenderness $\bar{\lambda}_{p,3}$ for Class 3 sections may differ substantially from country to country because of statistical variations in imperfections and in material properties, which are not sufficiently well known to be quantified accurately ; a review of the main national codes shows that it is varying from 0.6 up to 0.9 approximately. Eurocode 3 has adopted $\bar{\lambda}_{p3} = 0.7$ as limit plate slenderness of Class 3 compression elements, which are those for which the yield strength may be reached in the extreme fibre of the cross-section. In plate elements for which

$\bar{\lambda}_p < \bar{\lambda}_{p3}$, no plate buckling can occur before the maximum compressive stress reaches the yield strength.

In contrast to a Class 3 section whose behaviour is elastic, a Class 1 section must exhibit its full plastic capacity. To do that, it is necessary the section allows for rather large inelastic deformations and strain hardening in the most stressed fibres, without plate buckling occurs. That is only possible for compression elements the normalized plate slenderness of which does not exceed a limit value $\bar{\lambda}_{p1}$, that is much lower than $\bar{\lambda}_{p3}$. The values of $\bar{\lambda}_{p1}$ which are found in the main national codes range from 0.46 up to 0.60 ; this range is explained by the differences in the amount of rotation capacity required for platic design. Eurocode 3 has adopted the value $\bar{\lambda}_{p1} = 0.50$ for Class 1 sections.

A Class 2 section has still to exhibit its plastic capacity but may have rapidly a drop-off in capacity. The plate element is yielded and the material strained in the plastic range. The limit normalized plate slenderness $\bar{\lambda}_{p2}$ for Class 2 sections is obviously larger than $\bar{\lambda}_{p1}$ and lower than $\bar{\lambda}_{p3}$. Eurocode 3 has adopted the value $\bar{\lambda}_{p2} = 0.60$ for Class 2 sections.

When turned by 90° the σ - ϵ diagram is, as well known, a representation of the stress distribution over half the depth of the cross-section. Therefore the relation between the classes of cross-sections and the maximum compressive strain is represented by figure 4. The ranges of plate slenderness to which each of the models is applicable are represented schematically in figure 5.

As a result, the limit (maximum) b/t ratios for compression elements are listed in Tables 2 to 5. These limits are given for Classes 1, 2 and 3 plate elements, depending on the boundary conditions and the stress distribution. They are given by reference to a factor ϵ, which accounts for the yield strength of the material of which the plate element is made.

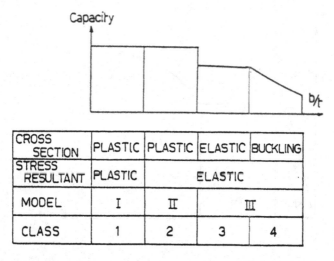

CROSS SECTION	PLASTIC	PLASTIC	ELASTIC	BUCKLING
STRESS RESULTANT	PLASTIC	ELASTIC		
MODEL	I	II	III	
CLASS	1	2	3	4

Figure 1 - Relation between models and classes of cross-sections

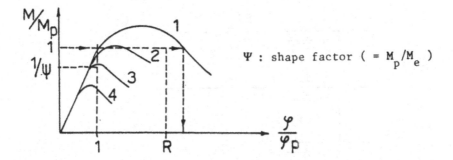

Ψ : shape factor ($= M_p/M_e$)

Figure 2 - Normalized moment-rotation curves for the four classes of cross-sections

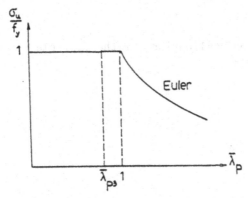

Figure 3 - Normalized load-slenderness curve for an ideal plate

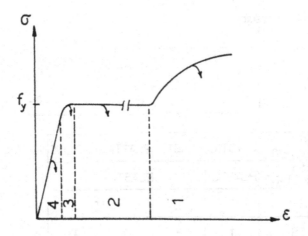

Figure 4 – Load-shortening curves for plate elements of different
b/t, ratios

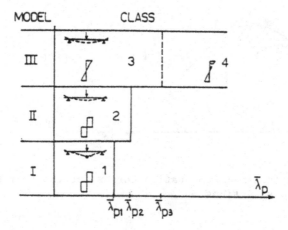

Figure 5 – Range of application for the four classes of cross-
sections

CHAPTER 3
TENSION FIELD MODELS FOR WEBS IN SHEAR

1. INTRODUCTION.

Modern steel construction makes an intensive use of built-up sections, more especially of plate girders. In contrast with universal beams, plate girders have a slender web (i.e. characterized by a large d/t ratio); the web, that is assembled to equal or unequal flanges by longitudinal fillet welds, belongs usually to class 4 (see Lecture 2). The main role of the web is to sustain shear; in this respect, a slender web can exhibit a large postbuckling strength, so that the shear resistance is normally much higher than the critical buckling resistance. When the shear capacity of the web exceeds notably the design shear load, the web material is likely to experience coincident direct stresses and shear stresses.

The shear buckling stress of a rectangular plate (a x d) writes :

$$\tau_{cr} = k_\tau \, \sigma_E$$

$$= k_\tau \ [\pi^2 E/12 \ (1 - \nu^2)] \ (t/d)^2 \tag{1}$$

Should this plate be the web of a plate girder, d represents the web depth while a is the spacing of transverse stiffeners. The buckling coefficient k_τ depends on the aspect ratio $\alpha = a/d$ and on support conditions. Usually the web of a plate girder is conservatively assumed as simply supported along its four edges; accordingly, k_τ writes:

$$k_\tau = 4 + 5.34/\alpha^2 \quad (\alpha \leq 1)$$

$$= 5.34 + 4/\alpha^2 \quad (\alpha \geq 1)$$

When the length of the plate is large compared to the width ($\alpha \gg 1$), the buckling coefficient approaches 5.34.

For a *perfect plate*, made with an elastic-perfectly plastic material, shear buckling should govern the design when the elastic critical shear buckling stress τ_{cr} does not exceed the material shear yield stress $\tau_{yw} = f_{yw}/\sqrt{3}$.
For a simply supported unstiffened plate ($\alpha \gg 1$), the limiting web depth-to-thickness ratio is :

$$d/t_w \geq 1.57 \ \sqrt{k_\tau} \ \sqrt{E/f_{yw}} \tag{2}$$

For plate slenderness d/t_w lower than the limiting value given by (2), yield should be governing. Actually the material behaves elastically up to a limit of proportionnality only; in addition account must be taken of the detrimental influence of residual stresses, which are liable to initiate prematurous yielding. There is thus a domain of inelastic plate buckling, where plate buckling interacts with yielding.
Accordingly, as soon as τ_{cr} exceeds $\beta \ f_{yw}/\sqrt{3}$, a reduced critical buckling stress τ_{cr} must be substituted for τ_{cr}. A lot of different expressions are suggested in this respect; that recommended by EC3 - 1989 uses $\beta = 0.67$ and writes, in normalized coordinates :

$$\tau_{cr}^{red}/\tau_{yw} = [1 - 0.8 \ (\bar{\lambda}_w - 0.8)] \ngtr 1 \tag{3}$$

$$\ngtr 1/\bar{\lambda}_w^2$$

with the normalized plate slenderness :

$$\overline{\lambda}_w = \sqrt{\tau_{yw}/\tau_{cr}} \tag{4}$$

Accordingly, above described shear resistance should be represented by the schematic plot of figure 1, as long as no postbuckling resistance is accounted for.

2. BACKGROUND OF ULTIMATE SHEAR MODELS.

The physical explanation of how a slender web can exhibit a postbuckling shear strength is developed here on base of a rectangular ideal slender plate, which is assumed to be perfectly flat and without residual stresses. These assumptions are only aimed at facilitating somewhat the reasoning and help in the understanding of the successive steps of the plate behaviour up to collapse ; they are indeed not at all compulsory because usual geometrical and structural imperfections are without significative effect on the ultimate shear strength.

Under increasing shear load a slender plate should buckle elastically at a stress level that can be predicted theoretically: the critical shear buckling stress τ_{cr}. Until this stage, the ideal plate shows only a beamaction strength, similarly to what happens in the web of any hot-rolled section. Pure shear in the plate is equivalent to equal but opposite principal direct stresses σ_1 and σ_2, acting at 45° with respect to shear; shear buckling at τ_{cr} may thus be understood as produced by the principal compressive stress $\sigma_2 = -\tau_{cr}$ (fig. 2.a). Once the plate has buckled, the web plate is no more able to transmit any additional direct stress in the direction of σ_2, while the perpendicular direction is still liable to sustain direct stresses. Henceforth the latter can still work like the tensile diagonal of a Pratt truss, the chords and the posts of which would be constituted by the flanges and the transverse stiffeners respectively. Otherwise speaking, the shear load can increase beyond the critical shear buckling load, provided there is a change of the stress distribution subsequent to buckling; an appreciable postbuckling strength may be mobilized because of the diagonal tensile stresses that develop. This second stage corresponds to the tension-field action strength (fig. 2.b).

Whereas the principal compressive stresses are still limited, the principal tensile stresses can increase furthermore up to yielding. That results in principal stresses of different magnitude, which are no more sloped at 45° on the plate edges (fig. 3); as a result, the edges are subject not only to shear stresses but also to direct stresses, which must be supported by edge members because of the equilibrium. If these edge members are rigid, the postcritical strength may develop uniformly over the whole area of the plate. Where, in contrast, the edge members are flexible, the direct stresses shall be limited and the direction of principal stresses shall change; only a part of the postbuckling strength reserve can be mobilised. The ultimate shear load experienced by a plate girder should thus appear as a combination of the carrying capacity of the web and that of the edge members respectively.

The assessment of the shear capacity of the web of a plate girder at the ultimate limit state requires consequently the knowledge of :
a) the magnitude of tensile and compressive principal stresses ;
b) the direction of the principal stresses ;
c) the contribution of edge members to the postbuckling strength.

In the extreme case of a *very slender* plate, the shear critical stress τ_{cr} is very small; as a first approximate, it may be disregarded. In accordance with what is explained above, the principal compressive stress σ_2 is zero. Shear load can thus be sustained by the postbuckling stress distribution, i.e. by tensile principal stresses in the web plate. This mode of resistance is termed *tension field* and the *tension field theories* are the methods used to describe the corresponding behaviour. The slender web of a plate girder should experience shear similarly to a Pratt truss, whose diagonals plays the role of the tension field in the web.

3. TENSION FIELD THEORIES.

3.1. The pioneer.

RODE (1916) was the pioneer of the tension field concept. His approach was rather rough; indeed he assumes that the diagonal tension field has a width equal to 50 times the web thickness.

3.2. Ideal tension field theory.

WAGNER (1929) developed the *ideal tension field theory*, which approaches the plate in shear as a system of ties covering the *whole* area of the web. This theory is quite appropriate for very thin webs (d/t_w = 600 up to 1000) connected to very rigid edge members. Prior to plate buckling, one has a pure shear stress distribution represented by a circle centered on the origin of the axes with a radius $\tau \leq \tau_{cr}$. According to WAGNER theory, one has $\tau_{cr} \approx 0$ and therefore σ_2 = 0 so that after web buckling, the circle shifts to the right, and is tangent to the τ axis (fig. 4.a).

At the ultimate load, the radius of the circle is τ_u. The tensile principal stress amounts :

$$\sigma_1 = 2 \ \tau_u \tag{5}$$

and develops over the whole web. The ultimate shear writes (fig. 5):

$$V_u = \sigma_1 \ t_w \ g \ \sin \theta \tag{6}$$

where θ is the slope of the tension field and g = d cos θ.

The maximum value of V_u corresponds to : i) the slope θ for which the first derivative $\partial V_u/\partial \theta$ vanishes, and ii) tensile yielding (σ_1 = f_{yw}) in the tension field. Then the principal tensile stresses are sloped 45° and the ultimate shear amounts :

$$V_{u,max} = 0.5 \ dt_w \ f_{yw} \tag{7}$$

That is 13 % less than the shear yield load :

$$V_y = dt_w \ \tau_{yw} = dt_w \ f_{yw}/\sqrt{3} \tag{8}$$

3.3. Complete tension field theory.

WAGNER theory had to be improved because plates of usual proportions - as used in civil engineering structures -, have edge members of finite bending rigidity, on the one hand, and have a non-negligible critical shear buckling resistance, on the other hand.   In the *complete tension field theory*, the edge members are assumed rigid but it is accounted for the critical shear resistance; therefore, one has $|\sigma_2| = \tau_{cr} \neq 0$. MOHR circle corresponding to this case is represented in figure 4.b, wherefrom, at the ultimate shear load :

$$\sigma_1 = 2\tau_u - \sigma_2 \tag{9}$$

The value of σ_1 at collapse is drawn from von MISES yield criterion :

$$\sqrt{\sigma_1^2 - \sigma_1\sigma_2 + \sigma_2^2} = f_{yw} \tag{10}$$

wherefrom :

$$\sigma_1 = 0.5 \; \sigma_2 + \sqrt{f_{yw}^2 - 0.75 \; \sigma_2^2} \tag{11}$$

and, account taken of (9) :

$$2 \; \tau_u = 0.5 \; \tau_{cr} + \sqrt{f_{yw}^2 - 0.75 \; \tau_{cr}^2} \tag{12}$$

The ultimate shear resistance is given as   :

$$V_u = \tau_u \; d \; t_w \tag{13}$$

The carrying capacity associated to above models is conveniently plotted in normalized coordinates $(V_u/V_y, \; \bar{\lambda}_p)$ as shown in figure 1. When the plate slenderness increases, the difference between both above tension field theories decreases, while that between critical shear theory and tension field theories is growing up.
An improvement has been brought to the tension field theories by developing the concept of *incomplete tension field*.

3.4. Incomplete tension field theory.

In the buckled plate, the principal tensile and compressive stresses

must be such that the deviation forces they produce at any point of the plate be in equilibrium (fig. 6). So the magnitude of the stresses shall depend on the transverse deflection of the buckled plate; it shall vary from point to point, so that the assumption of an homogeneous stress distribution, made as well in the ideal tension field theory as in the complete tension field theory, is no more valid. The principal compressive stresses, which are liable to maintain equilibrium, change in the postbuckling domain, so that
$|\sigma_2| > \tau_{cr}$; corresponding MOHR circle is shown in figure 4.c. The sole way for solving the problem would consist in integrating a set of two coupled 4th order differential equations (see Lecture 1), that account for large transverse displacements and the stabilizing effect of membrane forces. SKALOUD solved these differential equations for a plate in shear when there are no edge members, on the one hand, and when there are infinitely rigid edge members, on the other hand; computation efforts required accordingly are such that this approach is not at all appropriate for practice purposes.
Anyway, the model is still to be improved to account explicitly for the finite bending stiffness of the edge members.

4. INFLUENCE OF FLEXIBLE EDGE MEMBERS ON THE SHEAR CAPACITY.

Above tension field theories were actually based on tests conducted in the sphere of aircraft engineering, where extremely slender plates are used in conjunction with very strong ribs and stiffeners. Comparison of experimental and theoretical results were found in a very good agreement.

In the field of civil engineering, the assumption of undeformable edge members is no more acceptable. Let us consider a plate in the postbuckling range; its edges are subject to shear stresses and to additional direct stresses. In a plate girder, the plate is the web panel while the edge members are the flanges in the longitudinal direction and the transverse stiffeners in the perpendicular direction; these edge members experience the aforementioned direct stresses and consequently are prone to bend and displace in the plane of the web. The in-plane displacements of the edge members result first, in a drop

in magnitude of the membrane stresses, compared to what it would be with
very rigid edge members, and, second, in a change in direction of
principal stresses. When several web panels are continuously adjacent,
the direct stresses which develop in one panel can anchor strongly, due
to the sole presence of an adjacent panel; then, in-web plane
displacements of transverse stiffeners are *virtually* prohibited and
their effect can be disregarded. That is especially the situation of
intermediate transverse web stiffeners; in contrast, end stiffeners do
usually not benefit from such a favourable situation.

A step further in the improvement of tension field theories should be to
account properly of the ability to in-web plane displacements of the
edge members.

BASLER developed a tension field model, where the bending stiffness of
the flanges is fully neglected; the tension field anchors thus on the
sole transverse stiffeners (fig. 7). According to this approach, it is
assumed that the plate critical shear resistance :

$$V_{cr} = d \, t_w \, \tau_{cr} \qquad\qquad\qquad (14)$$

reached when the plate buckles, remains available at the ultimate limit
state ; in other words, the ultimate load is reached when the web
material yields in the tension field under coincident pure shear
stresses and direct stresses developed by the tension field action. In
addition, the frame composed of flanges and stiffeners contribute
partially the strength capacity. BASLER's theory leads to ultimate
shear loads, which are found in good agreement with experimental results
provided that the flange of the test specimens be thin and therefore
very flexible.

Tests conducted by ROCKEY and SKALOUD on plate girders with a same
geometry but flanges of different proportions, show that the bending
stiffness of the flanges is likely to influence appreciably the shear
capacity. Many researchers tried to improve BASLER's model accordingly;
it is not the place here to review all the work done in this respect.
Let us just point out the Cardiff model, which is probably the best

known. The Cardiff model is based on the onset of a plastic mechanism, which is determined by a yielded zone ECFB in the web and plastic hinges in the flanges (fig. 8). Similarly as in BASLER's model, it is assumed that the critical shear strength V_{cr} can be mobilised till the ultimate load is reached; thus, it superimposes the tension field resistance V_t. According to the Cardiff model, the tension field anchors not only on the transverse stiffeners but also on the flanges, in an extend which depends on the bending stiffness of these flanges. The tensile band is thus composed of a central portion, that anchors on the stiffeners, and of two additional lateral portions that anchor on the flanges. The larger the bending stiffness of the flanges, the larger the anchor lengths c_c and c_t on the flanges. The values of c_c and c_t are computed based on the consideration of a combined failure mechanism with plastic hinges in the flanges and a yielded band in the web (fig.8-10). Therefore, the effect of the flange stiffness is reflected through the plastic bending resistance of the flanges; as this resistance decreases in presence of an axial load in the flanges, the anchor lengths on the flanges shall be reduced according as the magnitude of direct stresses in the flanges.

The contribution of the tension band to the ultimate shear load is the vertical component of the resultant force in this band, i.e. :

$$V_t = g \, t_w \, \sigma_t \, \sin \theta \qquad\qquad (15)$$

where g is the width of the tension band, θ the slope of this band and σ_t the magnitude - at collapse - of the presumed uniform tensile stress distribution across this band.

The width g of the tensile band can be determined as a function of : i) the dimensions a and d of the web panel, ii) the inclination θ of this band; iii) the anchor lengths c_c and c_t on the flanges :

$$g = d \cos \theta - (a - c_c - c_t) \sin \theta \qquad\qquad (16)$$

Plastic hinges in the flanges are expected to occur in sections E and F, where flange shear force vanishes, and in sections C and B, where the flanges are considered as fully clamped.

Applying the principle of virtual work to the corresponding plastic mechanism in the upper flange gives (fig. 10) :

$$0.5 \; \sigma_t \; t_w \; c_c^2 \; \phi \; \sin^2 \theta = (M_{pE}^* + M_{pC}^*) \; \phi \tag{17}$$

where M_p^* is the plastic bending resistance of the flanges, with account taken of the possible axial load N_f in this flange :

$$M_p^* = M_{pf} \; [1 - (N_f/N_{pf})^2] \tag{18}$$

with, for a rectangular flange cross-section :

$$M_{pf} = 0.25 \; b_f \; t_f^2 \; f_{yf} \tag{19}$$

$$N_{pf} = b_f \; t_f \; f_{yf} \tag{20}$$

Length c_c is drawn from (17) :

$$c_c = (2/\sin \theta) \sqrt{(M_{pE}^* + M_{pC}^*)/2\sigma_t \; t_w} \; \nmid \; a \tag{21.a}$$

and c_t should be obtained similarly :

$$c_t = (2/\sin \theta) \sqrt{(M_{pB}^* + M_{pF}^*)/2\sigma_t \; t_w} \; \nmid \; a \tag{21.b}$$

Usually, M_{pB}^* and M_{pF}^*, on the one hand, and M_{pE}^* and M_{pC}^* , on the other hand, are not very different so that the anchor lengths write simply in the general following form :

$$c = 2/\sin \theta \sqrt{M_p^*/\sigma_t \; t_w} \tag{22}$$

where M_p^* is an *average* reduced bending resistance.

The stress σ_t in the tension band at collapse is derived from von MISES yielding criterion written with reference to the stress state related to directions u and v, respectively parallel and perpendicular to the tension field (fig. 11) :

$$\sqrt{\sigma_u^2 + \sigma_v^2 - \sigma_u \sigma_v + 3\tau_{uv}^2} = f_{yw} \tag{23}$$

where :

$$\sigma_u = \sigma_t + \tau_{cr} \sin 2\theta \tag{24.a}$$

$$\sigma_v = - \tau_{cr} \sin 2\theta \tag{24.b}$$

$$\tau = \tau_{cr} \cos 2\theta \tag{24.c}$$

Introducing (24) in (23) gives :

$$\sigma_t = \sqrt{f_{yw}^2 - \tau_{cr}^2 [3 - 2,25 \sin^2 2\theta]} - 1,5 \tau_{cr} \sin 2\theta \tag{25}$$

where τ_{cr}^{red} has to be possibly substituted for τ_{cr} when necessary.

At this stage, the sole remaining unknown is the slope θ of the tensile band. Its value should maximize V_t because of the static theorem of plastic design (indeed a state of stress has been defined which is in equilibrium with the shear load and complies with the yielding condition). Its search ought thus to proceed by trial and error ; however numerous computations have shown that the part of the curve $V_t = f(\theta)$ is very flat in the vicinity of its maximum. Thus any error on the assessment of θ_{opt} will only generate a slight error on V_t but on the safe side. ROCKEY observed on his numerous tests that θ_{opt} is always very close to 2/3 θ_d, where θ_d is the slope of the geometric diagonal of the web panel :

$$\theta_d = arctg (d/a) \tag{26}$$

This value can thus be used as a conservative approximate. Alternatively, iteration may be used to find the actual optimum value θ_{opt}.

Frame action is a third contribution V_f to the ultimate shear resistance. When the tensile diagonal of the pseudo-Pratt truss has exhausted its carrying capacity (because of yielding), the frame

constituted by the flanges and the transverse stiffeners is still able
to act as a mesh of a Vierendeel girder ; it can thus experience some
shear load till the onset of plastic hinges in this frame with the
result of a plastic collapse mechanism. Frame action is usually
disregarded because of the two main reasons : i) its contribution V_f is
usually much smaller than both other ones V_{cr} and V_t, and ii)
experiments and numerical analysis show that the real collapse of the
surrounding frame - which implies onset of plastic hinges - develops
after the maximum shear load has been overcome. The shear resistance V_u
would be given as :

$$V_u = V_{cr} + V_t \ (+ V_f) \tag{27}$$

DUBAS developed an ultimate shear model, which it is termed *simple
postcritical method* because of its simplicity. The basic idea regarding
the two successive steps of the behaviour is still the same; the
difference lies in the assessment of the postbuckling strength reserve.
Here the tension band is oriented according as the geometric diagonal of
the web panel (fig. 12); it is anchored on two rectangular gusset
plates, which have the same aspect ratio as the web panel and have
dimensions such that the critical shear stress of these gusset plates
reaches the material yield stress wherefrom :

$$c_s/d = \sqrt{\tau_{cr}/\tau_{yw}} \tag{28}$$

where τ_{cr} is the critical shear buckling stress of the web panel. The
additional shear contribution as a result of the tension field action
writes :

$$V_t = c_s t_w \ (\tau_{yw} - \tau_{cr}) \tag{29.a}$$

or, account taken of (28) :

$$V_t = dt_w \sqrt{\tau_{cr}/\tau_{yw}} \ (\tau_{yw} - \tau_{cr}) \tag{29.b}$$

The frame action is disregarded so that the shear buckling resistance
writes :

$$V_u = (V_{cr} + V_t) = dt_w \quad \sqrt{\tau_{cr}\tau_{yw}} \; [1 + \sqrt{\tau_{cr}/\tau_{yw}} - (\tau_{cr}/\tau_{yw})] \quad (30)$$

This expression is similar to the well known von KARMAN one, established for plates in uniaxial uniform compression. For large aspect ratios, the term between brackets is close to 1.0; then, the shear buckling resistance writes conservatively :

$$V_u = 0.9 \, d \, t_w \sqrt{\tau_{cr} \, \tau_{yw}} \tag{31}$$

which corresponds to the normalized average ultimate shear stress $\tau_u/\tau_{yw} = 0.9/\overline{\lambda}_w$.

To account for the detrimental effect of imperfections in the intermediate range of web slenderness values $\overline{\lambda}_w$, the simple postcritical normalized ultimate shear strength is determined as follows :

$$\tau_u/\tau_{yw} = 1 - 0.625 \, (\overline{\lambda}_w - 0.8) \not\geq 1 \tag{32}$$

$$\not\geq 0.9/\overline{\lambda}_w$$

which has a similar format as (3).

4. EXPERIMENTAL EVIDENCE FOR THE NEED OF BOTH APPROACHES.

All the test results dealing with plate girders subject to shear have been collected and compared with the theoretical ultimate shear forces, computed respectively in accordance with the Cardiff tension field model, on the one hand, and with the simple postcritical method, on the other hand. Plotting the values of the ratio V_{ex}/V_u against the values of aspect ratio $\alpha = a/d$ shows that (fig. 13-14) :

a) The tension field model is appropriate for transversely stiffened plate girders whose web panels aspect ratio is between 1 and 3; it is too conservative for large values of α and unsafe for values of α lower than 1 ;

b) The simple postcritical method is appropriate for large values of the aspect ratio (> 3) and too conservative for smaller values of α.

For large spacings of the transverse stiffeners ($\alpha > 3$), the tension band is found to develop with a slope higher than that computed based on the tension field approach and no more in relation with the aspect ratio.

It is also worthwhile examining whether the different parameters are appropriately accounted for in both methods. Some plots can be drawn which are not reproduced here; only the main conclusions are reported.

4.1. Comparison between test results and tension field approach.

A first question is dealing with the representativeness of these results. Therefore, diagrams of V_{ex}/V_u have been drawn against the relevant parameters: web depth, web thickness, web slenderness and material flange yield stress. With slenderness ranging from 60 up to 400 and yield stress up to 700 N/mm^2, it can be concluded that the data of the test specimens cover the range of values met in practice. In addition, because the average value and the standard deviation are approximately constant in the full range of these parameters, it may be concluded that the models are independant of these parameters.

The influence of the flange stiffness has been studied by plotting the value of the ratio V_{ex}/V_{pl} against the stiffness parameter :

$$V_f = I_f \cdot 10^6/a^3 t_w$$

where I_f is the moment of inertia of the flange. It has been observed that, for a specified web slenderness, the higher reduced experimental shear capacities are got for stiffer flanges. Then the question arises to know whether this influence of the flange stiffness is appropriately accounted for in the model. A slight tendency to have a lower average value is observed when the flange stiffness increases. A similar plot against the aspect ratio demonstrates a slight tendency to a larger average value when the aspect ratio increases. It can be concluded that the spacing of the transverse stiffeners is the reason of both tendencies ; indeed α is proportional to the panel length a while v_s is inversely proportional to a. To choose a slope of the tension band that is not depending on the spacing of the transverse stiffeners appears so as a simplification and is not rigourously justified.

Another parameter the influence of which would be worthwhile being examined is the critical shear contribution to the ultimate shear capacity.   The corresponding diagram shows that there is no dependency.

4.2. Comparison between test results and simple postcritical approach.

The plot of the reduced shear capacity against the plate slenderness shows that the model for τ_u, i.e. $\tau_u/\tau_{yw} = 0.9 /\overline{\lambda}_p$ is conservative.   The dependency of the experimental to theoretical ultimate shear ratio versus the aspect ratio shows a slight tendency to increase when α increases.

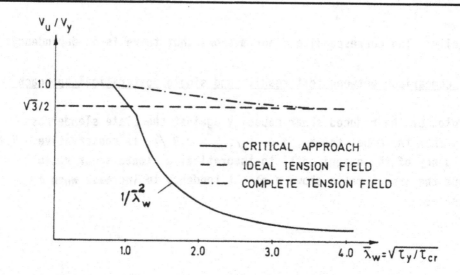

Fig. 1 - Ultimate shear stress according to different tension field
         theories.

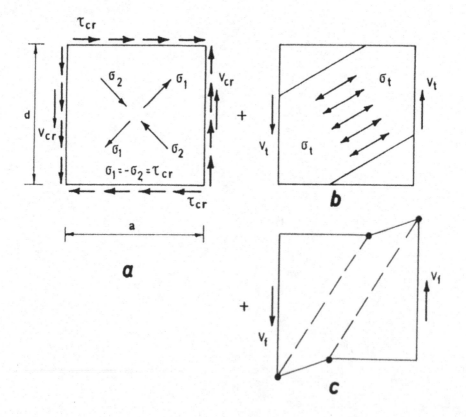

Fig. 2 - The several contributions to the ultimate shear resistance.

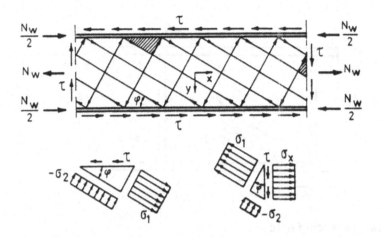

Fig. 3 - Components of the principal stresses at the boundary of a plate
         panel.

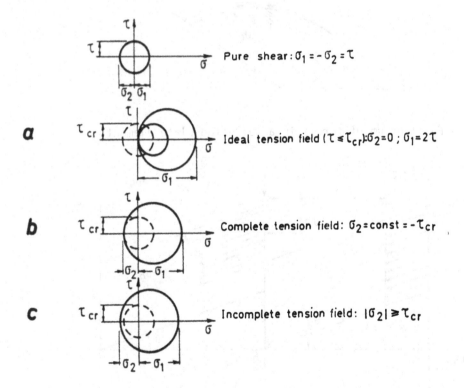

Fig. 4 - MOHR circles : a) ideal tension field ; b) complete tension
         field ; c) incomplete tension field.

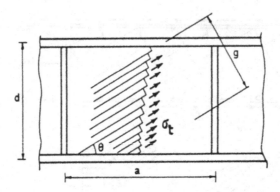

Fig. 5 - Ideal tension field.

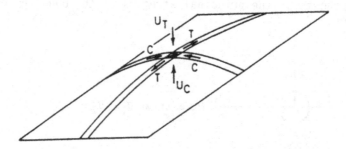

Fig. 6 - Equilibrium of forces in a deflected plate element.

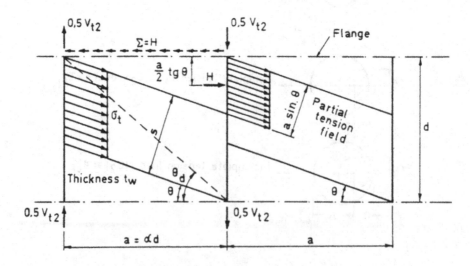

Fig. 7 - BASLER model for ultimate shear resistance.

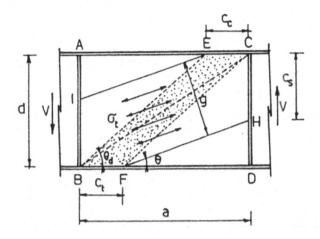

Fig. 8 - CARDIFF model for ultimate shear resistance.

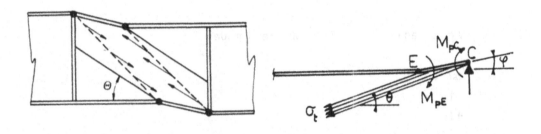

Fig. 9 - Combined failure mechanism.   Fig. 10 - Plastic mechanism in
                                                  the flange.

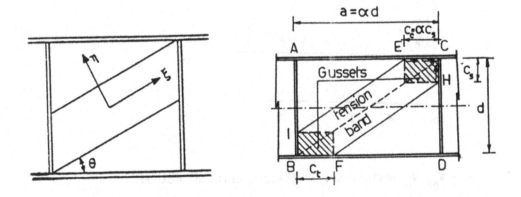

Fig. 11 - Stress state in the web plate. Fig. 12 - DUBAS model.

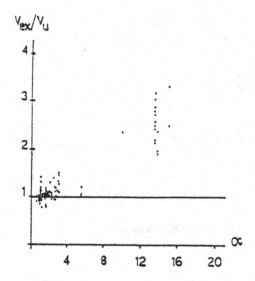

Fig. 13 - V_{ex}/V_u against α : Tension field model

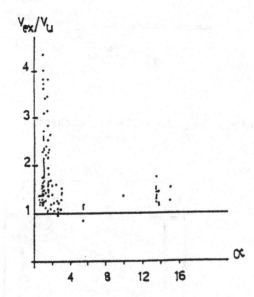

Fig. 14 - V_{ex}/V_u against α : Simple postcritical method.

CHAPTER 4
UNSTIFFENED AND TRANSVERSELY STIFFENED PLATE GIRDERS

1. INTRODUCTION.

For girders with large loads and long spans, it is often economical to use plate girders or truss girders. Plate girders have higher dead weight but lower fabrication costs. In more current conditions, plate girders may replace rolled sections when the reduced material price compensates a higher fabrication cost. The components of a plate girder look similar to those of a wide-flange section; they are termed as the top and bottom flanges, the web plate and the longitudinal fillet welds connecting the web to the flanges. These components, and more especially the web plate, are often slender, in contrast with universal beam sections, the proportions of which are governed by rolling requirements. The reduced web thickness allows for a higher girder depth giving lower flange forces and consequently smaller flange sections for a known bending moment.

The main function of a slender web is to resist shear forces; its contribution to bending capacity is very low. In this respect, a plate girder behaves nearly as a truss girder, the web plate playing the role of web members. The use of plate girders with slender longitudinally unstiffened webs is rendered possible because of the postcritical behaviour due to the longitudinal bending stress redistribution or to the tension field action in shear.

Because of the web slenderness, allowance must be made for the effects of local buckling when determining the moment resistance and/or the compression resistance of a plate girder cross-section, which belongs therefore to Class 4 sections. A first consequence is: for frames or girders made with fabricated slender I cross-sections, only an elastic global analysis is allowed for; a second one is: the cross-sections themselves should be checked elastically.

2. DESIGN CRITERIA FOR I PLATE GIRDER CROSS-SECTIONS.

Checking the ultimate carrying capacity of an I plate girder cross-section requires that local instability phenomena be accounted for in the design. Because instability is initiated by compressive stresses, the compression flange as well as the web plate are likely to be affected by this phenomenon.

When the plate girder is bent about its strong axis, - that is the regular loading condition - one flange is mainly subject to compression. When too slender, this flange can perish prematurously by *torsional buckling* in the elastic range, with the result of a possible limit for the maximum compressive stress.

The web experiences mainly shear stresses and subsidiarily direct compressive or tensile stresses; principal compressive stresses exist in some directions and are liable to initiate *web buckling*.

Bending curvature induces transverse stresses, i.e. stresses in the plane of the web but perpendicular to the longitudinal axis of the girder, which may cause a *vertical buckling of the flanges into the web* because of the slenderness of the web.

Last, the web may also buckle due to *patch loading*, when one or both flange(s) is (are) subject to point loads or loads acting over a rather short distribution length.

When members with slender sections are used as columns or as beam-columns, both flanges may experience compression with the result of possible plate buckling of both flanges.

Most of these design criteria are examined herebelow. For this purpose, it is assumed that :
a) the section is composed of steel plates having similar steel grades, thus excluding highly hybrid sections ;
b) the section is symmetrical with respect to the plane of the web ;
c) the section is bent about its strong axis ;
d) the flange widths are small compared to the girder span, so that shear lag is not at all significant ;
e) the flange thicknesses are small compared to the web depth, so that the gradient of direct stresses in the flanges can be neglected and reference be made to the stress at the flange centroids ;
f) the limit direct stresses are not governed by member instability, the latter being checked independently of the resistance of cross-sections.

3. CONCEPT OF EFFECTIVE WIDTH.

In contrast to rolled sections, plate girders with slender webs cannot exhibit neither the full plastic moment capacity, nor even the elastic yield moment. That is due to the local buckling of the web.

In accordance with the linear plate buckling theory, the bending capacity should be $M_{cr} = \sigma_{cr} W$, where W is the elastic section modulus. The ultimate bending strength is actually higher as a result of the direct stress redistribution which takes place in the postcritical range; a lower bound corresponds to the bending moment which causes yield in flange(s) - provided these are at least of Class 3 -, the contribution of the web being fully ignored, similarly to whar occurs in a truss.

To make the necessary allowance for reductions in resistance due to the
effects of local buckling, the most simple and most widely known manner
is to use the concept of effective width of compression elements.
*Compression elements include every element of a cross-section which is
either totally or partially in compression,* due to axial force or
bending moment, under the load combination considered.

3.1. Effective width of a perfect plate subject to uniaxial uniform compression.

Let us consider a rectangular plate which is assumed pefectly flat,
simply supported on its four edges and subject to longitudinal uniform
compression (fig. 1). The elastic plate critical buckling stress is :

$$\sigma_{cr} = k_\sigma \ [\pi^2 E/12(1 - \nu^2)](t/b)^2 \tag{1}$$

When the plate buckles, the bow effect results in non-uniform direct
stress distribution because the longitudinal fibres have no more the
same stiffness in the postbuckling range. The ultimate carrying
capacity of such a plate is generally assumed to be achieved when the
maximum stress σ_{max} (which occurs at both longitudinal edges) reaches
the material yield stress f_y. The ultimate compressive load is obtained
as :

$$N_u = 2 \int_0^{b/2} \sigma(y)tdy = b \ \bar{\sigma}_u \tag{2}$$

$\bar{\sigma}_u$ being the average stress at collapse. The use of equation (2) should
require the knowledge of the shape of the stress distribution at the
ultimate limit state. In order to by-pass this difficulty, von KARMAN
introduced the concept of an effective width b_e, smaller than b, such
that, when subject to uniform direct stress distribution of magnitude
$\sigma_{max} = f_y$, the width b_e transmits the same ultimate load N_u ; therefore:

$$N_u = b_e \ f_y \tag{3}$$

In addition, von KARMAN defined the effective width b_e as that of a
fictitious simply supported plate, which should have the same thickness

and aspect ratio as the actual plate and buckle for a plate critical
buckling stress equal to f_y. Accordingly, one can write :

$$\sigma_{cr} \ (b_e) = k_\sigma \ [\pi^2 E/12(1 - \nu^2)] \ (t/b_e)^2 = f_y \tag{4}$$

Comparing (1) and (4) yields :

$$b_e/b = \sqrt{\sigma_{cr}/f_y} \ \ngtr 1 \tag{5}$$

or account taken of (2) and (3) :

$$\bar{\sigma}_u = \sqrt{\sigma_{cr} \ f_y} \tag{6}$$

The normalized plate slenderness $\bar{\lambda}_p$ writes, with σ_{cr} given by (1) :

$$\bar{\lambda}_p = \sqrt{f_y/\sigma_{cr}} = (b/t)/(28,4 \ \epsilon \ \sqrt{k_\sigma}) \tag{7}$$

where k_σ is the buckling factor depending, in general terms, on the
aspect ratio, the boundary conditions and the stress ratio ψ and
$\epsilon = \sqrt{235/f_y}$ is a factor reflecting the steel grade.
The ratio b_e/b is obtained as the ratio between the ultimate load
$N_u = b_e \ f_y$ and the squash load $N_{pl} = b \ f_y$, termed as the normalized
ultimate load $\bar{N}_p = N_u/N_{pl}$. Thus the so-called von KARMAN formula for
plate buckling can be written:

$$\bar{N}_p = 1/\bar{\lambda}_p \ \ngtr 1 \tag{8}$$

and can be compared to EULER formula for column buckling $\bar{N} = 1/\bar{\lambda}^2$. Both
curves start at $\bar{N}$ or $\bar{N}_p = 1$ at $\bar{\lambda}$ or $\bar{\lambda}_p = 1$ and decrease when normalized
slenderness increase (fig. 2). The range between both curves represents
the postcritical strength reserve.

When the maximum calculated compressive stress σ_{max} in the plate element
is likely to remain significantly lower than the material yield stress
f_y, economy should command to compute the plate slenderness $\bar{\lambda}_p$ of this
element by substituting σ_{max} for f_y.

3.2. Effective width of an imperfect plate subject to uniaxial uniform compression.

Due to unavoidable imperfections - out-of-flatness and possible residual stresses - plate buckling is precipitated because of non-linear geometric effects and prematurous yielding. von KARMAN formula must be slightly modified in view to reflect the behaviour of an actual fabricated plate :

$$\overline{N}_p = b_e/b = [1 - 0.22/\overline{\lambda}_p]/\overline{\lambda}_p \nleq 1 \tag{9}$$

Formula (9) applies only when $\overline{\lambda}_p > 0.673$; for $\overline{\lambda} \leq 0.673$, $\overline{N}_p = 1$ which means a full efficiency.

3.3. Generalisation of effective width formula.

Though the effective width formula (9) was calibrated with test results on plates subject to uniaxial uniform compression, there is some evidence it can be extended to any kind of linear direct stress distribution provided - that the normalized plate slenderness be defined accordingly, i.e. by introducing the buckling coefficient k_σ related to the stress ratio $\psi = \sigma_2/\sigma_1$; σ_1 is the largest compressive stress and σ_2 is the lowest compressive or maximum tensile stress.

A similar formula is also used for shear loading provided that $\overline{\lambda}_p$ be defined in terms of shear stresses, i.e. $\overline{\lambda}_p = \sqrt{\tau_y/\tau_{cr}}$.

When there is a sign reversal in the direct stress distribution over the width b, only the portion b_c subject to compressive stresses has to be considered whereas the portion $(b - b_c)$ subject to tensile stresses is fully effective. Therefore expression (9) must be applied by substituting b_c for b:

$$b_e/b_c = [1 - 0.22/\overline{\lambda}_p]/\overline{\lambda}_p \nleq 1 \tag{10}$$

Once the effective width b_e of the compression zone is determined, it must be allocated by appropriate portions b_{e1} and b_{e2} respectively adjacent to the most and the least compressed fibres of the compression zone, when the plate is supported on both longitudinal edges (*internal compression elements*). These portions are however limited to $b_{e1} \ngtr 0,4\, b_e$ and $b_{e2} \ngtr 0.6\, b_e$. For *outstand compression elements*, which have one simply supported longitudinal edge and a free one, the whole effective width of the compression zone is adjacent either to the least compressed fibre or to the most compressed fibre, according as the stress gradient.

The procedure is summarized in Table 1 for internal compression elements and in Table 2 for outstand compression elements.

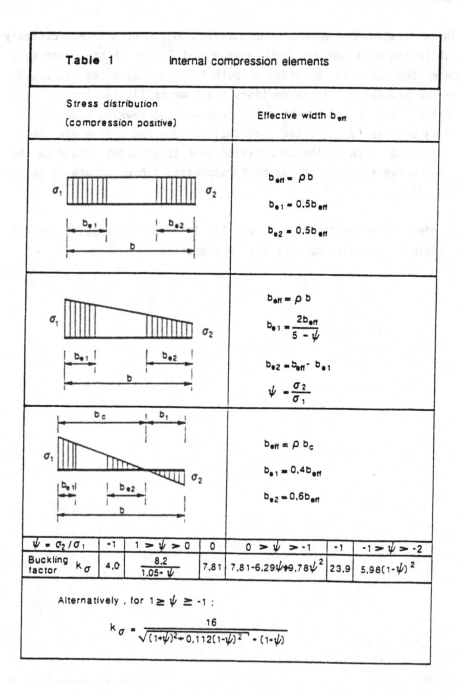

Table 1   Internal compression elements

Stress distribution (compression positive)	Effective width b_{eff}
	$b_{eff} = \rho b$ $b_{e1} = 0.5 b_{eff}$ $b_{e2} = 0.5 b_{eff}$
	$b_{eff} = \rho b$ $b_{e1} = \dfrac{2 b_{eff}}{5 - \psi}$ $b_{e2} = b_{eff} - b_{e1}$ $\psi = \dfrac{\sigma_2}{\sigma_1}$
	$b_{eff} = \rho b_c$ $b_{e1} = 0.4 b_{eff}$ $b_{e2} = 0.6 b_{eff}$

$\psi = \sigma_2/\sigma_1$	+1	$1 > \psi > 0$	0	$0 > \psi > -1$	-1	$-1 > \psi > -2$
Buckling factor k_σ	4,0	$\dfrac{8.2}{1.05 + \psi}$	7,81	$7.81 - 6.29\psi + 9.78\psi^2$	23,9	$5.98(1-\psi)^2$

Alternatively , for $1 \geq \psi \geq -1$:

$$k_\sigma = \frac{16}{\sqrt{(1+\psi)^2 + 0,112(1-\psi)^2} + (1+\psi)}$$

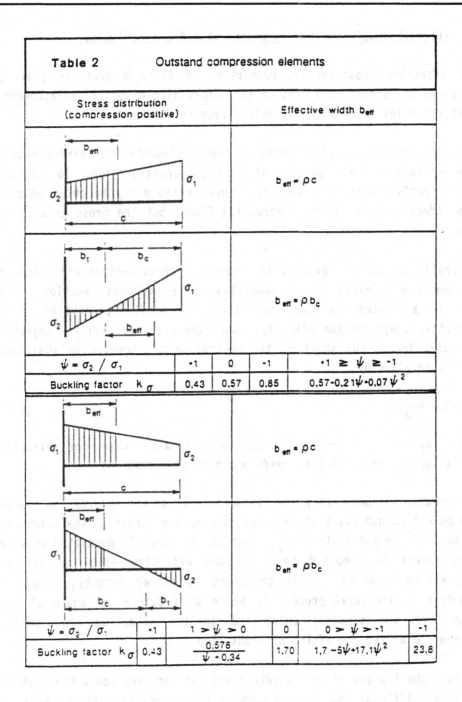

Table 2   Outstand compression elements

Stress distribution (compression positive)			Effective width b_{eff}	

$b_{eff} = \rho c$

$b_{eff} = \rho b_c$

$\psi = \sigma_2 / \sigma_1$	+1	0	-1	$+1 \geq \psi \geq -1$
Buckling factor k_σ	0,43	0,57	0,85	$0,57 - 0,21\psi + 0,07\psi^2$

$b_{eff} = \rho c$

$b_{eff} = \rho b_c$

$\psi = \sigma_2 / \sigma_1$	+1	$1 > \psi > 0$	0	$0 > \psi > -1$	-1
Buckling factor k_σ	0,43	$\dfrac{0,578}{\psi + 0,34}$	1,70	$1,7 - 5\psi + 17,1\psi^2$	23,8

4. EFFECTIVE CROSS-SECTION PROPERTIES OF CLASS 4 SECTIONS.

The effective cross-section properties of Class 4 sections shall be based on the effective widths of the compression elements, in accordance with the rules given in the previous section.

To determine the effective width of flange elements, the stress ratio ψ may be based on the properties of the gross cross-section. To determine the effective width of a web, the stress ratio ψ may be obtained using the effective area of the compression flange but the gross area of the web.

Generally the neutral axis of the effective cross-section will shift by an amount e compared to the neutral axis of the gross section. This must be accounted for when calculating the bending properties of the effective cross-section (fig. 3). When the cross-section is subject to an axial force, the shift of the neutral axis generates an additional moment ΔM given by :

$$\Delta M = Ne_N \tag{11}$$

where e_N is the shift of the neutral axis when the effective cross-section is subject to uniform compression (fig. 4).

For greater economy, the plate slenderness $\overline{\lambda}_p$ of any compression element may be determined using the maximum compressive stress in that element instead of the yield stress f_y. Such an allowance is permitted provided this stress be computed based on the effective width of all the compression elements. This procedure is rather lengthy; indeed it requires an iterative process in which ψ is determined again at each step from the stresses calculated in the effective cross-section determined at the end of the previous step.

Because the flanges of an I girder cross-section contribute the most the flexural stiffness and section moduli, the compression flange shall be proportioned to be fully effective ; therefore the normalized plate slenderness of the compression flange ought not exceed 0.673. Because

4. EFFECTIVE CROSS-SECTION PROPERTIES OF CLASS 4 SECTIONS.

The effective cross-section properties of Class 4 sections shall be based on the effective widths of the compression elements, in accordance with the rules given in the previous section.

To determine the effective width of flange elements, the stress ratio ψ may be based on the properties of the gross cross-section. To determine the effective width of a web, the stress ratio ψ may be obtained using the effective area of the compression flange but the gross area of the web.

Generally the neutral axis of the effective cross-section will shift by an amount e compared to the neutral axis of the gross section. This must be accounted for when calculating the bending properties of the effective cross-section (fig. 3). When the cross-section is subject to an axial force, the shift of the neutral axis generates an additional moment ΔM given by :

$$\Delta M = Ne_N \tag{11}$$

where e_N is the shift of the neutral axis when the effective cross-section is subject to uniform compression (fig. 4).

For greater economy, the plate slenderness $\bar{\lambda}_p$ of any compression element may be determined using the maximum compressive stress in that element instead of the yield stress f_y. Such an allowance is permitted provided this stress be computed based on the effective width of all the compression elements. This procedure is rather lengthy; indeed it requires an iterative process in which ψ is determined again at each step from the stresses calculated in the effective cross-section determined at the end of the previous step.

Because the flanges of an I girder cross-section contribute the most the flexural stiffness and section moduli, the compression flange shall be proportioned to be fully effective ; therefore the normalized plate slenderness of the compression flange ought not exceed 0.673. Because

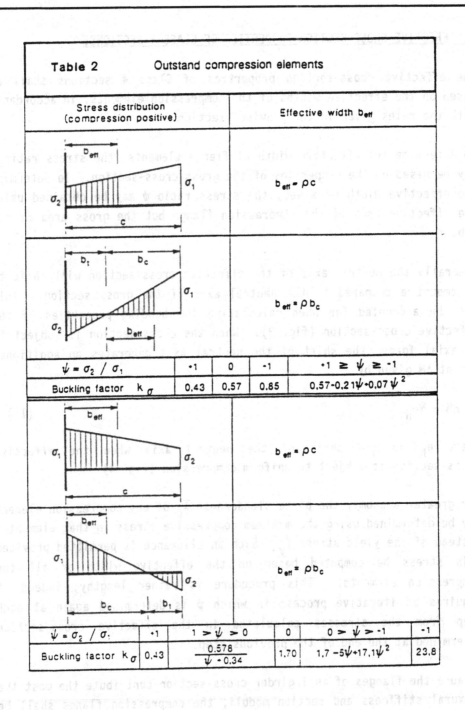

Table 2	Outstand compression elements			
Stress distribution (compression positive)		**Effective width b_eff**		
(diagram) σ_2, σ_1, b_{eff}, c		$b_{eff} = \rho c$		
(diagram) b_t, b_c, σ_1, σ_2, b_{eff}		$b_{eff} = \rho b_c$		

$\psi = \sigma_2 / \sigma_1$	+1	0	-1	$+1 \geq \psi \geq -1$
Buckling factor k_σ	0,43	0,57	0,85	$0,57 - 0,21\psi + 0,07\psi^2$

(diagram) b_{eff}, σ_1, σ_2, c		$b_{eff} = \rho c$			
(diagram) b_{eff}, σ_1, σ_2, b_c, b_t		$b_{eff} = \rho b_c$			

$\psi = \sigma_2 / \sigma_1$	+1	$1 > \psi > 0$	0	$0 > \psi > -1$	-1
Buckling factor k_σ	0,43	$\dfrac{0,578}{\psi + 0,34}$	1,70	$1,7 - 5\psi + 17,1\psi^2$	23,8

the buckling factor for an outstand compression element is k_σ = 0.43, this limit would correspond to a width-to-thickness ratio c/t_f = 12.5 and prevents from torsional buckling.

5. RESISTANCE OF CLASS 4 CROSS-SECTIONS.

5.1. Tension.

The design tension resistance N_{tRd} of a Class 4 cross-section is the smaller of :
a) the design plastic resistance of the gross section :

$$N_{pl,Rd} = Af_y/\gamma_{MO} \tag{12.a}$$

b) the design ultimate resistance of the net section at bolt holes:

$$N_{u,Rd} = 0,9 \; A_{net} \; f_u/\gamma_{M2} \tag{12.b}$$

where A is the cross-sectional area of the gross section, A_{net} that of the net section and f_u the ultimate tensile strength of the material. Coefficients γ_{MO} and γ_{M2} are partial safety factors ; they are respectively γ_{MO} = 1,00 and γ_{M2} = 1,25.

When a ductile behaviour is required, N_{pl} should be less than N_u, wherefrom:

$$0,9 \; (A_{net}/A) \geq (f_y/f_u) \; (\gamma_{M2}/\gamma_{MO}) \tag{13}$$

Each section shall fulfil :

$$N_{tSd} \leq N_{tRd} \tag{14}$$

where N_{tSd} is the design value of the tensile force.

5.2. Compression.

The design compression resistance of a Class 4 cross-section is :

$$N_{cRd} = A_{eff} \, f_y/\gamma_{M1} \tag{15}$$

where A_{eff} is the effective area of the cross-section determined in accordance with previous Sections 3 and 4 and $\gamma_{M1} = 1,10$. Usually fastener holes need not be allowed for in compression members.

When the Class 4 cross-section is not symmetrical, there is an additional moment ΔM due to the eccentricity e_N of the neutral axis of the effective section compared to that of the gross section ; then use must be made of the method for cross-sections subject to combined compression and bending.

Each section shall fulfil:

$$N_{cSd} \leq N_{cRd} \tag{16}$$

where N_{cSd} is the design value of the compressive force.

5.3. Bending.

In the absence of shear force, the design moment resistance of Class 4 cross-section *without bolt holes* is :

$$M_{Rd} = W_{eff} \, f_y/\gamma_{M1} \tag{17}$$

where W_{eff} is the effective section modulus computed in accordance with previous Sections 3 and 4 and $\gamma_{M1} = 1,10$.

Usually fastener holes in the compression zone of the cross-section need not be allowed for. Fastener holes in the tension flange need not be allowed for provided that the condition (13) on ductile behaviour be fulfilled ; if not, a reduced tensile flange area, derived from this condition, must be assumed for the computation of W_{eff}. Fastener holes in the tension zone of the web need not be allowed for provided that the condition (13) is satisfied for the complete tension zone comprising the tension flange plus the tension zone of the web.

Each cross-section shall fulfil :

$$M_{Sd} \leq M_{Rd} \qquad (18)$$

where M_{Sd} is the design value of the bending moment.

5.4. Bending and axial (compressive) force.

In the absence of shear force, Class 4 cross-sections are satisfactory when the stresses based on the effective widths of the compression elements satisfy the appropriate criterion, i.e. the elastic criterion. For cross-sections without holes, this criterion writes :

$$(N_{Sd}/A_{eff}) + (M_{Sd} + N_{Sd}\, e_N)/W_{eff} \leq f_y/\gamma_{M1} \qquad (19)$$

where A_{eff} = the effective area of the cross-section when subject to uniform compression ;

W_{eff} = the effective section modulus of the cross-section when subject only to moment about the relevant axis ;

e_N = the shift of the relevant neutral axis when the cross-section is subject to uniform compression.

5.5. Important remark.

In addition to the requirements given in this clause, the buckling resistance of the member shall also be checked according to appropriate clauses. Where necessary, frame stability should also be verified.

6. SHEAR BUCKLING RESISTANCE.

The limit depth-to-thickness ratio of the web, beyond which the shear resistance is governed by shear buckling, is derived from the condition of full efficiency $\overline{\lambda}_p > 0.673$, wherefrom :

$d/t_w > 30\, \epsilon\, \sqrt{k_\tau}$	for transversely stiffened web	(20.a)
$d/t_w \geq 69\, \epsilon$	for unstiffened web	(20.b)

where k_τ is the buckling coefficient for shear (see chapter 3). The limit is usually exceeded for webs of plate girders, the slenderness of which is generally larger, and sometimes much larger, than 100.

The shear buckling resistance of a slender web is presently evaluated by accounting for the postcritical strength reserve. It depends on the depth-to-thickness ratio d/t_w, on the spacing a of the possible transverse web stiffeners and on the possible anchor of tension fields to end stiffeners and flanges. The anchor provided by flanges is reduced by direct stresses in these flanges, due to bending moment and/or axial load acting on the girder.

In the present state of knowledge, it may be recommended to verify the shear buckling resistance using :
a) the simple postcritical method for unstiffened webs - provided that the web has transverse stiffeners at the supports - or for webs fitted with largely spaced transverse stiffeners (spacing-to-depth ratio a/d larger than 3);
b) the tension field method for closely spaced transverse stiffeners ($1 \leq a/d \leq 3$) provided adjacent web panels or stiff end posts provide anchor to tension fields.

The design value of the shear force V_{Sd} in each cross-section shall fulfil:

$$V_{Sd} \leq V_{Rd} \tag{21}$$

where V_{Rd} is the design buckling resistance.

The background of both methods is developed in previous chapter. It results however from a statistical evaluation of the test results that the appropriate specified partial safety factor $\gamma_{M1} = 1.10$ is reached provided that, the design buckling resistance be written as follows :
a) for the simple postcritical method :

$$V_{Rd} = dt_w \, \tau_u/\gamma_{M1} \tag{22}$$

b) for the tension field method :

$$V_{Rd} = (V_{cr} + 0.9\ V_t)/\gamma_{M1} \qquad\qquad (23)$$

with :

$$V_{cr} = dt_w\ \tau_{cr}\ \text{(or possibly } dt_w\ \tau_{cr}^{red}) \qquad\qquad (24)$$

$$V_t = g\ t_w\ \sigma_t\ \sin\theta \qquad\qquad (25)$$

End panels (fig. 5) are worthwhile being considered with a special care; indeed the web panels of a transversely stiffened web, which are adjacent to end posts, are not counterbalanced by adjacent web panels. Unless a suitable end post is supplied to anchor the tension field, end panels should be designed using the simple postcritical method. An end post which resists the tension field force is termed *stiff end post* and should satisfy the following criterion:

$$0.5\ H_\sigma\ c_s \le M_{pC}^* + M_{pH}^* \qquad\qquad (26)$$

H_σ is the horizontal component of the part of the resultant in the tension band, which anchors to the end post :

$$H_\sigma = \sigma_t\ (c_s\ \cos\theta - e\ \sin\theta)\ t_w\ \cos\theta \qquad\qquad (27)$$

where e is the distance between the flanges of the H section constituting the end post; when a single plate end post is used, e is equal to zero.
The distance c_s between points C and H is computed as :

$$c_s = d - (a - c_t)\ \text{tg}\ \theta \qquad\qquad (28)$$

Criterion (26) is derived from the principle of virtual work applied to a plastic beam mechanism in the end post, with plastic hinges at C - top of the end post - and H; it means that the end post is not allowed to reach its ultimate limit state before the end web panel does it. M_{pC}^* and M_{pH}^* are the plastic moments at points C and H, account taken of the influence of axial force existing at C and H respectively (see herebelow).

The shear buckling resistance of an end panel fitted with a stiff end
post is computed as for an intermediate panel. However, because there
is no possibility for self-equilibrium of the forces in the tension
band, the bending capacities M_p^* used in the expression of c_c must
account for axial forces N_f and N_s existing in the flange and in the
post respectively, as a result of the anchor of the tension band. At
point E, N_f is the horizontal component of the resultant force in the
tension band :

$$N_{f,E} = g \ t_w \ \sigma_t \ \cos \theta \tag{29}$$

At point C, the bending capacity is the lesser of two values: either the
plastic bending moment of flange, reduced by the axial force :

$$N_{f,C} = c_s \ t_w \ \sigma_t \ \cos^2 \theta \tag{30}$$

or the plastic bending moment of the end post, reduced by the axial
force :

$$N_{s,C} = c_c \ t_w \ \sigma_t \ \sin^2 \theta \tag{31}$$

At point H, the axial force in the end post is :

$$N_{s,H} = V_{Sd} - (d - c_s) \ t_w \ \tau_{cr} \tag{32}$$

When a single plate end post is used, the reduced plastic moments write:

$$M_{pE}^* = 0.25 \ b_f \ t_f^2 \ f_{yf} \ [1 - (N_{f,E}/b_f \ t_f \ f_{yf})^2] \tag{33}$$

$$M_{pC}^* = \text{Min} \begin{bmatrix} 0.25 \ b_f \ t_f^2 \ f_{yf} \ [1 - (N_{f,C}/b_f \ t_f \ f_{yf})^2] \\[2ex] 0.25 \ b_s \ t_s^2 \ f_{ys} \ [1 - (N_{s,C}/b_s \ t_s \ f_{ys})^2] \end{bmatrix} \tag{34.a,b}$$

$$M_{pH}^* = 0.25 \ b_s \ t_s^2 \ f_{ys} \ [1 - (N_{sH}/b_s \ t_s \ f_{ys})^2] \tag{35}$$

where b_f : flange width ;
     t_f : flange thickness ;
     b_s : single plate end post width ;
     t_s : single plate end post thickness ;

f_{yf} : yield strength of the flange material ;
f_{ys} : yield strength of the stiffener material.

A twin stiffener type of end post may be used as an alternative to the single plate type. It shall be noticed that the expressions of the reduced plastic moments of the end post shall have to be changed accordingly.

The search for the shear buckling resistance of end panels is an iterative procedure; indeed the anchor length c_c depends on the unknown σ_t, which affects directly M_p^* values. As a first approximate , it is often assumed that the effect of axial force can be disregarded as far as it does not exceed 30 % of the squash load of the flange.

7. DESIGN OF WELDS.

The forces used to check the web-to-flange welds should be compatible with the stress fields in the web panels according to the method used to determine the shear buckling resistance. The design of the web-to-stiffener welds should also be consistent with the design assumptions for the web panels. The welds connecting the end post to the top flange should be designed to resist M_{pC}^*, H_σ and N_{sC}.

8. INTERACTION BETWEEN SHEAR, BENDING AND AXIAL FORCE.

As the anchor lengths of the tension band to the flanges depend on the plastic bending resistance of these flanges, the theoretical shear resistance should depend on the direct stress level in the flange, i.e. on the bending moment and/or axial force. Computations demonstrate however that this interaction remains rather small; it is accepted that the design shear resistance of the web needs not be reduced to allow for the moment and/or axial force, provided that the flanges can resist the whole of the design values of the bending moment and axial force.

When the latter condition is not fulfilled, the web has to contribute supporting direct stresses. That results in changes of the stress state over the web depth and from point to point in the web, on the one hand,

and in a prematurous yielding of the web because of these additional
direct stresses, on the other hand. The interaction between shear and
direct stresses must thus be expected much more pronounced in this
range; that is indeed observed in the plot of the normalized
experimental shear resistance against the bending moment for
transversely stiffened webs.

Accordingly the interaction problem is summarized as follows (fig.
6.a,b):

- The cross-section may be assumed to be satisfactory when both
  following criteria are fulfilled :

$$M_{Sd} \leq M_{fRd} \tag{36}$$

and :

$$V_{Sd} \leq V_{Rd}^o \tag{37}$$

where M_{fRd} is the design plastic moment resistance of a cross-section
consisting in the flanges only (taking possibly account of the effective width of the compression flange), allowing for the presence of
the design axial force in the relevant flange. The design shear
resistance V_{Rd} is obtained from (22) or (23) depending on which one
of the tension field method or the simple postcritical method is
applicable; when use is made of (23),V_t is computed conservatively
with $c_c = c_t = 0$.

- Provided that V_{Sd} does not exceed 50 % of V_{Rd}^o, the design resistance
  of the cross-section to bending moment and axial force need not be
  reduced to allow for the shear force.

- When V_{Sd} exceeds 50 % of V_{Rd}^o, the following criterion should be
  satisfied :

$$M_{Sd} \leq M_{fRd} + (M_{NRd} - M_{fRd}) [1 - (2V_{Sd}/V_{Rd}^o - 1)^2] \tag{38}$$

where M_{NRd} is the reduced plastic resistance moment allowing for the
axial force.

- The cross-section is able to resist any combination of internal
  forces (M_{Sd}, V_{Sd}) when the representative point lies within the
  boundaries of the interaction curve defined accordingly.

9. FLANGE INDUCED BUCKLING.

The web of a plate girder does not contribute very significantly the bending resistance and is mainly designed for shear resistance with due account taken of the postcritical strength reserve; that results in possible very thin webs. Attention must however be paid to the danger of a collapse mode of too thin webs, which may buckle because of the transverse compression induced by the bending curvature.

Let us indeed examine the deformed configuration of a plate girder subject to bending. When curved, the flange which is subject to compression (tension) is in equilibrium provided radial transverse direct stresses develop at the web-to-flange junction (fig. 7); when the magnitude of the transverse compressive stresses is sufficient to produce buckling of the web, the compression flange looses the support provided by the web and may buckle as a compressed strut. This collapse mode, which is initiated by the web buckling, is termed *flange induced buckling*.

To prevent such a mode of collapse, the web slenderness d/t_w cannot exceed a limit value obtained as follows. Let us again assume a girder subject to pure bending ; the corresponding curvature results in uniformly distributed transverse direct stresses and constitutes the most detrimental loading case. Let us cut a vertical strip of unit width in the web, that is assumed to be pin-ended and to have an initial sinusoïdal out-of-straightness w_0 (fig. 8.a). Because of the axial load N, this strip shall shorten. The contributions to the shortening are due to the elastic linear behaviour on the one hand, and to the non-linear geometric behaviour (bow effect) on the other hand; it writes thus :

$$\delta = Nd/Et_w + 0.5 \int_0^d w'^2 \, dx \qquad (39)$$

The total deflection $w = w_0 + w_{add}$ is obtained as :

$$w(x) = [w_0/(1 - N/N_E)] \sin (\pi x/d) \qquad (40)$$

where N_E is the Eulerian column buckling load so that δ writes :

$$\delta = Nd/Et_w + (\pi^2/4d) [w_0/(1 - N/N_E)]^2 \tag{41}$$

The plot of δ against N/N_E shows (fig. 8.b) that for $w_0 \approx 0$, 1 t_w - which is a realistic amount of initial imperfection - the behaviour is quasi linear up to $N/N_E = 0.65$; it may be considered that the vertical buckling of the flange into the web is prevented as far as the response of the web to vertical load is linear, i.e. :

$$N/N_E \approx \sigma_v/\sigma_E < 0.65 \tag{42}$$

where the vertical direct stress in the unit web strip is deduced from the compression flange equilibrium according to :

$$\sigma_v = \sigma_f A_{fc}/Rt_w = 2 \sigma_f A_{fc} \epsilon_f/A_w \tag{43}$$

where :

A_{fc} : cross-sectional area of the compression flange ;

$A_w = dt_w$: cross-sectional area of the web ;

σ_f : axial stress in the compression flange ;

ϵ_f : axial strain in the compression flange ;

R : radius of bending curvature.

$$\sigma_E = [\pi^2 E/12 (1 - \nu^2)] (t_w/d)^2 \tag{44}$$

At the ultimate state σ_f is likely to reach the design yield stress f_{yd}, wherefrom $\epsilon_f = f_{yd}/E$. Account taken of (43) and (44), expression (42) yields :

$$d/t_w \le 0.55 (E/f_{yd}) \sqrt{A_w/A_{fc}} \tag{45}$$

This limit is too severe when the stress in the flanges is far from reaching the material yield stress. That is especially the case in the vicinity of the end supports, where shear is dominant while bending vanishes. Then it is allowed to substitute the actual design stress in the flange for the yield stress f_{yd}.

For the previous reasoning, it was assumed that there are no transverse stiffeners, which could constitute supports for the flange and therefore allow for larger slenderness limits. Such a beneficial effect is generally disregarded because small for usual spacings of transverse stiffeners.

When the girder is curved in elevation, with the compression flange on the concave side, the limit slenderness should be modified :

$$d/t_w = 0.55 \ (E/f_{yd}) \ \sqrt{A_w/A_{f,c}} \ / \ \sqrt{1 + dE/3rf_{yd}} \qquad (46)$$

where r is the radius of initial curvature of the compression flange.

Coefficient 0.55, which appears in (45) and (46) is only valid for sections which are checked elastically. When direct stress redistribution is allowed and when larger rotation capacity is required, this value must be reduced. More especially it becomes 0.3 for Class 1 sections and 0.4 for Class 2 sections.

10. WEB CRIPPLING.

It is often uneconomic to provide transverse stiffeners under concentrated loads; sometimes that is impossible, especially when a bridge girder is built by the launching erection procedure.

The problem of the postcritical behaviour of thin webs subjected to such a patch loading is still being considered with a view to establish a general expression of the patch load resistance, that would account for the influence of the numerous parameters.

The state of the art in this respect is presented elsewhere.

11. INFLUENCE OF POSTCRITICAL OUT-OF-PLANE DISPLACEMENTS.

Redistribution of longitudinal direct stresses as well as the formation of a tension field in the web require that appreciable out-of-plane displacements have occured. These might have detrimental influences on the global behaviour of plate girders with slender webs.

The serviceability criteria linked to the magnitude of the out-of-plane displacements are related to psychological considerations, i.e. to

possible limits for visible buckles which might induce fear regarding
the safety of the structure, and to potential snap-through phenomena,
when initial out-of-flatness of the web is in sympathy with the plate
buckling mode.   Such criteria are presently not well defined and are
seldom governing for the design.

On another hand, out-of-plane displacements may induce transverse
bending stresses in the web plate, the magnitude of which would be
likely to produce fatigue cracks at the web panel boundary, more
especially in the vicinity of the compression flange.   In contrast to
what is generally assumed for the computations, the web is not actually
hinged to the flanges so that the torsional stiffness of the latter
offers a partial flexural restraint with resulting bending stresses in
the fillet welds.   Even small ranges of such stresses may induce fatigue
cracks if the number of cycles is sufficiently high.   This problem is
presently not yet satisfactorily solved; a simple approach consists in
limiting the web slenderness, and thus the magnitude of postcritical
out-of-plane displacements together with that of the edge bending
stresses.   It is therefore suggested to limit the depth of the
compression zone of the web to 100 times the web thickness; for
symmetrical plate girders, that leads to $d/t_w \leq 200$.   Theoretical and
experimental work are presently in progress in several countries; their
results should provide with a better understanding of the phenomenon in
a very near future.

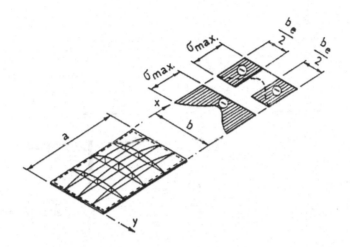

Fig. 1 - Concept of effective width of a compression plate

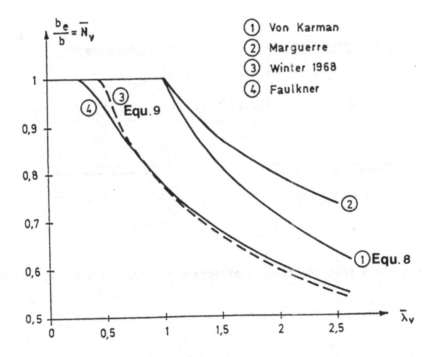

Fig. 2 - Normalized effective width versus the plate slenderness :
case of uniform compression of a simply supported plate.

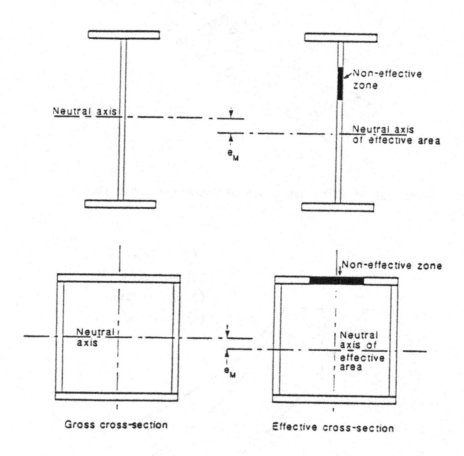

Fig. 3 - Class 4 cross-sections : effective section for bending moment.

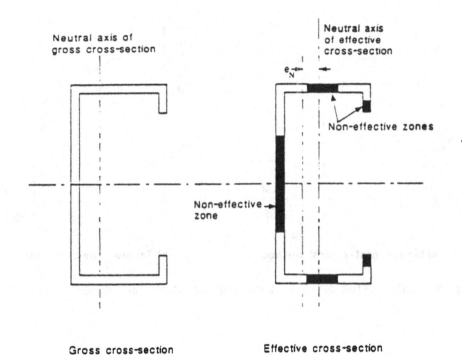

Fig. 4 - Class 4 cross-sections : effective section for axial force.

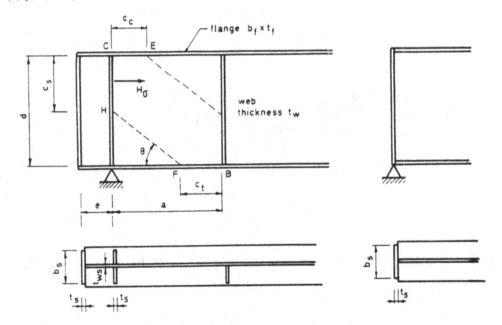

Fig. 5 - Tension field in end panels.

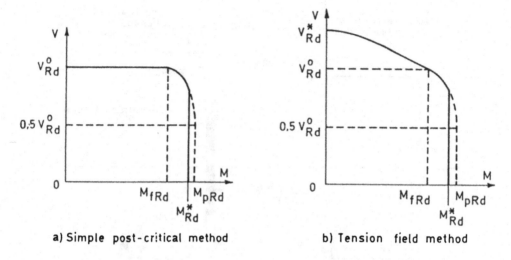

a) Simple post-critical method   **b) Tension field method**

Fig. 6 - Interaction of shear buckling resistance and moment resistance.

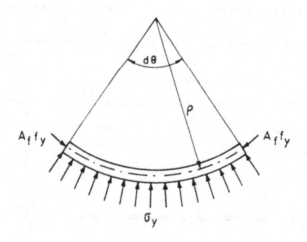

Fig. 7 - Transverse direct stresses in the web due to bending curvature.

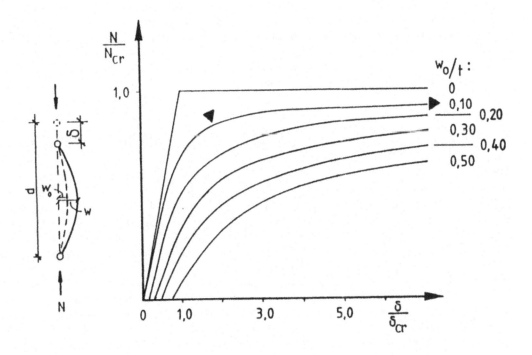

Fig. 8 - Non linear response of the web transverse strip.

CHAPTER 5
LONGITUDINALLY STIFFENED PLATE GIRDERS

1. INTRODUCTION.

Longitudinal stiffeners are used in addition to transverse stiffeners in some particular circumstances, especially when unusually deep girders with a very thin web are designed (fig. 1). The aim of such longitudinal stiffeners is twice : first, they strengthen the web with the result of an increase in buckling resistance, and second, they may reduce the transverse deflections of the web plate within the limits required by the serviceability limit states.

Longitudinal stiffeners contribute the bending stiffness of the cross-section if they are continuous and connected continuously to the web plate.

In addition tension field can only develop in the web if it can anchor on the surrounding frame, constituted by the transverse and the longitudinal stiffeners. Therefore transverse stiffeners are also aimed at being welded on the web plate. When both kinds of stiffeners are

located on the same side of the web plate, they must thus intersect ;
that does not go without increasing the fabrication cost because of
required cuttings and welding. A more simple and economical solution
consists in locating the transverse stiffeners on one side of the web
plate and the longitudinal ones on the other side; by doing so avoids
any crossing of both kinds of stiffeners.

Transverse and longitudinal stiffeners (or flanges) divide the web into
a series of unstiffened web portions, which are termed *web subpanels*.

The longitudinal stiffeners are liable to improve the moment and/or
axial force capacity, provided they are located and designed
appropriately. Not only they contribute an increase in section modulus
and/or cross-sectional area but also they may improve the efficiency of
the web plate subpanels with similar effects regarding the strength
properties of the girder cross-section. Moreover an increase in shear
capacity may result from the use of longitudinal stiffeners.

It is experimentally observed that the magnitude of the direct stresses
in the web at the location of the longitudinal stiffeners is
satisfactorily predicted by the linear distribution derived from the
elementary beam theory applied to an appropriate effective
cross-section. However reduction with respect to this linear
distribution occurs in the compression web subpanels as a result of
plate buckling when the loading is growing up. How to define the
effective cross-section of a longitudinally stiffened web is examined in
the following section.

2. EFFECTIVE CROSS-SECTION OF LONGITUDINALLY STIFFENED PLATE GIRDERS.

The determination of the effective cross-section of a longitudinally
stiffened web is conducted similarly as for an unstiffened web. However
one must consider here each web subpanel of depth d_i, the latter being
measured between two consecutive longitudinal stiffeners for the inner
web subpanels or between the flange and the nearest longitudinal
stiffener for both upper and lower web subpanels. Each web subpanel is
usually assumed to be simply supported along its four edges.

An efficiency η_i is computed for each web subpanel i, in accordance with von KARMAN formula introduced earlier :

$$\eta_i = [1 - 0.22/\overline{\lambda}_{p,i}]/\overline{\lambda}_{p,i} \not> 1 \qquad (1)$$

where $\overline{\lambda}_{p,i}$ is the normalized slenderness of the corresponding subpanel :

$$\overline{\lambda}_{p,i} = \sqrt{\sigma_{max,i}/\sigma_{cr,i}} \qquad (2.a)$$

$$= (d_i/t_w) \, (1.05/\sqrt{k_{\sigma,i}}) \, \sqrt{\sigma_{max,i}/E} \qquad (2.b)$$

It is implicitly assumed that the web has a constant thickness t_w over its whole depth; this assumption is in accordance with usual practice. The symbol i appearing as a subscript to some quantities means that these quantities are related to the subpanel under consideration. More especially, the buckling coefficient $k_{\sigma,i}$ shall depend on the stress ratio:

$$\psi_i = \sigma_{min,i}/\sigma_{max,i} \qquad (3)$$

in this subpanel, while $\sigma_{max,i}$ and $\sigma_{min,i}$ are respectively the largest compressive stress and the smallest compressive (or possibly tensile) stress in web subpanel i.

In contrast to what has been done for unstiffened webs, the maximum compressive stress is not taken equal to the material yield stress, in order not to be unduly conservative (especially for inner web subpanels where $|\sigma_{max,i}|$ can be appreciably lower than f_y). The stress ratios ψ_i may still be obtained using the linear stress distribution, in accordance with the elementary beam theory, and a section composed of the effective area(s) of the compression flange(s) but the gross area of the web.

Once computed according to (1), the efficiency η_i is applied to the depth $d_{c,i}$ of the compression zone of the subpanel i. Then the effective compression depth $d_{e,i}$ is deduced as :

$$d_{e,i} = \eta_i \, d_{c,i} \tag{4}$$

and allocated in parts to both extreme fibres of the compression zone in the subpanel i :

$$d'_{e,i} = [2/(5 - \psi_i)] \, d_{e,i} \nleq 0.4 \, d_{e,i} \tag{5.a}$$

$$d''_{e,i} = d_{e,i} - d'_{e,i} \ngeq 0.6 \, d_{e,i} \tag{5.b}$$

$d'_{e,i}$ is taken adjacent to the most compressed fibre and $d''_{e,i}$ adjacent to the least compressed fibre (or the neutral axis) of subpanel i.

The web portion subject to tensile stresses is taken fully effective.

Usually the flanges of plate girders are no so large that shear lag can occur. Then the tension flange is fully effective while for the compression flanges, some considerations developed for unstiffened plate girders are still valid here.

The effective girder cross-section, defined as explained above, looks like a section having as many holes as there are slender subpanels subject totally or partially to compressive stresses (fig. 2). The computation of the cross-sectional properties is then conducted as explained previously.

The concept of effective depth in each subpanel implicates that the longitudinal stiffeners are strong enough to mobilise direct stresses up to the ultimate capacity of the section. Therefore these stiffeners shall comply with some design requirements, which are discussed later on.

3. RESISTANCE OF CROSS-SECTIONS OF LONGITUDINALLY STIFFENED PLATE GIRDERS.

3.1. Tension.

See Section 5.1. of Chapter 4.

3.2. Compression.

See Section 5.2. of Chapter 4, under the reservation that the effective area of the cross-section be computed in accordance with previous Section 2 for $\psi_i = 1$.

3.3. Bending.

See Section 5.3. of Chapter 4, under the reservation that the effective section moduli of the cross-section be computed in accordance with previous Section 2 with the appropriate ψ_i values.

3.4. Bending and axial (compressive) force.

See Section 5.4. of Chapter 4.

4. SHEAR BUCKLING RESISTANCE OF LONGITUDINALLY STIFFENED PLATE GIRDERS.

The shear buckling resistance of a longitudinally stiffened web is computed according to the diagonal tension band model. It is however obvious that the postbuckling behaviour shall be significantly more complex than for unstiffened or transversely stiffened webs.

Ii is generally agreed that the initiation of shear buckling in the *weakest* web subpanel achieves the first stage of the web response, where a state of pure shear is assumed throughout the whole web. In addition it can be expected that longitudinal stiffeners contribute similarly an increase in ultimate shear load.

The results provided by four possible different approaches of the ultimate shear capacity of longitudinally stiffened webs have been evaluated by OSTAPENKO (fig. 3) by comparison with experimental values. These models differ from each other regarding the assumptions made about the number and properties of the tension band(s). The conclusion has been that one of them combines the advantages of great simplicity and satisfactory accuracy; therefore it has been selected as design model.

The design shear buckling resistance is still given as :

$$V_{Rd} = (V_{cr} + 0.9 \ V_t)/\gamma_{M4} \tag{6}$$

where V_{cr} is the beam-action strength and V_t the tension field-action strength; the frame-action strength is still neglected. γ_{M4} is a partial safety factor.

Beam-action strength V_{cr} is assessed using the critical shear buckling stress of the weakest subpanel :

$$V_{cr} = dt_w \ \text{Min} \ [\tau_{cr,i}^{red}] \tag{7}$$

where Min $[\tau_{cr,i}^{red}]$ is the smallest of the reduced critical plate buckling shear stresses relevant for the web subpanels. Once account is taken of the presence of the longitudinal stiffeners by increasing the critical shear stress, the procedure developed for transversely stiffened webs is pursued accordingly. Thus the *tension field-action* V_t is computed by introducing Min $[\tau_{cr,i}^{red}]$ in the expression of the tension field stress :

$$\sigma_t = \sqrt{f_{yw}^2 - \{\text{Min}[\tau_{cr,i}^{red}]\}^2 [3 - (1.5 \sin 2\theta)^2}$$
$$- 1.5 \ \text{Min} \ [\tau_{cr,i}^{red}] \sin 2\theta \tag{8}$$

wherefrom :

$$V_t = [d \ \text{cotg} \ \theta - a + c_c + c_t] \ t_w \ \sigma_t \ \sin^2 \theta \tag{9}$$

The determination of c_c, c_t, θ is conducted as already explained in previous lectures.

The tension field model, as used here, postulates that the tension field develops over the complete web depth, the influence of the longitudinal stiffeners upon the postbuckling action being totally neglected.

The plots of the ratios of experimental to theoretical ultimate shear loads against the aspect ratio of the longitudinally stiffened web panels, on the one hand, and against the flange stiffness factor, on the

other hand, demonstrate that tendancies similar to those observed for
transversely stiffened webs may be pointed out. A plot against the
number of longitudinal stiffeners demonstrates that the model is
conservative, due undoubtfully to the fact that the slope of the tension
field is not *directly* influenced by the presence of longitudinal
stiffeners.

A statistical evaluation procedure of the test results compared to the
results derived from the ultimate shear model yields the following
recommendations regarding the partial safety factor γ_{M4} :

γ_{M4} = 1.05 when one longitudinal stiffener only ;

γ_{M4} = 1.00 when at least two longitudinal stiffeners.

Because of the lack of sufficient experimental and/or theoretical
information dealing with the influence of longitudinal stiffeners on the
behaviour of end posts, it is suggested to disregard totally this
influence and to use the procedure developed earlier in Section 6 of
Chapter 4. For longitudinally stiffened webs, it seems appropriate, for
practical reasons, to design only rigid end posts.

5. INTERACTION BETWEEN SHEAR, BENDING AND AXIAL FORCE.

Provided the design resistance be computed in accordance with the
previous Sections, to account for the presence of longitudinal
stiffeners, the interaction between shear, bending and axial force is
treated by an interaction diagram, as already explained in Section 8 of
Chapter 4.

6. FLANGE INDUCED BUCKLING.

The limit d/t_w established for an unstiffened web is applied to each
subpanel by substituting the partial depth d_i of the subpanel for the
total web depth d. Doing so is probably conservative but does not seem
to be governing for usual proportions of longitudinally stiffened plate
girders.

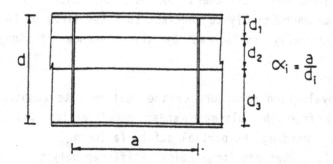

Fig. 1 - Stiffeners arrangement

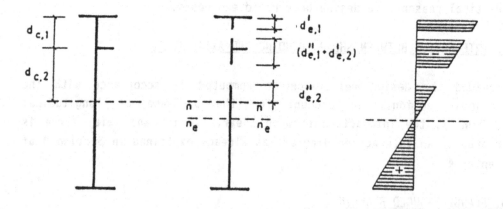

Fig. 2 - Full cross-section and effective cross-section for a
          longitudinally stiffened plate girder.

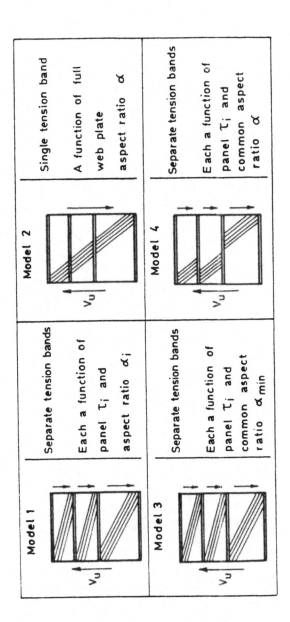

Figure 3 - Possible ultimate shear models

CHAPTER 6
DESIGN OF WEB STIFFENERS

INTRODUCTION.

Because the stiffeners have either to provide an anchor to the tension field or to mobilize direct stresses till the ultimate shear or bending capacity be reached, the stiffeners must be strong enough to resist all the parasitic forces which develop in the postbuckled range till collapse occurs. Otherwise speaking, the stiffeners have to constitute pseudo nodal lines for the plate buckling up to the ultimate limit load.

Therefore their relative flexural rigidity γ needs to be larger than the optimum flexural rigidity γ^* as defined in the linear theory of plate buckling :

$$\gamma \geq m\gamma^* \qquad (1)$$

Let us remind that the relative rigidity γ writes :

$$\gamma = EI_s/dD \qquad (2)$$

where d is the total web depth, D the rigidity $Et_w^3/12(1 - \nu^2)$ of the web plate of thickness t_w and I_s the moment of inertia of a fictitious strut composed of the stiffener properly plus an effective width of web plate.

This effective width is given as $b_e = t_w \sqrt{E/f_{yw}}$, and I_s is computed around the axis passing through the centroïd of the fictitious strut and parallel to the web. For an open section stiffener (angle, tee section, flat,...) the sole web of the stiffener is fitted with such an effective width of plate web; for a closed section stiffener, both webs are fitted with such an effective width of web plate (fig. 1) as far as their spacing does not exceed b_e, otherwise the total effective width shall be $(b + b_e)$.

The amplification factor m, used as a multiplyer for the optimum relative rigidity γ^*, reflects the detrimental effect of initial imperfections, on the one hand, and of the destabilizing forces experienced by the stiffeners in the postbuckle range, on the other hand. It should depend more especially on the web slenderness d/t_w.

It is also of paramount importance that the stiffeners do not exhibit a sudden loss of rigidity (because of local plate buckling of their wall elements) before ultimate capacity of the web plate be reached, either in shear or in bending. Therefore, it is required that the slenderness of the wall elements complies with the b/t limit values related to Class 2 sections.

2. DESIGN OF THE LONGITUDINAL STIFFENERS.

The value γ_L^* to be introduced in (1) depends on the kind of loading. Expressions of γ^* are existing which depend on the location of the stiffener under consideration and on the loading ; they are expressed in terms of the aspect ratio α and of the relative area δ. It is not the place here to list all these expressions. Let us just give some of them :

a) Stiffener at mid-depth for uniform compression :

$$\gamma_{LC}^* = \alpha^2 [8(1+2\delta)-1] - 0.5 \alpha^4 + 0.5(1+2\delta) \qquad \text{if } \alpha \leq \sqrt{8(1+2\delta)-1} \qquad (3.a)$$

$$= 0.5[8(1+2\delta)-1]^2 + 0.5(1+2\delta) \qquad\qquad \text{if } \alpha \geq \sqrt{8(1+2\delta)-1} \quad (3.b)$$

b) Siffener in the compression zone (at 1/5 of depth d) for pure bending:

$$\gamma^*_{Lb} = 25.\alpha - 6 + 78.\delta.\alpha^2 \qquad\qquad 0.5 \leq \alpha \leq 2 \quad (4)$$

c) Stiffener at mid-depth for pure shear :

$$\gamma^*_{Ls} = 5.4.\alpha^2.(2\alpha + 2.5\alpha^2 - \alpha^3 - 1) \qquad\qquad 0.5 \leq \alpha \leq 2 \quad (5)$$

Longitudinal stiffeners are checked regarding the required rigidity mainly; an additional strength requirement is only needed in special cases.

2.1. Rigidity requirement.

The amplification factor m_L will be reduced for hollow section stiffeners as a result of the beneficial effect of their large torsional rigidity. Thus, one has :

- open section stiffeners :

$$m_L = 0.023d/t_w - 1.5 \geq 1.25 \qquad\qquad\qquad (6)$$
$$\leq 4.00$$

- hollow section stiffeners :

$$m_L = 0.0105d/t_w \qquad\qquad \geq 1.25 \qquad\qquad\qquad (7)$$
$$\leq 2.50$$

The limit values 1.25 and 2.50 are obtained for $d/t_w = 120$ and 240, respectively.

Because of the approximative character of the values of m_L, a rough determination of the optimum rigidity γ^* is sufficient in most cases. An interpolation between the γ^* values for panels under bending and shear is usually not justified: the γ^* value corresponding to the determining stress case (usually bending) is used for the stiffener design.

For webs subjected to shear, a constant value m_L = 1.25 seems to be appropriate.

2.2. Strength requirement.

In addition to rigidity requirements, the longitudinal stiffeners may have to transmit wind loads and other similar forces to the transverse stiffeners.   The corresponding bending moment are however usually negligible; for horizontally curved girders, lateral bending due to the radial forces needs a careful consideration.

When a strength check is required, the stiffener is understood as the fictitious strut defined above in Section 1.   It is checked as a beam-column with an initial bow of a/500 with respect to the theoretical shape of its longitudinal axis and reference will be made to column buckling curce c (with appropriate α factor for aforementioned initial bow) and to buckling length equal to the transverse stiffeners spacing.

A similar check is recommended when the longitudinal stiffeners sustain an average direct compressive stress exceeding 2/3 of the direct stress in the nearest compression flange.

3. DESIGN OF THE TRANSVERSE STIFFENERS.

Transverse stiffeners act as posts of a Pratt truss during the postbuckling behaviour of a web subject to shear.   They are thus aimed at supporting compressive forces resulting from the anchor of the tension field.   Therefore both rigidity and strength requirements must be fulfilled.

3.1. Rigidity requirement.

A condition similar to (1) is usually to fulfil for transverse stiffeners in order to maximize the ultimate strength and to reduce the amplitude of the web deformation. The amplification factor is obtained as follows :

$$m_T = 0.027d/t_w - 1 \geq 1 \qquad (8)$$
$$\leq 3$$

The limit values 1 and 3 are obtained for d/t_w = 75 and 150, respectively.

It is usual to consider the value of γ_T^* related to the case of a very long web in shear, fitted with equally spaced transverse stiffeners, i.e. :
a) for an unstiffened web :

$$\gamma_{To}^* = 21/\alpha - 15.\alpha \geq 6 \qquad (9)$$

b) for a longitudinally stiffened web :

$$\gamma_T^* = \gamma_{To}^* \ [k_\tau(\alpha_i)/k_\tau(\alpha)].(d/d_{i,max})^2 \qquad (10)$$

where α is the aspect ratio of the web plate of depth d (α = a/d), α_i that of the subpanel of depth $d_{i,max}$ which is the most critical in shear (a_i = a/$d_{i,max}$). Shear buckling factors $k_\tau(\alpha)$ et $k_\tau(\alpha_i)$ are respectively related to the web plate and to the most critical subpanel. This value (10) is deduced by equating the shear critical load of the longitudinally stiffened web to that of an equivalent unstiffened web the thickness of which is $t_{w,equ}$. Thus, one has :

$$k_\tau(\alpha).t_{w,equ}^3/b^2 = k_\tau(\alpha_i).t_w^3/d_{i,max}^2$$

wherefrom :

$$(t_{w,equ}/t_w)^3 = [k_\tau(\alpha_i)/k_\tau(\alpha)].(d/d_{i,max})^2 \qquad (11)$$

3.2. Strength requirement.

Transverse stiffeners may be classified in *load bearing stiffeners*, which are located upright the supports of the beam, and in *intermediate stiffeners*.

An intermediate stiffener is adjacent to two web panels; its maximum loading is determined by assuming that any of these panels achieves its ultimate shear capacity. Actually the partial anchor of the tension band on the flange results in a compression force component acting at one end of the stiffener while the anchor on the transverse stiffener induces compressive forces components distributed over the anchor depth (fig. 2.a). As a result a transverse stiffener should be designed for a total compressive force cqual to the lesser of the vertical components of the stress resultants in the tension bands adjacent to the stiffener.

For a load bearing stiffener located upright intermediate supports, the compressive force components resulting from the adjacent tension bands cumulate because the sign of shear changes (fig. 2.b).

An *end load bearing stiffener* experiences the compressive forces induced by the tension band in the sole adjacent web panel.

As these compressive forces are induced by the postbuckling behaviour of the web plate, they are assumed to be applied in the middle plane of the web plate, with the result of a possible eccentricity with respect to the centroïd of the stiffener strut (stiffener properly plus effective width of web plate). Because of the normally large value of the compressive force in the load bearing stiffeners, these are preferably made by stiffening the web plate symmetrically - i.e. with stiffeners located on both sides of the web - with a view to nullify the aforementioned eccentricity of the compressive force. In contrast, intermediate transverse stiffeners usually experience smaller compressive forces and are therefore located on one side only of the web plate; consequently they are usually subject to compression and bending.

The stiffeners shall thus behave, generally speaking, as beam-columns.

Some comments are however useful in this respect :
a) The way the compressive forces are applied on the stiffeners influences the buckling length ;
b) Because compression struts are usually non-symmetrical welded sections, it is appropriate to account for an initial out-of-straightness somewhat larger than the basic value of 1/1000 of the actual length, on which the European buckling curves are based ;
c) Because ultimate shear implies yielding of a non-negligible portion of the web plate, the amount of effective width of web plate to be accounted for when defining the stiffener strut has to be reduced compared to the value $b_e = t_w \sqrt{E/f_y}$ used for longitudinal stiffeners;
d) Any longitudinal stiffener located in the compression zone of the web is liable to exert a destabilizing force on the transverse stiffener.

The design procedure of the transverse stiffeners has been simplified by formulating some assumptions the consequences of which have been validated by a lot of experiments :
1. An intermediate stiffener is subject to end compressive loads acting in the middle plane of the web :

$$N_s = V_{Sd} - dt_w \tau_{cr}^{red} \qquad (12)$$

   where V_{Sd} is the design shear load. The lesser value for the two panels adjacent to the stiffener should be used.
2. An end post is subject to end compressive loads acting in the middle plane of the web :

$$N_s = R_{Sd} \qquad (13)$$

   where R_{Sd} is the design bearing reaction.
3. Account must be taken of possible transverse loads acting on the stiffeners , perpendicular to the web plate, and equal to 1 % of the axial load in any longitudinal stiffener located in the compression zone of the web.
4. A reduced value of the effective width of web plate should be used, which amounts $b_e = 0.67 t_w \sqrt{E/f_y}$.
5. Account must be taken of the possible eccentricities of the compressive loads N_s.
6. Only column buckling perpendicular to the web plane has to be considered; the buckling length is taken as 0.7 times the web depth d

to account for the actual load distribution and the restraint provided by the web plate when the stiffener tends to buckle.

7. Use is made of European buckling curve c; however a modified value of the curve parameter α_c is used in order to account for an initial out-of-straightness d/500 of the stiffener strut.

8. A stiff end post (end load bearing stiffener) must fulfil additional requirements, as already explained in Chapter 4.

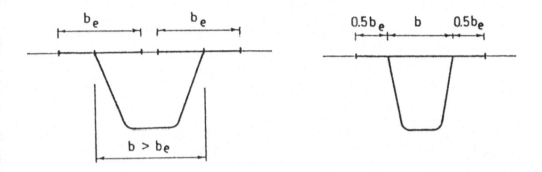

Fig. 1 - Effective width of web plate for a closed section stiffener.

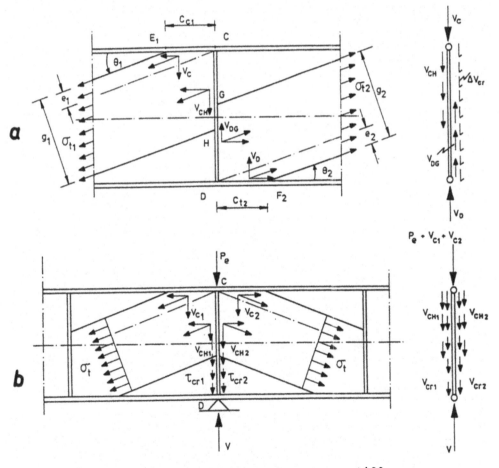

Fig. 2 - Compressive forces acting on transverse stiffeners.
        (a) intermediate stiffener
        (b) internal load bearing stiffener

ULTIMATE LOAD BEHAVIOUR OF WEBS SUBJECT TO PATCH LOADING

and

INTERACTION BETWEEN SHEAR LAG AND PLATE BUCKLING IN LONGITUDINALLY STIFFENED COMPRESSION FLANGES

M. Skaloud

Czechoslovak Academy of Sciences, Prague, Czechoslovakia

ABSTRACT

This part of the monograph, corresponding to seven hours of lecturing at the related International Advanced School, deals with two chapters on advanced analysis of steel plate and box girders; viz. (i) the behaviour of the slender webs of steel plate girders subject to partial edge loading and (ii) the interaction that exists between shear lag and plate buckling in the longitudinally stiffened compression flanges of steel box girder bridges and similar structures. Both chapters are mostly based on the results and conclusions of the research undertaken by the author and his associates in Prague during the last years, or (see one section of Chapter 2) in close cooperation with Prof. Maquoi's team at Liège University in Belgium.

1. ULTIMATE LOAD BEHAVIOUR OF WEBS SUBJECT TO PATCH LOADING

1.1. Definition of Patch Loading

The notation "patch load" (or "partial edge load" or "discrete edge load") is used for a load with a short distribution length along the girder, applied to the flange in the plane of the web (see Fig. 1.1).

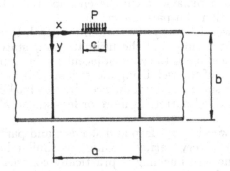

Fig. 1.1. A web panel between vertical stiffeners subjected to a partial edge load.

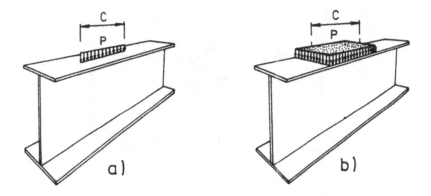

Fig. 1.2. Two kinds of partial edge load acting on a web panel: a) load uniformly distributed over a portion of the flange axis, b) load uniformly distributed over a portion of the flange area.

Of course, in reality, such a partial edge load can be materialized in two ways, as shown in Figs 1.2 a and b.

This type of loading is frequently encountered in constructional steelwork. In buildings, there is a general tendency to locate transverse stiffeners at positions where there are large concentrated forces, but even there transverse stiffeners are sometimes avoided because of their cost. However, in the case of structures subject to moving loads (such as for crane girders, or bridge girders erected by "launching"), the use of transverse stiffeners for the same purpose is not possible. Then the web of the girder is necessarily under the action of patch loading.

1.2. Yielding, Buckling and Crippling – Three Phenomena Connected with the Behaviour of Webs Subject to Patch Loading

Three types of web behaviour can be encountered when a steel plate girder is under the action of a partial edge load.

First, when the web is stocky (i.e., when its depth-to-thickness ratio is low), no web buckling occurs, and the ultimate limit state of the girder is determined by web yielding in the zone adjacent to the applied patch load.

Second, in the case of the web being more slender, it buckles, so that the ultimate limit state is substantially influenced by this buckling. If the web is not very deep, the buckled pattern is more or less spread all over the whole surface of the web.

Third, when the web is very deep and slender, and particularly so when the partial edge load is very narrow ("knife" or similar loadings) and the loaded flange thin, the web buckling is practically confined to the very neighbourhood of the load, with other parts of the web being practically unaffected by buckling. Even the failure mechanism of the girder develops just in this portion of the web (a segmental line plastic hinge) and in the adjacent part of the loaded flange (a set of plastic hinges).

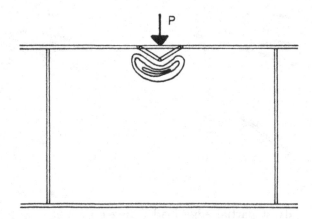

Fig. 1.3. Web "crippling".

But, on the whole, phenomenon 3 is not very different from phenomenon 2, since both of them are governed by web buckling. Hence, in the following, we shall distinguish merely two fundamental cases, viz.: (i) yielding and stress state in non-buckling webs subject to partial edge loading and (ii) buckling of webs subject to partial edge loading.

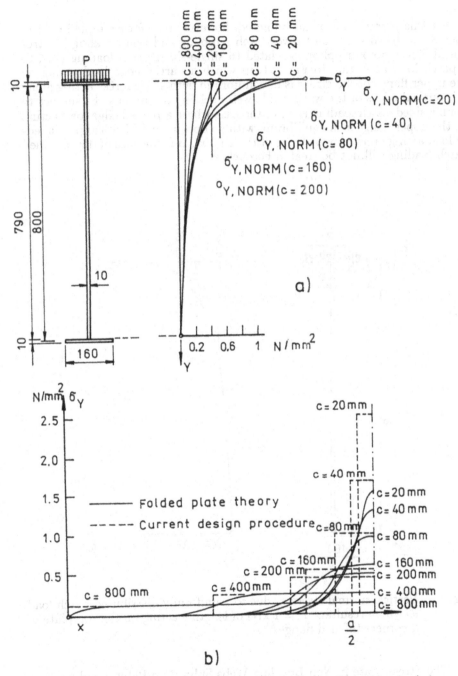

Fig.1.4. Vertical normal strees σ_y due to a uniformly distributed load p (whose resultant $P = 1$ kN) acting on a portion c of the area of the upper flange having thickness $t_f = 10$ mm: a) distribution over the web depth under the load centre, b) distribution over one half of the web width at its connection to the flange.

At this juncture it should be noted, however, that for example in Eurocode 3 another distinction between web crippling and web buckling is introduced. There web crippling is related to single-sided patch loading (i.e., to
a patch load according to Fig. 1.1, in which a partial edge load acting on
the upper flange of the girder is in equilibrium with shear forces acting along
the vertical edges of the web), while web buckling occurs in webs subject to
double-sided patch loading (i.e., in the case where a partial edge load acting
on the upper flange is in equilibrium with an oposite and identical partial edge load acting on the lower flange of the girder). But the case of double-sided
patch loading will not be treated herebelow.

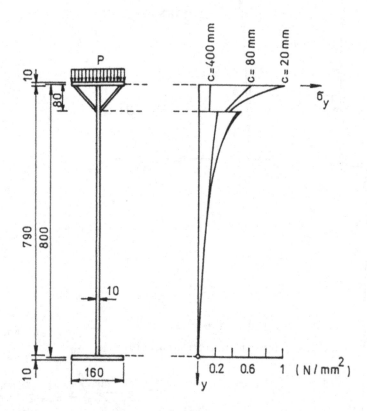

Fig. 1.5. Vertical normal stress σ_y due to a uniformly distributed patch load
   p (whose resultant $P = 1$ kN) of length c acting on the top plate of
   a corner-stiffened flange.

1.3. The Stress State in Non-Buckling Webs Subject to Patch Loading

In this section, we shall deal with the elastic stress state that develops in
an unbuckled web under the action of a patch load.

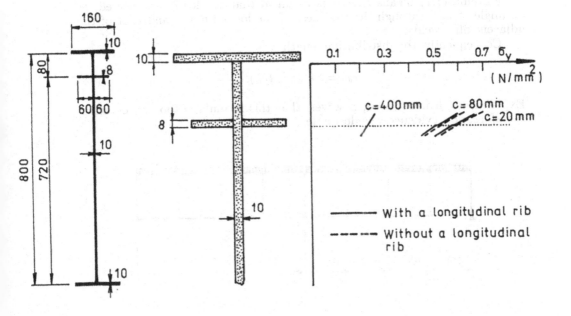

Fig.1.6. Effect of a horizontal rib (located at the distance of one fifth of the web depth from the loaded flange) upon the vertical normal stress σ_y due to a uniformly distributed patch load p (whose resultant $P = 1$ kN) of length c acting on the upper flange of thickness $t_f = 10$ mm.

1.3.1. Simple design formula

For a long period of time, the following simple formula was used in this respect:

$$\sigma_{y,norm} = \frac{P}{[c + 2(t_f + t_{f,w})]t_w}. \tag{1.1}$$

P denotes the resultant of the uniformly distributed partial edge load p, t_f the flange thickness, $t_{f,w}$ that of the fillet weld, and t_w that of the web.

Formula (1.1) means that it is assumed that the load is dispersed, at an angle of 45° , through the thickness of the loaded flange and that of the adjacent fillet weld.

Consequently, the distribution length

$$c_{dis} = c + 2(t_f + t_{f,w}). \tag{1.2}$$

Experiments demonstrated, however, that this formula is too conservative; therefore, new evidence was desirable.

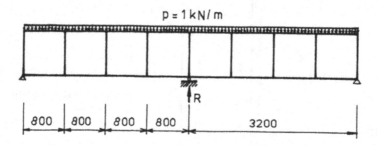

Fig.1.7. A two-span continuous beam with a vertical stiffener subjected to a compressive reactive force.

1.3.2. Analytical solution to the problem

Using the apparatus of folded plate theory, V. Křístek and the author (see [1.1]) carried out an analytical investigation into the state of stress of steel plate girders subject to patch loading. The results are presented in the following figures. They were calculated for a unit load, which is such that, whatever the distribution of the loading, its resultant $P = \int_{A_p} p \, dA_p = 1kN$. So the curves given in the figures represent a kind of influence lines for the stress state in the web.

Let us start with an unstiffened web attached to ordinary flat flanges (see Fig. 1.4). As it became manifest during the analysis that the vertical normal stresses, σ_y , predominated over all other components of the stress state, the figure (and the following ones) concentrate on these stresses only.

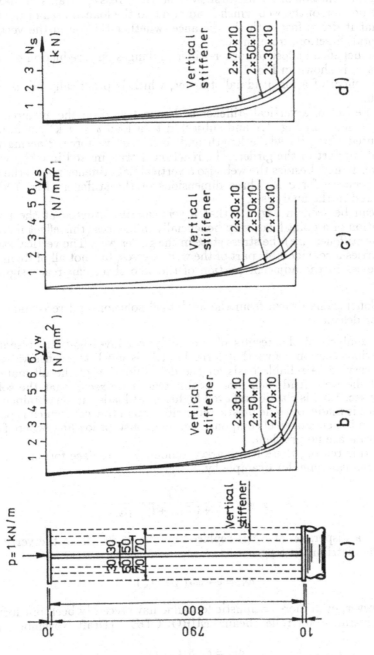

Fig.1.8. Effect of a vertical stiffener upon the stress state: a) geometrical characteristics and loading, b) stress $\sigma_{y,w}$ in the web at its connection to the stiffener, c) stress $\sigma_{y,s}$ in the stiffener, d) normal force N_s in the stiffener.

An examination of the figure shows that the stress σ_y attains its maximum in that portion of the web which is adjacent to the loaded area of the flange, and that it drops fast with the distance, whether this be in the vertical or horizontal direction, from there.

The beneficial effect of corner-stiffened flanges on a reduction of vertical stresses σ_y is shown in Fig. 1.5.

The effect of a longitudinal stiffener, which is practically nil, is seen in Fig. 1.6.

The effect of a vertical stiffener is demonstrated on the example of the girder depicted in Fig. 1.7 and subjected to a load $q = 1$ kN/m uniformly distributed along its whole length and to a reactive force R acting at the internal support of the girder. The reaction is transmitted into the web via the lower flange; besides the web also a vertical flat stiffener resists the action of the reactive force R. Three dimensions of the stiffener (Fig. 1.8) were considered in the analysis.

It can be seen in Fig. 1.8 that a vertical rib, situated at the point of application of a point load, very beneficially influences (this effect increasing with the stiffener size) the stress state in the girder web. The vertical stiffener then carries a considerable part of the web stresses, but not all of them. Even the stresses in the adjacent portion of the web sheet can reach significant values.

1.3.3. Conclusions drawn from the analytical solution and recommendation for design

An analysis of the results of the analytical investigation presented in the previous section shows that formula (1.1) is much too conservative and cannot serve as a reliable basis for the determination of the ultimate limit state of the web. And even more is this true if we recall that the solution given in sec. 1.3.2 is elastic and even higher limit loads can be obtained when allowance is made for some plastic redistribution in the web, which is possible at least in the case of steel components under constant loading, where fatigue phenomena are no problem.

For this reason, there is a marked tendency to replace formula (1.1) by a more realistic one, for example by

$$\sigma_{y,norm} = \frac{P}{[c + 3.5\,(t_f + t_{f,w})]\,t_w}, \tag{1.3}$$

in which a dispersion at 45° is replaced by one at 60° (from the vertical). * Then the distribution length

$$c_{dis} = c + 3.5\,(t_f + t_{f,w}). \tag{1.4}$$

However, even more optimistic formulae have recently been put forward.

For instance, for rolled beams EUROCODE 3 (Draft November, 1983) gives

$$c_{dis} = c + 5\,(t_f + r), \tag{1.5}$$

---

* i.e., at an inclination of 30° from the horizontal

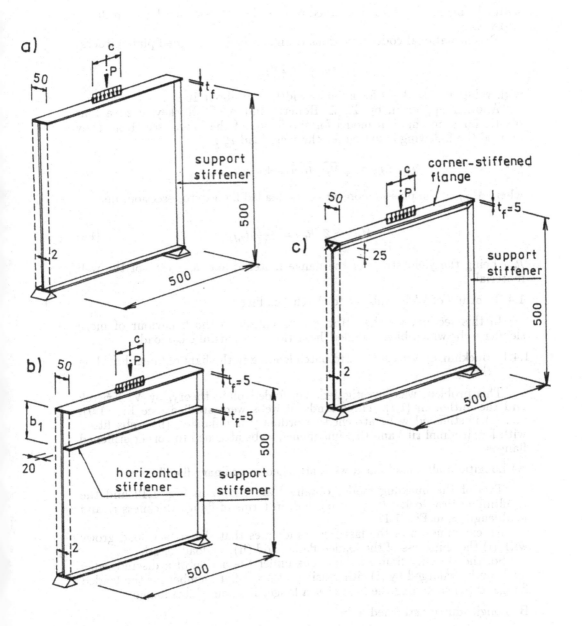

Fig.1.9. Panels studied by V. Křístek and the author : a) an unstiffened
web attached to flat flanges, b) a web fitted with a longitudinal
stiffener, c) a web attached to corner-stiffened flanges.

with r being the depth of the curved part of the cross-section between flange and web.

Recent national codes of various countries give for welded plate girders

$$c_{dis} = c + 2\beta t_f, \tag{1.6}$$

with values of $\beta = 3 - 4$ for a flange width b_f of about $10\ t_f$.

Another approach, by T. M. Roberts and K. C. Rockey, used a four plastic hinge mechanism model for the flange of the beam; see [1.2]. They derived the following equation for the yield load P_y :

$$P_y = 4\sqrt{M_{pl,f}\, t_w\, R_{y,w}} + c t_w\, R_{y,w}, \tag{1.7}$$

where $M_{pl,f}$ = the plastic moment of the loaded flange cross-section, i.e.,

$$M_{pl,f} = Z_f\, R_{y,f} = \frac{b_f\, t_f^2}{4} R_{y,f}, \tag{1.8}$$

$R_{y,f}$ being the yield stress of the flange material and $R_{y,w}$ that of the web material.

1.4. Buckling of Webs Subject to Patch Loading

In this section, we shall turn our attention to the behaviour of more slender webs which buckle under the action of a partial edge load.

1.4.1. Buckling of webs subject to patch loading in the light of linear buckling theory

This problem was investigated, via folded plate theory, by V. Křístek and the author in [1.1]. Three kinds of webs were studied (see Fig. 1.9), viz. (i) unstiffened webs attached to ordinary flat flanges, (ii) webs fitted with longitudinal ribs and (iii) unstiffened webs attached to corner-stiffened flanges.

A) Longitudinally unstiffened webs attached to ordinary flat flanges

Two of the buckling modes obtained are shown in Fig. 1.10 and the resulting critical loads, P_{cr} , are plotted, in terms of flange thickness t_f and load length c, in Fig. 1.11.

An examination of the last figure indicates that the critical load grows with (i) the thickness of the loaded flange and (ii) the load length.

So, the stability limit of thin webs under the action of a discrete edge load can be enlarged by (i) either using a thicker flat member for the loaded flange or (ii) spreading the load over a longer interval of this flange.

B) Longitudinally stiffened webs

It is of some importance to check whether the stability limit of webs subjected to a patch load can be increased by reinforcing the web by a longitudinal stiffener. For this reason, V. Křístek and the author studied the buckling problem of the web given in Fig. 1.9 b and fitted with a longitudinal rib situated at various distances b_1 from the loaded flange. The size of

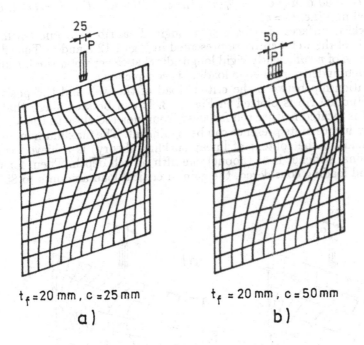

$t_f = 20$ mm, $c = 25$ mm $t_f = 20$ mm, $c = 50$ mm

a)   b)

Fig.1.10. Buckled pattern of an unstiffened web attached to flanges of thickness $t_f = 20$ mm and subjected to a partial edge load P of length
c: a) $c = 25$ mm, b) $c = 50$ mm.

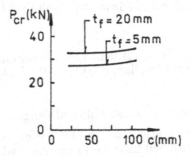

Fig.1.11. Critical load P_{cr} , in terms of flange thickness t_f and load length c,
of an unstiffened web subjected to a partial edge load.

the stiffener was the same in all cases; the distance b_1 assuming the following values : (i) $b_1 = 50$ mm (i.e. $b_1 = b/10$), (ii) $b_1 = 100$ mm (i.e. $b_1 = b/5$), and (iii) $b_1 = 250$ mm (i.e. $b_1 = b/2$).

The buckled surfaces of webs with a longitudinal rib a) at one tenth, b) in the middle of the web depth are presented in Figs 1.12 a and b. The effect of the presence of a sufficiently rigid longitudinal stiffener upon the buckling mode of a web under partial edge loading is conspicuous.

The relationship between the critical load of the web and the position of the longitudinal rib is plotted in Fig. 1.13. It can be seen there that, if the stiffener is located neither too far away from, nor too close to the loaded flange, the influence of its presence can be significant. The maximum increase in critical load (in the context of linear buckling theory) is achieved when the longitudinal stiffener is at (about) one fifth of web depth. Then, for the geometry and loading considered, the gain in critical load amounts to 85%.

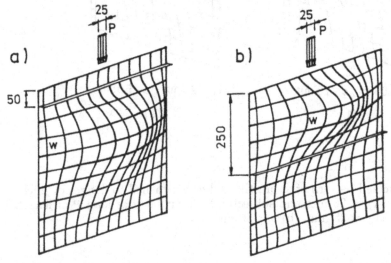

$t_f = 5$ mm, $b_1 = b/10 = 50$ mm, $c = 25$ mm $t_f = 5$ mm, $b_1 = b/2 = 250$ mm, $c = 25$ mm

Fig. 1.12. Buckled pattern of a web stiffened by a longitudinal rib situated at
the distance b_1 from the flange a) $b_1 = b/10$, b) $b_1 = b/2$.

C) Unstiffened webs attached to corner-stiffened flanges

V. Křístek and the author also looked into the effect of a corner-stiffened flange (Fig. 1.9 c). The obtained increase in critical load amounted, in the case studied with a relatively small corner-stiffened flange, to 35% and would be even more significant if a more rigid flange were used.

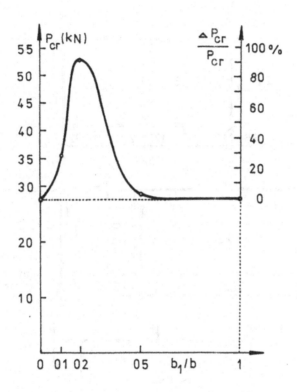

Fig.1.13. Critical load, in terms of the distance b_1 of the longitudinal rib from the loaded flange, of a longitudinally stiffened web subjected to a partial edge load.

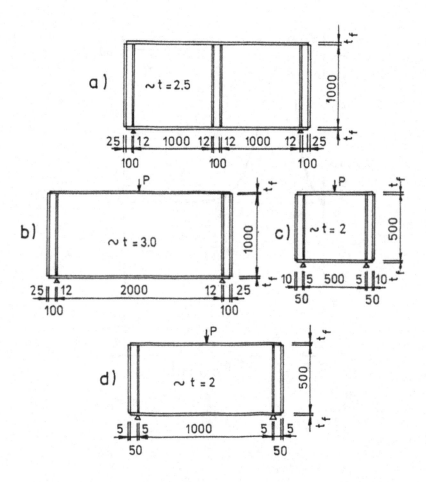

Fig.1.14. General details of the test girders : a) series TG 1 – TG 5', b) series TG 6 – TG 10' and TG 11 – TG 15', c) series STG 1 – STG 6 , and d) STG 7 – STG 11.

Table 1.1. Geometrical Characteristics of the Test Girders

Girder	α	λ	$b_{f[mm]}$	$t_{f[mm]}$	I_f/a^3t
TG 1			160	5.5	0.887
TG 1'			160	5.42	0.849
TG 1"			160	5.42	0.849
TG 2			200	10.09	6.85
TG 3	1	400	200	16.24	28.55
TG 4			200	20.17	54.70
TG 5			250	30.88	245.39
TG 5'			250	30.50	236.44
TG 6			160	6.29	0.138
TG 6'			160	6.38	0.144
TG 7			200	10.00	0.694
TG 8			200	16.55	3.148
TG 9			199	19.78	5.347
TG 10			251	29.9	23.297
TG 10'			250.9	29.95	23.405
TG 11	2	333	160	6.32	0.140
TG 11'			160	6.28	0.138
TG 12			200	10.34	0.768
TG 13			200	16.38	3.0523
TG 14			199	19.68	5.267
TG 15			250.8	30.05	23.670
TG 15'			250.9	29.95	23.405
STG 1			50	5.95	3.48
STG 2			50	5.97	3.50
STG 3			45	16.21	63.89
STG 4	1		45	16.25	64.36
STG 5			50	24.80	254.22
STG 6		250	50	24.40	242.11
STG 7			50	4.98	0.257
STG 8			50	4.96	0.254
STG 9			45	15.88	7.51
STG 10	2		45	15.88	7.51
STG 11			60	24.80	38.13
STG 12			60	24.80	38.13

1.4.2. Ultimate load behaviour of webs subject to patch loading

As theoretical evidence about this problem, based on an application of the non-linear theory of large deflections, is to date very scarce, design approaches using formulae for the ultimate loads of webs under the action of an in-plane partial edge load are usually based on experimental results. Also in Prague, numerous series of tests on plate girders under the above kind of loading have been under way for fifteen years now, girders both without and with longitudinal stiffeners and subject both to constant and repeated loadings being studied.

A) Tests on plate girders without longitudinal stiffeners

This campaign of tests comprised 34 experiments and was carried out by P. Novák and the author (see [1.3]).

In its framework, several series of experimental girders (see Fig. 1.14 and Table 1.1) were tested. In each of them, the web dimensions (i.e. the web width a, web depth b, web thickness t, and, consequently, also the side ratio $\alpha = a/b$ and the depth-to-thickness ratio $\lambda = b/t$) were constant, whereas the flange dimensions varied from girder to girder, in order that the effect of flange rigidity upon the behaviour of the web and the whole girder could be studied. In one test series, the experimental girders were subjected to partial edge loads of two different widths c, so that also the effect of loading width upon web performance could be investigated.

The whole experimental investigation and the results obtained are in detail described in [1.3]. Here, we shall confine ourselves to only very briefly presenting the main conclusions.

As, besides web thickness and load length, it is particularly the size of loaded flange that considerably influences the ultimate load behaviour of plate girders subjected to patch loading, this effect will now be illustrated in the light of the data obtained.

First, let us study the influence of flange dimensions upon the buckled patterns of the webs studied. This is shown in Fig. 1.15 (related to girder TG 6' having a thin flange with $t_f = 6.3$ mm) and Fig. 1.15 b (associated with girder TG 10 having a thick flange – $t_f = 30$ mm). A comparison of both figures indicates that the effect of the size of the loaded flange on the web buckled pattern is very significant : in the latter case (girder TG 10 with bulky flanges), the buckled surface is much more distributed over the web panel, so that the behaviour of the web is much more homogeneous.

Let us now perform a similar comparison for the patterns of vertical membrane stress σ_{my} in buckled webs, which play an important role in the plastification process of the web and in the failure mechanism of the whole girder. This is done (for membrane strains ϵ_{my}) in Fig. 1.16, again for a girder with thin flanges (this time it is TG 1) and one with heavy flanges (TG 5).

The width of the intensively stressed part of the web depended on the flange thickness. In the case of a girder with flexible flanges (such as TG 1, whose strain plots are depicted in Fig. 1.16 a), the membrane stress pattern, like the buckled surface w of the web discussed above, is localized in the vicinity of the partial edge load. On the other hand, for a girder with rigid

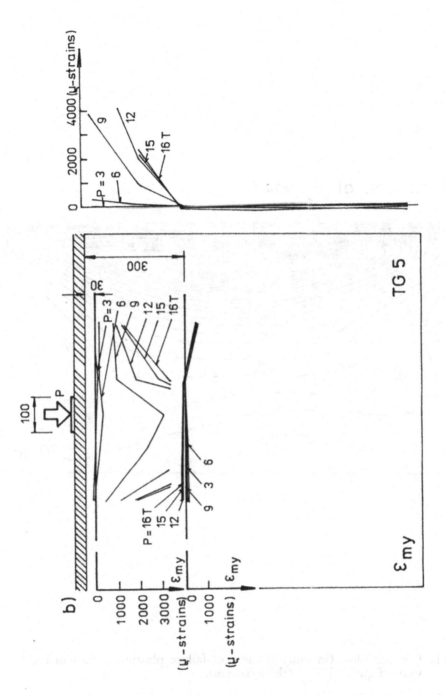

Fig.1.16 b. Membrane strains $\epsilon_{m,y}$ in the web of girder TG 5 (thick flanges).

After failure at P_u $=14$ T

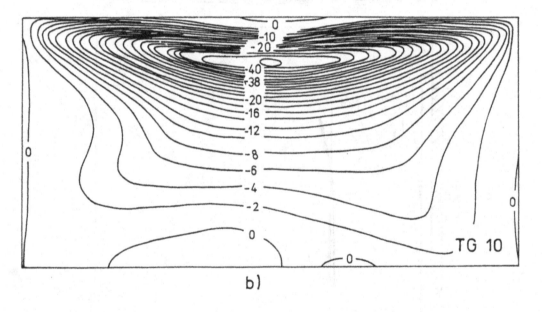

b)

Fig.1.15 b. Contour plots (in mm) of the post-failure plastic residues in the
webs of girder TG 10 (thick flanges).

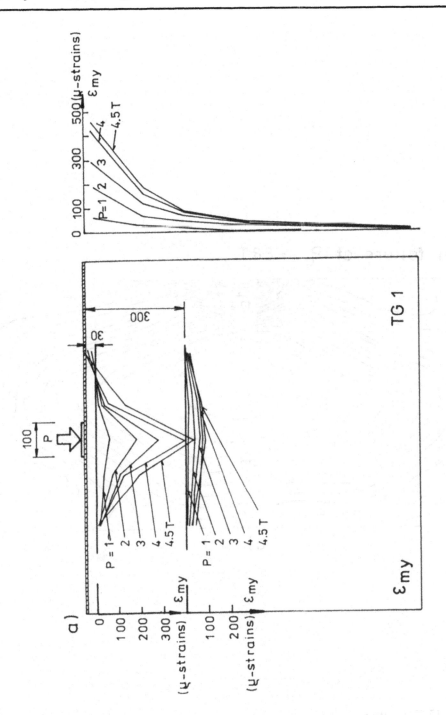

Fig.1.16 a. Membrane strains $\epsilon_{m,y}$ in the web of girder TG 1 (thin flanges).

After failure at P_u = 8.8 T

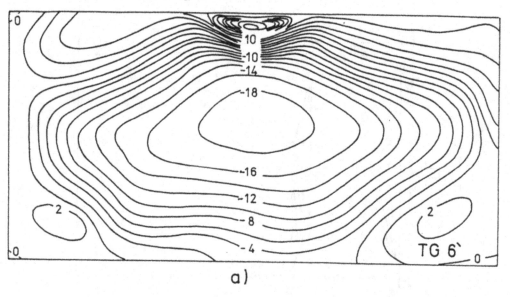

a)

Fig.1.15 a. Contour plots (in mm) of the post-failure plastic residues in the
webs of girder TG 6' (thin flanges)

flanges (such as TG 5, studied in Fig. 1.16 b), the stress pattern, again like the buckled surface w, is wider and more distributed over the width of the web panel. The behaviour of the web is then more homogeneous; this affecting – as will be demonstrated below – very beneficially the ultimate load of the girder.

It also follows from the figures that in the case of flexible flanges are the strains ϵ_{my} still small under a load amounting to 90% of the experimental load-carrying capacity. For this reason it seems very likely that the loaded flange and the adjacent portion of the web buckled inwardly (and the girder failed) without any pronounced plastification in a significant portion of the web.

The situation is different with girders having bulky flanges, where the vertical membrane strains ϵ_{my} in the upper part of the web are larger. They indicate that at higher loads the aforementioned (fairly wide and deep) area of the web has already yielded, without this bringing about failure of the test girder. Such a girder can therefore sustain further load – thanks to the rigidity of the boundary framework consisting of the flanges and vertical stiffeners – even after the web has plastified in a considerable portion.

Let us now pay attention to the failure mechanism of plate girders subject to patch loading.

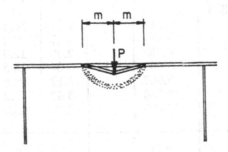

Fig. 1.17. Plastic hinges in the loaded flange.

To start with, we can say that it follows from experimental observations that a girder subjected to a discrete edge load fails when the web buckles to the extent of not being capable to provide the loaded flange with sufficient support, and the flange, under the action of the partial edge load and an axial force induced by the flexure of the girder, buckles inwardly.

Then it was concluded during the tests that in the case of girders with flexible flanges there was a tendency for a set of more or less developed plastic hinges to form in the loaded flange. One or two formed under the partial edge load itself, one occurring when the load was short (similar to a knife edge load) and two developing – at the ends of the loaded interval – when the load

Table 1.2. Experimental Ultimate Loads of the Test Girders

Girder	α	λ	I_f/a^3t	c/a	P_{cr}^{RB} (T)	P_u (T)	$\frac{P_u}{P_{cr}^{RB}}$
TG 1			0.887		2.18	5.0	2.29
TG 1'			0.849		2.18	5.0	2.29
TG 1"			0.849		2.18	5.25	2.41
TG 2	1	400	6.85	0.1	2.24	6.5	2.90
TG 3			28.55		2.36	7.0	2.97
TG 4			54.70		2.48	9.0	3.63
TG 5			245.39		2.78	18.0	6.47
TG 5'			236.44		2.78	18.5	6.65
TG 6			0.138		2.80	7.9	2.82
TG 6'			0.144		2.80	8.8	3.14
TG 7			0.694		2.85	10.0	3.51
TG 8			3.148	0.05	2.96	12.0	4.05
TG 9			5.347		3.05	12.8	4.20
TG 10			23.297		3.33	14.0	4.20
TG 10'			23.405		3.34	16.0	4.79
TG 11	2	333	0.140		2.87	10.0	3.48
TG 11'			0.138		2.87	9.0	3.14
TG 12			0.768		2.92	12.0	4.11
TG 13			3.052	0.1	3.04	13.5	4.44
TG 14			5.267		3.12	15.5	4.97
TG 15			26.630		3.42	17.2	5.03
TG 15'			23.405		3.42	15.0	4.39
STG 1			3.48		2.18	3.6	1.65
STG 2			3.50		2.2	4.0	1.82
STG 3	1		63.89		2.53	5.5	2.17
STG 4			64.36		2.53	5.5	2.17
STG 5			254.22		2.81	7.5	2.64
STG 6			242.11		2.85	8.0	2.81
STG 7		250	0.257	0.1	1.6	3.75	2.34
STG 8			0.254		1.6	3.5	2.76
STG 9	2		7.51		1.74	4.8	2.76
STG 10			7.51		1.74	5.25	3.02
STG 11			38.13		1.98	6.0	3.03
STG 12			38.13		1.98	5.5	2.78

length c was larger. Two other plastic hinges occurred at a certain distance from there. It was observed that this distance grew with the size of the flange.

However, it was also seen that an increase in the flange size led to a reduction of the "hinge character" of the plastic hinges. When the flange was very flexible, the plastic hinges were well developed. With increasing the flange thickness, the hinges became less conspicuous; and for bulky flanges they completely vanished. Only monotonous plastic flexure of the loaded flange was detected in the latter case. This conclusion is similar to the observation the author made in his shear girder tests [1.4], when studying the effect of flange inertia upon the character and location of the plastic hinges in the flanges of a plate girder loaded in shear.

Besides the partial plastic hinges in the loaded flange, the author also detected yielded zones in the web. They were segmental and tended to emanate from the flange hinges (in the case of thin flanges) or from the web panel corners (when the flanges were thick). It was noted that they coincided with the areas of sudden change in the curvature of the buckled surface of the web; therefore, they represented a kind of line plastic hinges in the web. In the test, they were marked either by peeling off of lime paint or steel scale.

Schematically, the failure mechanism is shown in Fig. 1.17.

To conclude, let us present in Table 1.2 the ultimate loads of the test girders – in comparison with the critical loads $P_{cr}^{R,B}$, evaluated by means of the charts given in [1.5].

An analysis of the table makes it possible to study the effect of (i) web geometry and (ii) flange geometry upon the load-carrying capacities of the test girders. In particular, for a given web thickness and a given load length, it is flange size that very considerably influences the ultimate load of the girder. This is fully compatible with what was said above about the effect of the flange size upon (i) web buckled pattern and (ii) web stress state.

B) Tests on plate girders with longitudinal stiffeners

Test girders

This experimental investigation was carried out by K. Januš, I. Kárníková and the author. It comprised 180 tests, sub-divided into four stages, the following quantities being varied in them:
   - the distance of the longitudinal stiffener from the loaded flange,
   - the size (and flexural rigidity) of the longitudinal stiffener,
   - the character (single- or double-sided) of the longitudinal stiffener,
   - the depth-to-thickness ratio of the web,
   - the aspect ratio of the web, and
   - the size of the loaded flange.

In all experiments, the load length, c, was constant, $c = a/10$. The general details of the test girders can be seen in Fig. 1.18.

The objective of the first test series was to obtain some general information about the effect of all main parameters influencing the ultimate load behaviour of longitudinally stiffened plate girders under the action of a discrete edge load, viz. of aspect ratio of web panel, size of longitudinal stiffener and its distance from loaded flange. Then, in the second and third stages of the research, the longitudinal rib was located at a constant distance, $b_1 = 0.2b$,

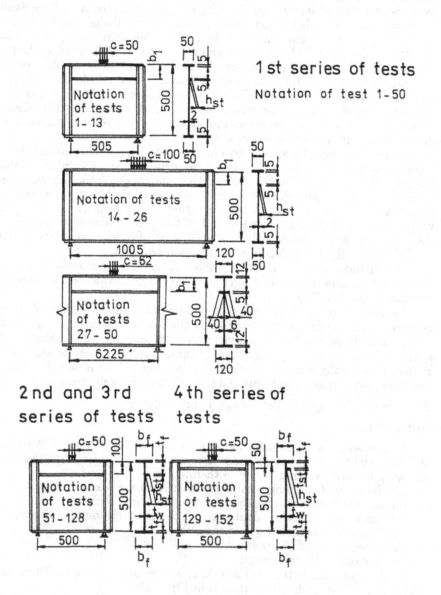

Fig.1.18. General details of the longitudially stiffened test girders.

from the loaded flange, the aim of the tests being focused on studying profoundly the role of stiffener rigidity (or flexibility). In these experiments, also web thickness was varied, web plates being fabricated from sheet of thickness $t_w = 2$, 4 and 6 mm. The tests of the following (i.e., fourth) stage were conducted on girders having a constant web thickness, $t_w = 4$ mm. Apart from that, the test panels were practically identical with those of the second and third series, but the longitudinal rib was moved nearer to the loaded flange, viz. to the distance of $b_1 = 0.1\,b$.

The whole experimental investigation and its results are in detail described in [1.6] and [1.7]. Therefore, here only the main conclusions are given.

Ultimate loads of test girders

Fig. 1.19 a shows the relationship between the experimental ultimate loads, P_u^{exp} , of the test girders belonging to the first research stage and having a longitudinal stiffener of (i) 12 × 5 mm and (ii) 30 × 5 mm (which was (i) the largest and (ii) the smallest size tested) and the distance, b_1 , of the stiffener from the loaded flange. In the relationship, the dimensionless quantity P_u^{exp}/P_u is used, where P_u is the transformed (with due allowance to the fact that the material characteristics of the longitudinally stiffened test girders were different) ultimate strength of the experimental girders without longitudinal stiffening.

Fig. 1.19 b then shows the same relationship, but plotted for the average values of the experimental ultimate loads of all test girders.

Failure mechanisms of test girders

All test girders collapsed by the formation of a failure mechanism which consisted of (i) a segmental line plastic hinge in the web sheet and (ii) three point plastic hinges in the loaded flange. The segmental line plastic hinge in the web formed in the vicinity of the loaded flange, the size of this flange considerably influencing the configuration of the line hinge. The larger the size of the flange, the longer and deeper the segment. The effect of the longitudinal rib was reflected in shortening the length and depth of the arc of the line hinge, if compared with the situation in a web having the same dimensions but not being reinforced by a longitudinal rib (see Fig. 1.20).

The plastic hinges in the loaded flange occurred (i) at mid-span, i.e. under the narrow partial edge load, and (ii) at those flange sections from which the segmental line plastic hinge in the web emanated. The distance of the outer hinges from the central one was influenced by the flange size; i.e. the larger the size of the flange, the greater was the distance of the outer hinges.

Effect of the size of the loaded flange

An examination of the obtained results indicates that the effect of the size of the loaded flange upon the load-bearing capacities of the webs (and the whole plate girders) is very significant. It considerably influences the dispersion of the load in the web, its buckling and even the collapse mechanism of the plate girder.

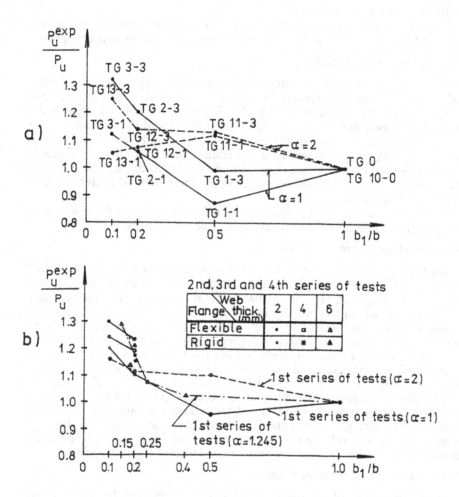

Fig.1.19. Relationships between the experimental ultimate loads, P_u^{exp}, and
the distance, b_1, of the longitudinal stiffener from the loaded flange:
a) individual ultimate loads for some girders of the lst test series,
b) average values for all test series.

Effect of the position and size of the longitudinal stiffener

An analysis of the results reveals that a longitudinal rib can substantially improve the behaviour of a plate girder web subject to partial edge loading only when the rib is located in the vicinity of the loaded flange, i.e., if $b_1 < b/4$. This conclusion follows from the observation that, in the case of a web under the action of a discrete edge load, the buckling is pronounced merely in the neighbourhood of the applied patch load. If the longitudinal stiffener is positioned off this region, it is not able to influence the web buckling and, consequently, the ultimate load behaviour of the girder.

Fig. 1.20. Collapse mechanism.

As for the effect of the size (and rigidity) of the rib, this was pronounced only in the range of flexible ribs, where the curve plotting the relationship between ultimate load and stiffener rigidity exhibited a conspicuous rising tendency. In the domain of larger stiffeners, the curve was flat, so that the growth of ultimate strength with a further increase in stiffener size and rigidity was very slow (see Figs 1.21 a – c).

The quantity γ_r^{opt} appearing in the figures is the optimum rigidity of

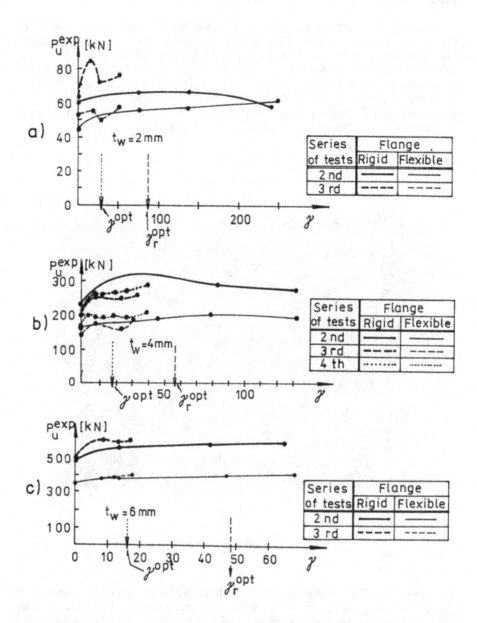

Fig.1.21. Relationship between the experimental ultimate loads, P_u^{exp} , and the flexural rigidity, γ , of the longitudinal stiffener.

perfectly rigid longitudinal stiffeners, for which K. Januš, I. Kárníková and the author established, by means of their experimental results, in [1.7] a formula. An inspection of the figures shows, however, that, at least as far as the effect of stiffener size on ultimate strength is concerned (i.e., without caring whether the stiffener deflects or not), considerably weaker stiffeners (e.g., those with $\gamma = \gamma^{opt} = \gamma_r^{opt}/3$, or even smaller ones) will do.

Effect of the aspect ratio of the web and of the character of the longitudinal stiffener

Neither the aspect ratio of the web, nor the character (single- or double-sided) of the stiffener influenced in a significant way the ultimate load behaviour of the webs of the test girders.

Effect of the longitudinal stiffener on web deflection

One of the objectives of the tests was to study how the presence of a longitudinal stiffener can help to reduce the deflections of webs under partial edge loading if compared with the buckled patterns of unstiffened webs.

An analysis of the obtained data indicates that, in the case of the writer's experiments, this reduction was pronounced and amounted on an average to 42%.

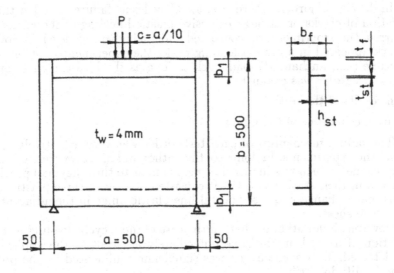

Fig. 1.22. Test girders.

C) Teste on plate girders subject to repeated loading – web "breathing"

This is the most recent stage of experiments conducted (again by K. Januš, I. Kárníková and the author) in Prague on plate girders under the action of a partial edge load.

Test girders

80 test girders subject to variable repeated patch loading have been tested to date, their webs being fitted with a longitudinal stiffener positioned at (i) one tenth, (ii) one fifth of the web depth. For comparison, a few girders without stiffening were tested, too.

The test girders had the same dimensions, and were fabricated from the same material, as those used in the static patch loading tests, this being so as to enable the author and his associates to compare the results (ultimate loads, onset-of-yielding loads, etc.) of both respective experimental series.

In all experiments, the load length, c, was constant, $c = a/10$. The general details of the test girders can be seen in Fig. 1.22.

Test set-up

The load P cycled between (i) zero and (ii) a value P_{max} , which in turn was varied between α) the statical ultimate load P_u^{exp} and β) the onset-of-surface-yielding load, detected also in the related static test. Thus, under the above loading, the webs of the girders tested behaved in the elasto-plastic range and, consequently, their performance was expected to be governed by low-cycle fatigue. Therefore, the basic number of loading cycles was chosen so as to be equal to 5×10^4 .

In the case of girders where after 5×10^4 cycles no failure (whether through initiation of cracks or through excessive plastic buckling of the girder web) occurred, the experiment was continued under a higher load level. If, however, a crack appeared in a certain loading cycle, the experiment went on, under the same (i.e., unchanged) load level, as long as the load-carrying capacity of the test girder was exhausted.

Main results of the tests

Failure mechanisms of the girders

The failure mechanism of each test girder was very attentively studied during the experiments. In doing so, the author and his co-workers concluded that this mechanism was in principle very similar to that they had previously observed in their static tests, viz. it consisted of a set of three plastic hinges in the loaded flange and a segmental line plastic hinge in the adjacent zone of the web sheet.

However, in addition to that, in the case of most cyclic loading tests, the initiation of a crack in the zone immediately adjacent to the applied load was detected. This phenomenon was significianty influenced by the position of the longitudinal rib.

With girders having a rib at the distance $b_1 = 0.1\,b$ from the loaded flange, such a crack appeared (with the exception of two tests) as late as the girder was collapsing, the crack initiating in the web sheet either in the segmental line hinge or close to the weld underneath the loaded flange. In the case of some girders, no crack developed at the collapse of the girder.

Conversely, with girders having a stiffener at $b_1 = 0.2\,b$, a crack practically always (i.e., barring one case) initiated close to the weld, but the test girder

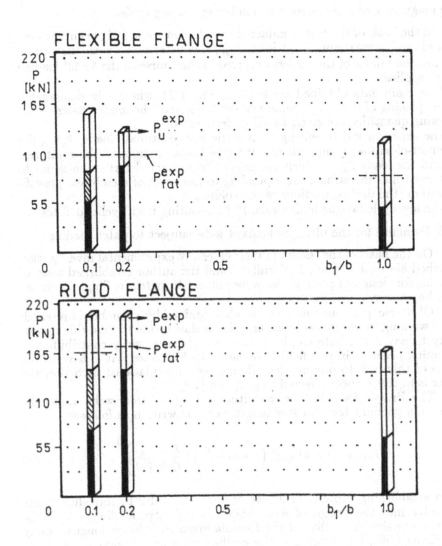

Fig.1.23. Main data obtained.

concerned was able to sustain thereafter a high number (of the order of 10^5) of loading cycles before its load-bearing capacity was exhausted.

The magnitude of load versus the number of loading cycles

On the basis of the tests conducted to date, low-cycle fatigue curves were plotted, and a maximum limit load, P_{fat}^{exp}, was determined such that under this load no failure of the girders occurred in the course of the 5×10^4 loading cycles applied.

The main data obtained are listed in Fig. 1.23, where – in dependence on the placing of the longitudinal rib and the size of the loaded flange – the following quantities are given for all girders tested:

a) the experimental (resulting from static tests) ultimate loads, P_u^{exp} , the onset-of-surface-yielding loads, P^{su} (the black portions of the columns), and the loads $P_{3\epsilon}^{me}$, which are related to unitary deformation equal to three times the strains corresponding to the onset of membrane plastification (the dashed portions of the columns),

b) the low-cycle fatigue limit loads, P_{fat}^{exp} , resulting from cyclic load tests.

1.4.3. Formulae for the ultimate loads of webs subject to patch loading

On the basis of the results of their extensive experimental investigation described above, K. Januš, I. Kárníková and the author established a set of formulae for design of plate girder webs subject to patch loading, which are given herebelow.

Of course, precious evidence was also obtained by a number of research teams working in other countries, and the author wants to take this opportunity to pay his tribute to all of those who have recently contributed to promoting progress in this field of research. Unfortunately, it is beyond the means of this text to present their design recommendations; therefore, the reader is in this respect referred to [1.2] and [1.8].

The Prague formulae for the ultimate loads of steel webs under the action of a partial edge load P of length c can be written as follows:

$$P_u(x = b_1) = AR_{d,w}t_w^2 \left(1 + 0.004\frac{c}{t_w}\right) \left(\frac{I_f}{t_w^4}\sqrt{\frac{R_{d,f}}{210}}\right)^B . \qquad (1.9)$$

When establishing formula (1.9), we strove – in order to introduce some continuity into the design of webs under partial edge loading – to make its structure similar to that of the formula given in [1.9] for longitudinally unstiffened webs; but, of course, the coefficients A and B, derived via non-linear regression, were different from their counterparts in [1.9] in order to allow for the beneficial presence of the longitudinal stiffener. Moreover, the effect of flange material was also inserted into the formula.

Formula (1.9) is written – as is frequent practice in the currently held Limit State Design philosophy – not as a function of yield stresses, R_y , but it is formulated in terms of so-called design strengths, R_d (which represent a kind of guaranteed – with a given probability – minimum yield stress; usually $R_d \doteq 0.85 R_y$, R_y being the normative yield stress of the steel concerned). Then,

$R_{d,w}$ is the design strength of the web material and $R_{d,f}$ that of the flange material.

But this formula can be easily transformed for yield stresses $R_{y,w}$ and $R_{y,f}$; viz. by multiplying the coefficient A by the numerical value of the ratio $R_{d,w}/R_{y,w}$, by replacing $R_{d,w}$ by $R_{y,w}$ and by writing the square root as $\sqrt{R_{y,f}/240}$.

b_1 is the distance of the longitudinal rib (if there is any) from the loaded flange.

I_f is the second moment of area (with respect to the centroidal axis perpendicular to the web plane) of the loaded flange. A and B are experimentally verified coefficients, derived via non-linear regression, which reflect (i) the absence or presence of a longitudinal stiffener, (ii) its position and size, if such a stiffener does exist, and (iii) the kind of loading (i.e., its constant or repeated character).

On the basis of the Prague results, the coefficients A and B assume (in a formulation with design strengths, R_d; in one with yield stresses R_y, the coefficients would be transformed as indicated above) the following values:

A) Constant loading

Webs with no longitudinal ribs (or when their distance $b_1 \geq 0.25b$):

$$A = 13.8, \quad B = 0.153. \tag{1.10a}$$

Webs with one longitudinal rib at $b_1 = 0.2b$:

$$A = 15.9, \quad B = 0.153. \tag{1.10b}$$

Webs with one longitudinal rib at $b_1 = 0.1b$:

$$A = 17.1, \quad B = 0.153. \tag{1.10c}$$

As can be seen, in this case the optimum position of the rib is one tenth of web depth, i.e. the stiffener is situated close to the loaded flange. For intermediate positions it can be linearly interpolated.

B) Repeated loading

Webs with no longitudinal ribs (or when their distance $b_1 \geq 0.25b$):

$$A = 10, \quad B = 0.132. \tag{1.11}$$

A comparison with Eq.(1.10 a) reveals that a marked reduction in ultimate strength follows in this case from the mechanism of damage cumulation in webs "breathing" under repeated patch loading.

Let us now pass to the case of longitudinally stiffened webs.

An examination of the Prague results shows that with repeated loading the optimum position of the longitudinal rib is rather at one fifth of web depth (i.e., for $b_1 = 0.2\ b$). Then formulae (1.10 b) can again be employed, which means a considerable increase in ultimate load when compared with the previous case of longitudinally unstiffened webs.

C) Combined action of a partial edge load and a bending moment

If the web under study is subject to the combined action of a patch load and a bending moment, the ultimate loads resulting from the formulae given above in secs (A) and (B) are multiplied by a reduction factor

$$\eta_b = \sqrt{1 - \left(\frac{\sigma_{b,w}}{R_{d,w}}\right)^2},$$ (1.12)

where $\sigma_{b,w}$ is the bending stress at the extreme (i.e., adjacent to the loaded flange) fibres of the web and $R_{d,w}$ is again the design strength of the web material.

References

1.1. Škaloud, M., Křístek, V.: Stability problems of steel box girder bridges, Transactions of the Czech. Academy of Sci., Series of Techn. Sci., No.1, 1980, 1–122.
1.2. Roberts, T.M., Rockey, K.C.: A mechanism solution for predicting the collapse loads of slender plate girders when subjected to in-plane patch loading, Proc. of Instn. Civ. Engrs, Part 2 (1979), 155–175.
1.3. Škaloud, M., Novák, P.: Post-buckled behaviour of webs under partial edge loading, Transactions of the Czech. Academy of Sci., Series of Techn. Sci., No.3, 1975, 1–94.
1.4. Škaloud, M., Zörnerová, M.: Post-buckled behaviour of webs in shear, Transactions of the Czech. Academy of Sci., Series of Techn. Sci., No.3, 1972, 1–96.
1.5. Rockey, K.C., Bagchi, D.K.: Buckling of plate girder webs under partial edge loading, Int. J.Mech. Sci., Pergamon Press, 12 (1970).
1.6. Škaloud, M., Kárníková, I.: Experimental research on the limit state of the plate elements of steel bridges, Transations of the Czech. Academy of Sci., Series of Techn. Sci., No.1, 1985, 1–141.
1.7. Januš, K., Kárníková, I. and Škaloud, M.: Experimental investigation into the ultimate load behaviour of longitudinally stiffened steel webs under partial edge loading, Acta technica ČSAV, 33 (1988), 2, 162–195.
1.8. Behaviour and design of steel plated structures (Edited by P.Dubas and E.Gehri), Applied Statics and Steel Structures, Swiss Federal Institute of Technology, Zurich, January 1986, 1–247.
1.9. Drdácký, M.: Steel webs with flanges subjected to partial edge loading, Chalmers University of Technology, Göteborg, 1982.

2. INTERACTION BETWEEN SHEAR LAG AND PLATE BUCKLING IN LONGITUDINALLY STIFFENED COMPRESSION FLANGES

2.1. Introduction

The design of the longitudinally stiffened compression flanges of steel box girder bridges and similar structures is governed by the interaction that in such flanges usually occurs between shear lag and flange buckling. Several approaches to analysis of this problem have been established to date. Here we shall deal with three of them (the first two having also been included into the document "Stiffened Compression Flanges of Box Girders", edited by the ECCS Technical Working Group 8/3 "Plated Structures"), viz.
(i) the British approach,
(ii) the Liège – Prague approach, and
(iii) the Prague approach.

2.2. The British Approach

This approach is based on an extensive study of the problem carried out recently at (i) the Imperial College of Science and Technology, London (by Prof. Dowling et al.) and (ii) the University of Surrey, Guildford (by Prof. Harding et al.).

It consists of two steps.

In the first one, the ultimate load $\bar{\sigma}_u$ of the compression flange is calculated by taking account of flange buckling only. This is achieved via the so-called strut approach (or the bar-simulation analogy), which means that the load-carrying capacity of the flange concerned is determined as the sum of the ultimate strengths of the longitudinal stiffeners taken as struts. The effective cross-section of each stiffener is considered to be composed of the cross-section of the stiffener proper and of the effective portions (calculated by means of a formula for the effective width of a compression plate) of the adjacent sheet panels. The effective portions of the boundary sheet panels, i.e., of the panels next to the longitudinal flange edges – at the junction of the flange with the girder webs – can also be accounted for. This is usually materialized approximatelly by enlarging the number of the longitudinal ribs by one.

In the second step, the analysis is modified so as to also take the effect of shear lag into consideration. This consists in multiplying the aforesaid flange ultimate loads $\bar{\sigma}_u$, determined with due regard to flange buckling but regardless of shear lag, by a factor ψ of shear lag effect.

In the context of the first version of the British approach, this factor
(i) is taken as unity for $b/L \leq 0.2$ (where L is the girder span in the case of simply supported beams or the distance between points of contraflexure for continuous spans and b is the overall flange width).
(ii) For $b/L > 0.2$, it is given by the following equation

$$\psi = \varrho_s^{(b/2L)}. \tag{2.1}$$

There ϱ_s is the elastic shear lag effective width ratio, which in various Codes of Practice is given by formulae, charts or tables.

In the latest version of the British approach, which is due to J. Harding, ψ is again taken as unity for $b/L \leq 0.2$, but for $b/L > 0.2$

$$\psi = \varrho_s \left(2.26 - \frac{\delta^2 \lambda^2}{13\ 000} \right) < \varrho_s, \qquad (2.2)$$

where $\delta = mA_s/bt$ (m being the number of longitudinal stiffeners, A_s the cross-sectional area of one stiffener without effective width of flange sheet and t the thickness of flange sheet), $\lambda =$ the slenderness ratio of the stiffener with account being taken of its effective cross-section. ϱ_s has the same meaning as above and can be easily calculated using the following formulae established by J.Harding:

$$\varrho_s = e^{-(b/L)^{0.75}(1.4+\delta)} \qquad (2.3a)$$

for a girder under point load and for $b/L \leq 1$;

$$\varrho_s = e^{-(b/L)^{1.75}(1.0+\omega\delta)}, \qquad (2.3b)$$

where $\omega = 1.75 - 1.5\frac{b}{L}$, for a girder under uniformly distributed loading and for $b/L \leq 1$.

For $b/L > 1$(but such cases will be rare), ϱ_s should be taken as the value for $b/L = 1$ divided by b/L.

2.3. The Liège – Prague Approach

This approach, based on a solution via the theory of large deflections, is more complex. In its context, the analysis is divided into two stages: in the first one, the interaction between global flange buckling and shear lag is dealt with, while in the other stage the solution is modified so as to take also account of local flange sheet buckling between longitudinal ribs. The solution was carried out in cooperation between (i) the Department of Mechanics and Stability of Structures of the University of Liège (Ph. Jetteur and R. Maquoi) and (ii) the Stability Department of the Institute of Theoretical and Applied Mechanics of the Czechoslovak Academy of Sciences (M. Zörnerová and the author), see [2.1].

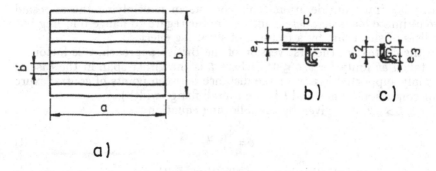

Fig. 2.1 a. Geometry of longitudinally stiffened compression flanges.

2.3.1. Interaction of shear lag with global buckling

Let us study the post-buckled behaviour of a longitudinally stiffened compression flange as depicted in Fig. 2.1 a.

In our analysis, the discretly stiffened flange plate is substituted by two continuous structural systems, viz. (i) the isotropic flange sheet and (ii) a fictitious plate that idealizes the longitudinal ribs, and the geometrical characteristics of which are as follows:

$$D_x = \frac{EI}{b},$$
(2.4a)

$$D_y = 0,$$
(2.4b)

$$H = 0.$$
(2.4c)

As the longitudinal stiffeners are one-sided, the centroids of both continuous systems do not coincide. The eccentricity of the centroid of the sheet will be denoted as e_1 (Fig. 2.1 b), that of the fictitious plate as e_2 ; both are measured from the centroid of the sheet plus stiffeners system.

The interaction between the compression flange and the vertical webs of the bridge girder is disregarded; hence, the flange is assumed to be simply supported along its four edges. The dimensions of the longitudinal ribs are such that the local instability of stiffeners is no problem. However, the analysis takes account of the shear lag phenomenon present in the flange. Then the distribution of the longitudinal normal stress, σ_x , across the flange width is not uniform.

The formulation is based on a generalized Reissner's principle that takes account of the non-linear geometrical effects. Reissner's principle is a variational principle with two independent fields, σ and u , about which assumptions are made a priori. Both fields are bound to be continuous, but besides that they can be arbitrary. However, it is in the nature of things that the quality of the obtained results significantly depends on whether or not a realistic choice for the two fields have been made.

The mathematical analysis starts from the functional

$$W = \int \left[N_{x1}(u_x - e_1 w_{xx} + \frac{1}{2}w_x^2 + w_x w_{0x}) + \right.$$

$$+ N_{y1}(v_y - e_1 w_{yy} + \frac{1}{2}w_y^2 + w_y w_{0y}) +$$

$$+ N_{xy1}(u_y + v_x + w_x w_y + w_x w_{0y} + w_y w_{0x} - 2e_1 w_{x,y}) -$$

$$- \frac{1}{2Et}(N_{x1}^2 - 2\nu N_{x1}N_{y1} + N_{y1}^2 + 2(1+\nu)N_{xy1}^2) +$$

$$+ N_{x2}(u_x + e_2 w_{xx} + \frac{1}{2}w_x^2 + w_x w_{0x}) - \frac{b}{2EA}N_{x2}^2 +$$

$$+ \frac{D}{2}(w_{xx} + w_{yy})^2 - D(1-\nu)(w_{xx}w_{yy} - w_{xx}^2) + \frac{EI}{2b}w_{xx}^2 \right] dxdy +$$

$$+ \int \left[N_{x1}(\bar{U} - u + e_1 w_x) + N_{x2}(\bar{U} - u - e_2 w_x) \right] dy \Big|_{x/a=0}^{x/a=1},$$
(2.5)

where u, v and w are the displacements measured with respect to the horizontal plane passing through the centroid of the sheet plus stiffeners system.

The following assumptions for the force and displacements fields are introduced into the analysis:

$$N_{xt} = t\left\{\sigma_0\left(1 - \epsilon\sin\frac{\pi y}{b}\right) - \sigma_1\sin^2\frac{m\pi y}{b} - e_1\sigma_f\sin\frac{m\pi y}{b}\sin\frac{\pi x}{a}\right\}, \quad (2.6a)$$

$$N_{y1} = 0, \quad (2.6b)$$

$$N_{xy1} = 0, \quad (2.6c)$$

$$N_{x2} = \frac{A}{b}\left\{\sigma_0\left(1 - \epsilon\sin\frac{\pi y}{b}\right) - \sigma_1\sin^2\frac{m\pi y}{b} + e_2\sigma_f\sin\frac{m\pi y}{b}\sin\frac{\pi x}{a}\right\}, \quad (2.6d)$$

$$u = u_0\left(1 - \epsilon\sin\frac{\pi y}{b}\right)\frac{x}{a}, \quad (2.7a)$$

$$\bar{U} = \bar{U}_0\left(1 - \epsilon\sin\frac{\pi y}{b}\right), \quad (2.7b)$$

$$v = 0, \quad (2.7c)$$

$$w = f\sin\frac{\pi x}{a}\sin\frac{m\pi y}{b}, \quad (2.7d)$$

$$w_0 = f_0\sin\frac{\pi x}{a}\sin\frac{m\pi y}{b}. \quad (2.7e)$$

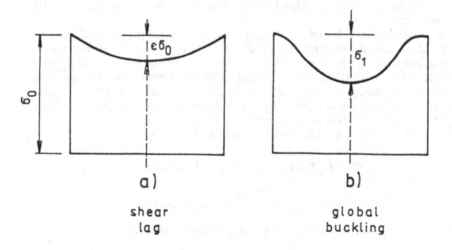

Fig. 2.2. Effect of shear lag and flange buckling.

The meaning of σ_0, σ_1 and ϵ follows from Fig. 2.2. There it can be seen that σ_0 is a compression stress field that acts in the flange plate in the absence of any shear lag and buckling phenomena. $\epsilon\sigma_0$ denotes the loss of efficiency that occurs in the flange owing to shear lag, whilst σ_1 takes account of the loss of flange efficiency due to buckling. σ_f is a parameter of the bending stress in the sheet plus stiffeners system; m designates the number of half-waves (in the transverse direction, y) of the buckled surface of the flange.

A is the area of the cross-sections of all longitudinal stiffeners.

Bearing in mind that the distribution of the longitudinal normal stress, σ_{xs}, that exists in the flange in the absence of buckling, but takes account of the effects of shear lag, can be written as follows:

$$\sigma_{xs} = \sigma_0 \left(1 - \epsilon \sin \frac{\pi y}{b}\right),\tag{2.8}$$

the shear lag factor, ϱ_s, can be defined in the following way:

$$\varrho_s = \frac{\int_0^l \sigma_0 \left(1 - \epsilon \sin \frac{\pi y}{b}\right) dy}{\sigma_0 b} = 1 - \frac{1\epsilon}{\pi}.\tag{2.9}$$

The condition of the minimum of the functional W, i.e. $\partial W = 0$, leads to the mathematical solution of the problem (see [2.1]), which then forms a basis for the subsequent static solution.

To start with, let us confine our analysis to global overall buckling of longitudinally stiffened compression flanges. Even then several approaches to the definition of the ultimate limit state are possible, each of them reflecting a different level of plasticity in the structure.

A) Definition assuming a high level of plasticity in the flange

A very optimistic definition of the ultimate limit state of longitudinally stiffened flanges is achieved when the limit state is defined by the condition that the average longitudinal normal stress, $\bar{\sigma}_{stiffener}$, in the stiffeners attains the yield stress, R_y, or the design stress, R_d, of the flange material.

It is, perhaps, worthwhile to note at this juncture that in the Limit State Design philosophy the s.c. design strength, R_d, describes a certain (defined via a certain degree of probability) minimum yield stress of the material of the structural element under consideration. Then $R_d = kR_y$, where, for instance, in Czechoslovakia the coefficient k is about 0.85. Thus, for most elements made of Steel 37 one has $R_d = 210 \ N/mm^2$ and for those fabricated from Steel 52 $R_d = 290 \ N/mm^2$.

So,

$$\bar{\sigma}_{stiffener} = -R_d.\tag{2.10}$$

It is in the nature of things that the above limit state criterion allows a high level of plastic deformations in the flange. For example, it permits a complete redistribution of stresses when shear lag is present and when there is no flange buckling.

The results of the calculations (performed under the assumption of $m = 1$) of the corresponding ultimate loads, $\bar{\sigma}_u$, of longitudinally stiffened compression flanges are plotted, for $R_d = 210$ and $290 \ N/mm^2$, for various values

---

*   In those countries where the concept of design strength, R_d, is not introduced into analysis, but one works with yield stress, R_y, the quantity R_d in the following equations is to be replaced by R_y.

of e_2/a and in terms of the parameter $\sqrt{R_d/\sigma_{cr}}$ (σ_{cr} being the critical load of the stiffened flange – see formula (2.11)), in Fig. 2.3.

$$\sigma_{cr} = \frac{\pi^2}{a^2} \frac{\left[bD(1+m^2\alpha^2)^2 + EI + E(e_1^2 bt + e_2^2 A)\right]}{bt + A} .$$

$$(2.11)$$

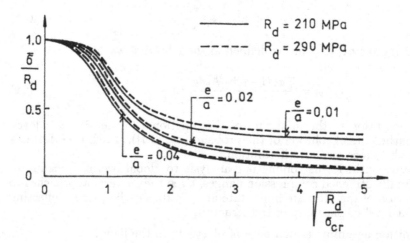

Fig. 2.3. Ultimate loads of longitudinally stiffened compression flanges – limit state criterion (2.10).

The numerical calculations were carried out for $f_0 = a/500$, this value having been chosen on the basis of the conclusions of the work of the Task Group "Tolerances in Steel Plated Structures", which operated within the framework of IABSE [2.2].

An analysis of the equations and the results obtained shows that, if the above plastic definition of the ultimate limit state is used, the ultimate load of the flange is not influenced by the shear lag factor, ϱ_s . Then the effect of shear lag can be completely neglected, at least as far as ultimate limit state considerations are concerned.

B) Definition assuming a lower level of plasticity in the flange

The above definition of the ultimate limit state operated with the average stress $\bar{\sigma}_{stiffener}$ related to the whole surface of the flange. A more severe definition of the ultimate limit state is adopted when the analysis is based on the quantity $\bar{\sigma}_{stiffener}(x = a/2)$, i.e. on the average longitudinal stress calculated for the most loaded transverse section of the flange($x = a/2$). Then the limit state criterion may be written as

$$\bar{\sigma}_{stiffener}(x = a/2) = R_d.$$

$$(2.12)$$

The resulting ultimate loads, $\bar{\sigma}_u$, calculated again for $f_0 = a/500$ and plotted in the same way as above in sec. A., are shown in Fig. 2.4. As expected, they are lower than those given in Fig. 2.3, which are related to a higher degree of plasticity in the flange.

It again follows from an examination of the governing equations and the results obtained that, even in the context of the "less plastic" definition of the ultimate limit state, the resulting ultimate load of the flange is not affected by shear lag.

C) Elastic (or quasi-elastic) definition

In most countries the design of steel bridges and similar structures is based on elastic or quasi-elastic considerations. This means that the ultimate limit state is based on criteria that operate with the onset of yielding or a well-limited degree of plasticity in the bridge structure.

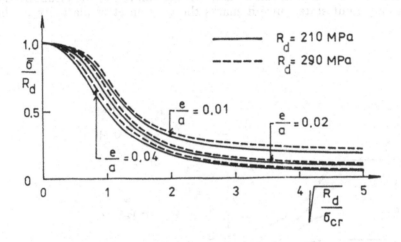

Fig. 2.4. Ultimate loads of longitudinally stiffened compression flanges –
limit state criterion (2.12).

Then, of course, the definition of the ultimate limit state of compression flanges must follow the same approach. Experiments on large-scale test girders show (see, for instance, [2.3]) that yielding in longitudinally stiffened compression flanges can commence either in the flange sheet or in the ribs. Consequently, the limit state is defined by the three following criteria:
a) onset of membrane yielding in the flange sheet at the longitudinal edges of the flange,
b) onset of membrane yielding in the flange sheet in the middle of the flange,
c_1) onset of yielding at the centroid of the central longitudinal stiffener (i.e., of the stiffening element proper – without any effective portion of the flange sheet).
Condition (a) of the ultimate limit state is influenced both by shear lag

and the membrane stress state due to flange buckling, conditions (b) and (c) are in addition affected by the bending stresses occurring in the flange as a result of its buckling. Then decisive is that of them which furnishes the lowest ultimate load of the flange. This is dependent on the geometrical parameters of the flange and on the intensity of the shear lag effect.

It is obvious that the application of the aforesaid criterion c_1 (i.e., condition of the onset of yielding at the centroid of the central longitudinal rib) involves the occurrence of some plastic deformations at least in a part (between the stiffener outstand and the stiffener centroid) of the most loaded (i.e., that which is situated in the middle of the flange, which is most influenced by bending stresses due to flange buckling) longitudinal rib. For this reason an even more severe criterion of the limit state may be preferable in some cases, such as

c_2) onset of yielding at the extreme fibres (located at a distance e_3 – see Fig. 2.1 c – from the sheet + stiffeners centroid) of the stiffener outstand.

Of course, this criterion can be used with advantage for determining the serviceability limit state, since it marks the very onset of plasticity in the flange.

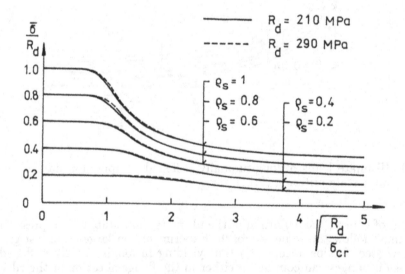

Fig. 2.5 a. Ultimate loads of longitudinally stiffened compression flanges limit state criterion (a).

With there being three criteria of the ultimate limit state, we obtain three values of $\bar{\sigma}$. The lowest among them, $\bar{\sigma}_{min}$ is decisive for the analysis. If the dimensions of the partial flange panels are such that, for the corresponding

out in Prague by M. Zörnerová and the author, that was already mentioned in sec. 2.3.3 and which comprised 2240 cases. An analysis of the conclusions of the study * indicates that, in comparison with the two large-deflection theory orthotropic plate approaches, the British strut approach furnishes ultimate loads that,for practically important geometrics, are

(i) practically always on the safe side,

(ii) frequently rather conservative, with the difference being 20 and even more per cent.

2.6. The Serviceability Limit State of Longitudinally Stiffened Compression Flanges

The analysis described in the foregoing sections was concentrated on the ultimate limit state of the longitudinally stiffened compression flanges of steel bridges and similar structures. So, it remains to explain here whether or not it is desirable to introduce also a limit state of serviceability into the design of the compression flanges; and, if so, to say how this limit state should be defined.

It is obvious that it would not be realistic to determine the limit state of serviceability by means of some aesthetic considerations, i.e. for instance on the basis of the maximum flange deflection. Nor would it be very pertinent to relate this limit state to the danger of snap-through phenomena in the buckled surface of the flange, since in the case of ordinary flanges such phenomena happen rather scarcely.

More rational does it appear to base the definition of the serviceability limit state on the requirement that under service loads any kind of plastification (i.e., not only membrane but also surface plasticization) of the compression flange be avoided. The reason for this is as follows: Every day during their long "service lives", the compression flanges are subject to many times repeated loading cycles and, consequently, "breathe" under them. Then flange buckling interacts with the phenomenon of cummulation of damage, i.e. with fatigue phenomena in the flange. More evidence is needed about the "breathing" of plate elements under repeated loading and about their failure mechanism; therefore, extensive research in this line is being conducted by the author and his associates in Prague at the moment. Until more accurate and reliable enough data about the aforesaid phenomenon are available, it seems to be advisable to define the limit state of serviceability of longitudinally stiffened compression flanges by the onset of yielding at the extreme fibres of the cross-section of the most heavily loaded longitudinal rib, i.e. with the aid of criterion (c_2) of sec. 2.3. Then the serviceability limit load of the flange can be easily determined by means of charts given in Figs 2.6 a – d in sec. 2.3 or in Figs 2.12 a – c in sec. 2.4, provided that quantily e_3 (i.e., the distance of the stiffener extreme fibres from the centroid of the sheet plus stiffeners system) is inserted for e.

It could be commented at this juncture that the criterion of the onset of yielding at the extreme fibres of the most heavily loaded longitudinal stiffener does not satisfactorily safeguard against the occurrence of surface

---

* It is beyond the means of this publication to include the numerous graphs obtained, but the reader will be made acquainted with them in another paper.

loading, local buckling of the sheet panels is no problem, this lowest value already determines the ultimate load of the flange:

$$\bar{\sigma}_{min} = \bar{\sigma}_u.$$

A numerical examination of the individual limit state criteria reveals that
(i) the first of them (onset of membrane yielding in the sheet at the longitudinal edges) is decisive when the effect of shear lag is very pronounced, and that of flange buckling is weak.
(ii) When, on the contrary, the size of the flange elements is such that flange buckling plays an important role (and, consequently, the effect of the bending stresses in the central zone of the flange is great), and this phenomenon is not overwhelmed by the redistribution of membrane stresses due to shear lag, it is either the second (onset of plastification in the sheet in the middle of the flange) or – and more frequently – the third (onset of yielding in the central stiffener) criterion that governs the analysis.

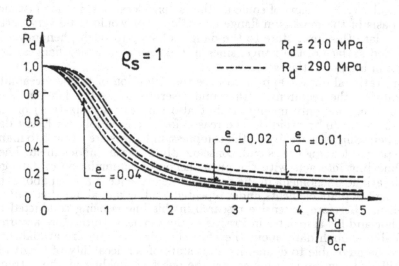

Fig. 2.6 a.

In order to facilitate the work of the designer, it was desired to present the results of the analysis in a way that the designer should not need to solve the system of non-linear equations. This is done in Figs 2.5 and 2.6; the former figure concerns the limit state criterion (a) and the latter criteria (b, c). Both e_1 and e_2 are to be substituted for e in Fig. 2.6. If the designer wishes to replace criterion (c_1) by the more cautious one (c_2), which leads to avoiding any plastic deformations in the longitudinal ribs, Fig. 2.6 can again be used, but on the condition that the parameter e_3 is inserted for e. Each of figures (2.6 a – f) is plotted for a particular value of the shear lag factor, ϱ_s.

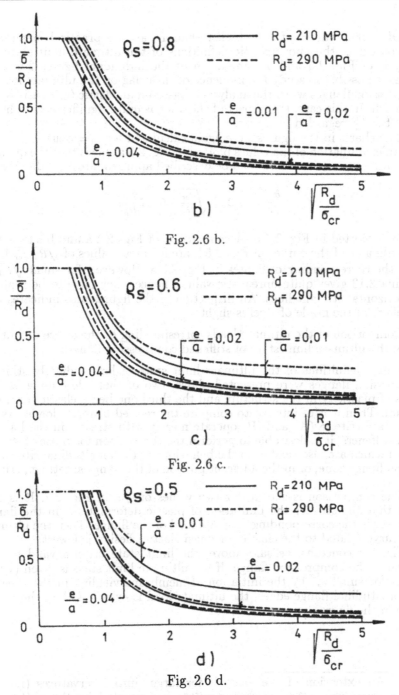

b)

Fig. 2.6 b.

c)

Fig. 2.6 c.

d)

Fig. 2.6 d.

Fig. 2.6. Ultimate loads of longitudinally stiffened compression flanges limit state criteria (b, c_1 or c_2).

Of course, values of $\bar{\sigma}/R_d$ higher than 1 have no practical meaning in the context of the above elastic definitions of the ultimate limit state. An analysis of Fig. 2.5 and a comparison of the individual figures 2.6 a – d makes it possible to study the influence of shear lag on the ultimate loads of compression flanges when the analysis is based on an elastic (or quasi-elastic) approach. In this case, the impact of shear lag is quite significant and, hence, cannot be disregarded.

The charts in the figures were again calculated for $f_0 = a/500$. *

An examination of the values plotted in Fig. 2.5 a shows that the quantity $\bar{\sigma}$ resulting from criterion (a) can be described approximately by the formula

$$\frac{\bar{\sigma}}{R_d} = 1.052\sqrt{\varrho_s}\sqrt{\frac{\sigma_{cr}}{R_d}} - 0.277\frac{\sigma_{cr}}{R_d}, \tag{2.13}$$

which is plotted in Fig. 2.5 b. A comparison of Figs 2.5 a and b shows that the ordinates of the curves in Fig. 2.5 b are for most values of $\sqrt{R_d/\sigma_{cr}}$ lower than the corresponding ordinates in Fig. 2.5 a. However, for small $\sqrt{R_d/\sigma_{cr}}$ formula 2.13 gives quite unrealistic values; hence, they have to be replaced by horizontal straight lines. An inspection of the figures also indicates that the effect of the grade of steel is slight.

D) Comparison of the ultimate loads corresponding to the above definitions of the ultimate limit state of stiffened compression flanges

Three definitions of the ultimate limit state of longitudinally stiffened compression flanges were presented above, two of them describing various levels of plasticity in the structure and the third one being elastic (or quasi-elastic). Then it is of interest to compare the related ultimate loads. As the limit state criteria (A) and (B) operate merely with stresses in the longitudinal stiffeners, it is desirable to perform the comparison for those of criteria (a – c) which are also related to the behaviour of the longitudinal ribs in the central flange zone, or in the adjacent portion of the flange sheet; i.e., criteria (b – c_2).

The comparison is shown, for two values of e/a, in Fig. 2.7. It is seen there that the stricter the limitation of plastic deformations in the flange, the lower is the corresponding ultimate load of the flange. Then the ultimate load curve related to the elastic (or quasi-elastic) design is lowest.

Fig. 2.7 concerns, as said above, the limit state criteria valid for the middle of the compression flange. If the ultimate limit state is given (which occurs for small ϱ_s) by the initiation of membrane yielding in the sheet at the longitudinal flange edges, the ultimate load is even less than the lowest curve in the figure.

---

* An extension of the analysis for larger initial cervatures (i.e., for $f_0 = a/100$, $a/200$, $a/300$, $a/400$ and $a/500$) is presented by the writer and M. Zörnerová in [2.4], where the reader can also find all corresponding charts for design.

2.3.2. Effect of local sheet buckling

A) Analytical solution

When the geometry of partial flange sheet panels, of width b', between the longitudinal ribs is such that local sheet buckling occurs, the above analysis of the interaction between global flange buckling and shear lag must be completed so as to take this local buckling into consideration.

For the sheet panels of ordinary longitudinally stiffened plates, presenting the normally encountered magnitudes of initial imperfections, this can be done with sufficient accuracy by means of the well-known Faulkner formula, where the maximum stress σ_{max} is substituted by the stress $\bar{\sigma}_{sheet}$, acting in the flange sheet and related to the attainment of the ultimate limit state.

The effective width, b'_e, of the sheet panels reads

$$b'_e = \varrho_{b,l} b',$$ (2.14)

where

$$\varrho_{b,l} = 1.052\sqrt{\frac{\sigma'_{cr}}{\bar{\sigma}_{sheet}}} - 0.277\frac{\sigma'_{cr}}{\bar{\sigma}_{sheet}}.$$ (2.15)

σ'_{cr} is the critical stress of the sheet panel, hence

$$\sigma'_{cr} = \frac{\pi^2 E}{12(1-\nu^2)}\left(\frac{t}{b'}\right)^2 k,$$ (2.16)

where k is the coefficient of the critical load for an unstiffened compression plate simply supported on all four boundaries, i.e., $k = 4$. Thus after substitution we obtain

$$\sigma'_{cr} = 3.61E\left(\frac{t}{b'}\right)^2.$$ (2.17)

After inserting this value into (2.15), one has

$$\varrho_{b,l} = 2\frac{t}{b'}\sqrt{\frac{E}{\bar{\sigma}_{sheet}}}\left(1 - 0.5\frac{t}{b'}\sqrt{\frac{E}{\bar{\sigma}_{sheet}}}\right).$$ (2.18)

Then the ultimate load of a longitudinally stiffened compression flange, determined with due regard to (i) shear lag, (ii) global flange buckling and (iii) local buckling of partial sheet panels, is written as follows:

$$\bar{\sigma}_{u,l} = \frac{\varrho_{b,l} bt\bar{\sigma}_{sheet} + A\bar{\sigma}_{stiffener}}{bt + A},$$ (2.19)

where A denotes the total area of all longitudinal stiffener cross-sections

$$A = mA_{st},$$ (2.20)

m being their number and A_{st} the cross-sectional area of one stiffener.

For practical reasons it is advantageous to introduce the following parameter, K, of the effect of local sheet buckling:

$$K = \frac{\bar{\sigma}_{u,l}}{\bar{\sigma}_u} , \tag{2.21}$$

with $\bar{\sigma}_u$ designating the ultimate load of the same flange, but worked out regardless of the influence of local buckling.

Let us further define the parameter

$$\delta_A = \frac{A}{bt + A} , \tag{2.22}$$

which measures the area of the cross-sections of all longitudinal ribs with respect to the cross-sectional area of the whole stiffened flange.

Then formula (2.19) can be rewritten in this form

$$\bar{\sigma}_{u,l} = \varrho_{b,l}(1 - \delta_A)\bar{\sigma}_{sheet} + \delta_A\bar{\sigma}_{stiffener}. \tag{2.23}$$

At this juncture it is also useful to note that the requirement that the statical moment of the cross-section of the sheet + stiffeners system, with respect to its centroidal axis, be nil gives the following relationship between e_1, e_2 and δ_A :

$$\frac{e_1}{e_2} = \frac{\delta_A}{(1 - \delta_A)}. \tag{2.24}$$

M. Zörnerová and the author strove to solve the whole problem of the effect of local sheet buckling in a way that the designer could obtain the values of K, for all flange geometries encountered in ordinary steel box girder bridges, in the form of a suitable graphical representation. Therefore, an extensive parametric study of this problem was performed by them.

The results are given in Figs 2.8 – 2.10 herebelow.

For intermediate values of ϱ_s, δ_A and e/a, it can be linearly interpolated.

B) A simplified analysis of the effect of local sheet buckling

It was seen above that the analytical approach to local sheet buckling is rather time-consuming. In particular, this is due to the necessity to work out painstakingly the values of $\bar{\sigma}_{sheet}$ and $\bar{\sigma}_{stiffener}$.

For this reason let us check whether it is not possible to simply put

$$\bar{\sigma}_{sheet} = \bar{\sigma}_{stiffener} = \bar{\sigma}_u , \tag{2.25}$$

where $\bar{\sigma}_u$ is the ultimate load of the flange determined with due regard to the interaction between shear lag and global flange buckling (which can easily be done with the aid of the graphs presented above), but regardless of the buckling of the partial sheet panels.

Then formula (2.23) takes this form

$$\bar{\sigma}_{u,1} = \bar{\sigma}_u[\varrho_{b,l}(1 - \delta_A) + \delta_A] \tag{2.26}$$

and

$$K = \frac{\bar{\sigma}_{u,1}}{\bar{\sigma}_u} = \varrho_{b,l}(1 - \delta_A) + \delta_A. \tag{2.27}$$

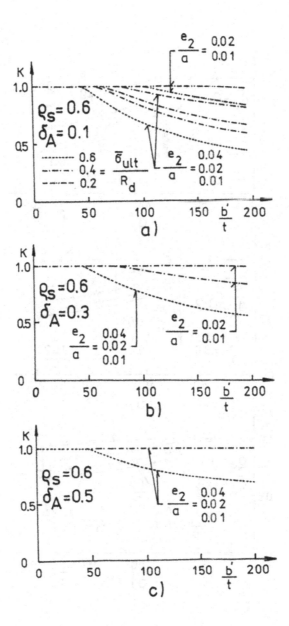

Fig. 2.8. Effect of local sheet buckling for a longitudinally stiffened flange,
ultimate limit state criterion related to the onset of membrane yielding at the longitudinal edges,
$\varrho_s = 0.6$;   a) $\delta_A = 0.1$,   b) $\delta_A = 0.3$,   c) $\delta_A = 0.5$.

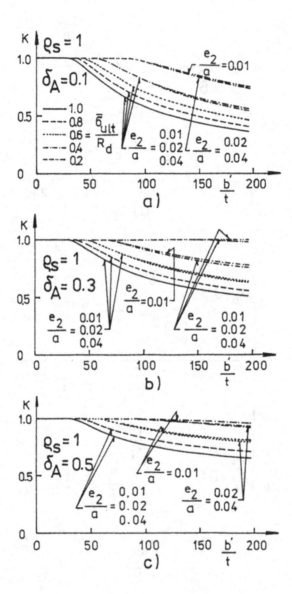

Fig. 2.9. Effect of local sheet buckling for a longitudinally stiffened flange,
ultimate limit state criterion related to the onset of yielding at the
centroid of the central longitudinal stiffener,
$\varrho_s = 1.0$; a) $\delta_A = 0.1$, b) $\delta_A = 0.3$, c) $\delta_A = 0.5$.

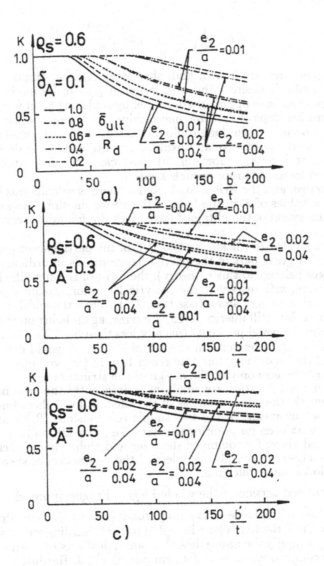

Fig. 2.10. Effect of local sheet buckling for a longitudinally stiffened flange, ultimate limit state criterion related to the onset of yielding at the centroid of the central longitudinal stiffener,
$\varrho_s = 0.6$; a) $\delta_A = 0.1$, b) $\delta_A = 0.3$, c) $\delta_A = 0.5$.

Of course, the Faulkner formula can then also be rewritten in the following more straightforward form:

$$\varrho_{b,l} = 2\frac{t}{b'}\sqrt{\frac{E}{\bar{\sigma}_u}}\left(1 - 0.5\frac{t}{b'}\sqrt{\frac{E}{\bar{\sigma}_u}}\right), \qquad\qquad (2.28)$$

where all quantities are known and, therefore, may be simply substituted into the formula. If desired, the calculation of $\varrho_{b,l}$ can be made even more accurate via successive approximations, i.e. by replacing $\bar{\sigma}_u$ in formula (2.27) by the successive steps of that maximum value of the longitudinal membrane stress which occurs at the longitudinal edges of the sheet panel and which corresponds to the successive approximations obtained in the determination of the ultimate load $\bar{\sigma}_{u,l}$. However, in most cases the first approximation already provides an accurate enough solution.

M.Zörnerová and the writer conducted numerous calculations in order to ascertain the values of K in the light of the above simplified approach.

An examination of the obtained data gives the following very important conclusion.

When the width-to-thickness ratio, b'/t, of the partial sheet panels is less than 50 (and a great part of the partial flange panels in ordinary steel box girder bridges belongs to this category), the error of the simplified approach described above, with respect to the analytical solution explained in sec. A, is on the safe side and very small (less than 2%). When the width-to-thickness ratio is greater, but still inferior to 80, the error, again being on the safe side, does not surpass several per cents (usually less than 6 – 7%). For higher b'/t-ratios the error again remains on the safe side (i.e., it would not jeopardize the safety of the structure), but may reach 10 – 15 (or even more) % (i.e., it would imperil the economy of the design of the structure).

Hence, if the b'/t-ratios are small, as is frequently the case in ordinary steel bridgework, the whole analysis of the effect of local sheet buckling can be carried out by means of the two simple relationships (2.26) and (2.27). This means that then no additional charts (i.e., in addition to those which were presented above for combined shear lag and global flange buckling) are needed to analyse the complex problem of the interaction of shear lag with global and local flange buckling.

2.3.3. A simplified version of the whole Liège – Prague approach

The authors of the Liège – Prague approach (i.e., Dr. Jetteur, Prof. Maquoi, Dr. Zörnerová and the writer) are indepted to Prof. Harding for his striving to significiantly simplify the above described analytical version of the approach.

In his recommendation (see, for example, [2.5]), J. Harding

(i) keeps formula (2.13), related to limit state criterion (a), i.e. to the onset of membrane plastification at the longitudinal flange edges, but replaces there ϱ_s by ψ according to (2.2).

(ii) As for criteria (b) and (c), i.e. those related to the onset of yielding in the central portion of the flange, he keeps only the charts given in Fig. 2.6 a and being valid for $\varrho_s = 1$. This means that he uses them for all values of ϱ_s. By comparing the curves given in Figs 2.6 a – d, the reader will see that the aforesaid assumption is conservative.

(iii) As for the effect of local sheet buckling, he uses the simple formula (2.26) for all width-to-thickness ratios, b'/t, of the partial sheet panels. After some transformations, he writes the formula in this form:

$$\bar{\sigma}_{u,l} = \bar{\sigma}_u \frac{mA_{be} + b'_e t}{A_f}, \tag{2.29}$$

where $A_{be} = A_{st} + b'_e t$ is the effective area of a stiffener(A_{st} denoting the cross-sectional area of one stiffener without effective width of sheet panel) and $A_f = mA_{st} + bt$ is the full area of the flange (i.e., of flange sheet + flange longitudinal stiffeners). And, as above, $b'_e = \varrho_{b,l} b'$ and m is the number of the stiffeners.

The reader will recall that M. Zörnerová and the writer have found out (see sec. 2.3.2, part B above) that this simplified treatment of local sheet buckling is always conservative, that for b'/t-ratios ≤ 50 the involved error is very small (inferior to 2%), but that for larger values of b'/t the error can rearch 10, 15 or even more %.

With the view to compare the simplified version by J. Harding with the original "pure culture" of the Liège – Prague approach by Ph. Jetteur, R. Maquoi, M. Zörnerová and the writer, and extensive parametric study was carried out in Prague by the author in cooperation with M. Zörnerová, in which (i) the aspect ratio of the flange panel investigated, (ii) the flange sheet thickness, (iii) the number, (iv) configuration and (v) size of the longitudinal ribs stiffening the flange were varied from case to case. On the whole, about 2240 cases were studied, thereby covering a great majority of flange geometries of practical interest.

It follows from the conclusions * of the study among other things that for practically important flange geometries the simplified version gives, in comparison with the analytical version of the Liège – Prague approach, results that are

(i) almost always on the safe side,

(ii) rather conservative and frequently very conservative, the difference amounting to 20, 30 and even more percent.

2.4. The Prague Approach

This approach, established by M. Zörnerová and the author, is a more accurate version of the Liège – Prague approach, described in the previous sec. 2.3, the higher accuracy being achieved by using more complex assumptions for displacement and stress (force) fields.

These assumptions are as follows:

$$N_{x1} = t \left\{ \sigma_0 \left(1 - \epsilon \sin \frac{\pi y}{b} \right) - \sigma_1 \sin^2 \frac{\pi y}{b} - \sigma_2 \sin^2 \frac{2\pi y}{b} - \right.$$

$$-\sigma_3 \sin^2 \frac{3\pi y}{b} - \sigma_4 \sin^2 \frac{4\pi y}{b} - \sigma_5 \sin^2 \frac{5\pi y}{b} -$$

---

* It is beyond the means of this publication to present the corresponding numerous graphs, but the reader will soon be made acquainted with them in another paper.

$$-e_1\sigma_{1f}\sin\frac{\pi x}{a}\sin\frac{\pi y}{b}-e_1\sigma_{3f}\sin\frac{\pi x}{a}\sin\frac{3\pi y}{b}-$$

$$-e_1\sigma_{5f}\sin\frac{\pi x}{a}\sin\frac{\pi y}{b}\bigg\}\,;\tag{2.30a}$$

$$N_{y1}=0,\tag{2.30b}$$

$$N_{xy1}=0,\tag{2.30c}$$

$$N_{x,2}=\frac{A}{b}\left\{\sigma_0\left(1-\epsilon\sin\frac{\pi y}{b}\right)-\sigma_1\sin^2\frac{\pi y}{b}-\sigma_2\sin^2\frac{2\pi y}{b}-\right.$$

$$-\sigma_3\sin^2\frac{3\pi y}{b}-\sigma_4\sin^2\frac{4\pi y}{b}-\sigma_5\sin^2\frac{5\pi y}{b}+$$

$$+e_2\sigma_{1f}\sin\frac{\pi x}{a}\sin\frac{\pi y}{b}+e_2\sigma_{3f}\sin\frac{\pi x}{a}\sin\frac{3\pi y}{b}+$$

$$\left.+e_2\sigma_{5f}\sin\frac{\pi x}{a}\sin\frac{\pi y}{b}\right\},\tag{2.30d}$$

$$u=u_0\left(1-\epsilon\sin\frac{\pi y}{b}\right)\frac{x}{a}-\frac{1}{2}\left(\frac{1}{2}\frac{\pi x}{a}+\frac{a}{4\pi}\sin\frac{\pi x}{a}\right)\frac{\pi^2}{a^2}\left(w_{11}^2\sin^2\frac{\pi y}{b}+\right.$$

$$+2w_{11}w_{13}\sin\frac{\pi y}{b}\sin\frac{3\pi y}{b}+2w_{11}w_{15}\sin\frac{\pi y}{b}\sin\frac{5\pi y}{b}+$$

$$\left.+w_{13}^2\sin^2\frac{3\pi y}{b}+2w_{13}w_{15}\sin\frac{3\pi y}{b}\sin\frac{5\pi y}{b}+w_{15}^2\sin^2\frac{5\pi y}{b}\right)-$$

$$-\left(\frac{1}{2}\frac{\pi x}{a}+\frac{a}{4\pi}\sin\frac{\pi x}{a}\right)\frac{\pi^2}{a^2}\left(w_{11}w_{01}\sin^2\frac{\pi y}{b}+\right.$$

$$\left.+w_{13}w_{01}\sin\frac{\pi y}{b}\sin\frac{3\pi y}{b}+w_{15}w_{01}\sin\frac{\pi y}{b}\sin\frac{5\pi y}{b}\right)\,,\tag{2.31a}$$

$$\bar{U}=\bar{U}_0\left(1-\epsilon\sin\frac{\pi y}{b}\right),\tag{2.31b}$$

$$v=0,\tag{2.31c}$$

$$w=w_{11}\sin\frac{\pi x}{a}\sin\frac{\pi y}{b}+w_{13}\sin\frac{\pi x}{a}\sin\frac{3\pi y}{b}+$$

$$+w_{15}\sin\frac{\pi x}{a}\sin\frac{5\pi y}{b},\tag{2.31d}$$

$$w_0=w_{01}\sin\frac{\pi x}{a}\sin\frac{\pi y}{b}.\tag{2.31e}$$

The meaning of the unknown quantities σ_0,σ_1 and ϵ again follows from Fig. 2.2. The other unknown quantities are w_{11},w_{13} and w_{15} (i.e., the three parameters of the buckled surface); $\sigma_2,\sigma_3,\sigma_4$ and σ_5 (i.e., the other four parameters of the distribution of the longitudinal membrane stress over the flange breadth; consequently, playing a similar role to parameter σ_1 in the Liège – Prague approach shown in Fig. 2.2 b); σ_{1f},σ_{3f} and σ_{5f} (i.e., the three parameters of the distribution of the longitudinal bending stress over the flange breadth; consequently, playing a similar role to parameter σ_f in the Liège – Prague approach).

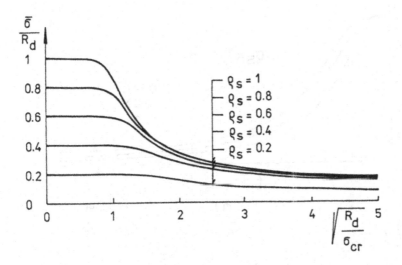

Fig. 2.11. Ultimate loads of longitudinally stiffened compression flanges   –
limit state criterion (a).

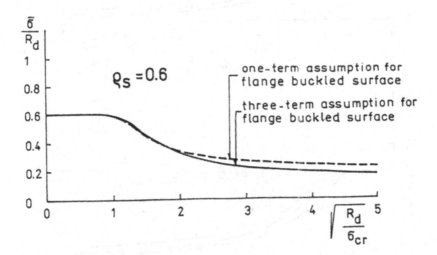

Fig.2.13. Difference between the results obtained via (i) the Liège – Prague
approach and (ii) the Prague one.

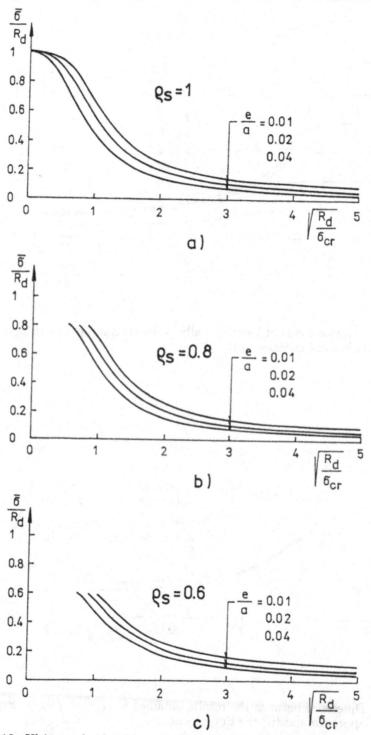

Fig. 2.12. Ultimate loads of longitudinally stiffened compression flanges limit state criterion (b, c_1 or c_2).

A is again the total cross-sectional area of all stiffeners, taken without any effective portion of the flange sheet.

The analytical model used for the solution is the same as in the Liège – Prague approach (see above sec. 2.3), viz. a generalized Reissner's principle. However, it is in the nature of things that the Prague solution, using more complex assumptions for displacement and stress fields, proved to be much more laborious and time-consuming than that in the Liège – Prague analysis. Also the same definitions of the ultimate limit state can be used as those applied in the Liège – Prague approach, whether they be their plastic or quasi-elastic variants. Then we obtain flange ultimate load curves $\bar{\sigma}$ qualitatively similar to those we had above in sec. 2.3.

Again, first the problem of global flange buckling is dealt with.

It is beyond the means of this publication to include all the curves obtained; therefore, we will present here merely the curves following from the quasi-elastic solution (see criteria a, b, c_1 and c_2 in sec.2.3).

For example, curves $\bar{\sigma}/R_d$ related to limit state criterion (a), i.e. to the onset of membrane plastification at the longitudinal flange edges, are given in Fig. 2.11, again in terms of shear lag factor ϱ_s and flange geometry parameter $\sqrt{R_d/\sigma_{cr}}$. Curves $\bar{\sigma}/R_d$ associated with limit state criteria (b, c_1 or c_2), i.e. with plasticization of the central part of the flange, are shown – for various $\sqrt{R_d/\sigma_{cr}}, e/a$ and ϱ_s – in Figs 2.12 a, b and c.

At this juncture, the reader can ask about the difference between the $\bar{\sigma}/R_d$ - curves yielding from (i) the Liège – Prague approach and (ii) the Prague one. For one particular case (viz., for criterion (a) and $\varrho_s = 0.6$), this is demonstrated in Fig. 2.13, from where it is seen that the more accurate Prague solution gives lower (as is usual when a variational principle combined with more realistic assumptions for displacement and stress fields is applied) ultimate loads, but the difference is not great. And a similar conclusion results even from other comparisons. Of course, in some cases the difference is larger; but for $\sqrt{\sigma_{cr}/R_d} < 2.5$ (and the geometries of the compression flanges of ordinary bridges are such that $\sqrt{\sigma_{cr}/R_d}$ is almost always within this range) is less than 20% and usually smaller.

Next, we can proceed to analysis of the effect of local flange sheet buckling. Again, the line of solution is similar to that explained above in sec. 2.3 in connection with the Liège – Prague approach, but the analysis involved is much more time-consuming. In the end, we obtain charts for local sheet buckling coefficient K, similar to those depicted in Figs 2.8 and 2.9 of sec. 2.3. Analysing the "new" and "old" values of K, M. Zörnerová and the author came to the conclusion that the "old" values of K, i.e. those given in Figs 2.8 and 2.9, can be employed with sufficient accuracy (i.e., with an error usually inferior to 5%) even in analysis via the Prague approach.

2.5. Comparison between (i) the British Approach and (ii) the Liège – Prague and Prague Approaches

It remains to comment on the difference between the ultimate loads calculated via (i) British strut approach (see sec.2.2 above) on the one side and (ii) the Liège – Prague approach and the Prague one, based on a non-linear modified orthotropic plate concept (see secs 2.3 and 2.4 above) on the other. This was one of the objectives of the extensive parametric study carried

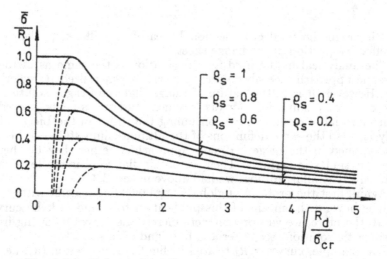

Fig. 2.5 b. Values of $\bar{\sigma}$ following from formula (2.13).

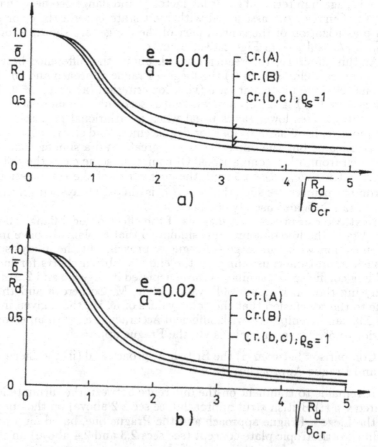

Fig. 2.7. Comparison of the ultimate loads corresponding to different limit state criteria.

plastification in other parts of the flange. But, on the other hand, it should be borne in mind that, for example, potential surface plastification near the longitudinal flange edges is safely checked by ultimate limit state criterion (a), i.e., by the beginning of membrane yielding there, since bending stresses in this area are small and, hence, the difference between membrane and surface plastification is slight (i.e., very much inferior to the impact of the difference between (i) ultimate design loads and (ii) service ones). Moreover, it follows from a parametric study carried out by the authors that, for ordinary flanges, surface plasticization in the sheet in the central flange zone practically never arises prior to the beginning of yielding at the outmost fibres of the most intensively loaded longitudinal rib.

One of the objectives of the parametric study carried out recently in Prague by M. Zörnerová and the author (see above) was to look into the impact of the aforementioned definition of serviceability limit state upon the design of longitudinally stiffened compression flanges. It was concluded that the above introduction of the limit state of serviceability had some practical meaning only when a plastic criterion for the limit state of resistence was employed. If, on the contrary, a quasi-elastic criterion was used for the ultimate limit state (for example, applying criteria (a), i.e. onset of membrane yielding at the longitudinal edges of the flange, (b), i.e. onset of membrane yielding in the flange sheet in the middle of the flange, and (c_1), i.e. onset of yielding at the centroid of the central longitudinal stiffener), it was always the ultimate limit state check that governed the design. This was due to the fact that the difference between the limit load curves coresponding to yielding at (i) the stiffener centroid and (ii) the stiffener extreme fibres was less than the impact of the difference between (α) ultimate design loads and (β) service design loads.

References

2.1. Jetteur, Ph., Maquoi, R., Škaloud, M. and Zörnerová, M.: Interaction of shear lag with plate buckling in longitudinally stiffened compression flanges. Acta technica ČSAV, 29(1984), 3, 376-397.

2.2. Massonnet, Ch. et al.: Tolerances in steel plated structures. IABSE Surveys S - 14/80, August, 1980.

2.3. Škaloud, M., Kárníková, I.: Experimental research on the limit state of the plate elements of steel bridges, Transactions of the Czech. Academy of Sci., Series of Techn. Sci., No. 1, 1985, 1-141.

2.4. Škaloud, M., Zörnerová, M.: The role of larger initial curvatures and of local sheet buckling in the interaction of shear lag with buckling in longitudinally stiffened compression flanges, Acta technica ČSAV, 29 (1984), 5, 623-647.

2.5. Europian recommendations for the design of longitudinally stiffened webs and of stiffened compression flanges. Document edited by the ECCS Technical Working Group 8/3: Plated Structures, 1990.

PART 2

SHELLS

THE BEHAVIOUR AND DESIGN OF
CYLINDRICAL SHELL STRUCTURES

J. E. Harding
University of Surrey, Guildford, UK

Abstract

The buckling and collapse behaviour of shells is a complex subject about which numerous books have been written.

This chapter is intended to serve as a simple introduction to the behaviour and design of a restricted class of shells, namely cylindrical shells stiffened by ring or orthogonal stiffeners.

The treatment is non-mathematical with an emphasis on explaining the behaviour of the components. This is demonstrated by reference to the results of numerical analyses.

The geometries considered are typical of the main leg components of offshore rig legs and where specialist design rules are mentioned they are taken from the area of offshore design.

Introduction

The attractiveness of cylindrical shells comes partly from the fact that the cylindrical shape is the most efficient for resisting axial compressive loading and partly because it also acts efficiently when loaded by uniform internal or external pressure. The former pressure is resisted by membrane tension stresses while the latter by membrane compression in the circumferential direction. Under axial loading the curvature adds significantly to its local buckling resistance. While many cylindrical shell structures are stiffened, perhaps by ring stiffeners or by longitudinal stringer stiffeners, the inherent load resisting properties of the shell plating remain. The basic purpose of the stiffening,as with all plated structures, is to enhance the local buckling performance of the cylinder so that more slender geometries can be utilised while maintaining reasonable collapse stress values under the relevant applied loading. The governing slenderness values of the structure also govern the type of buckling of the shell, for example the relationship between overall column buckling and local buckling of the shell plating for uniaxially loaded cylindrical shells.

Cylindrical shells are used in such structures as offshore oil platforms and transmission towers where the local slenderness is modest (radius/shell thickness ratio: R/t of the order of 300 or less for the main legs of oil platforms and 50 or less for use as column members such as in transmission tower elements or bracing members of offshore structures).

Appreciably more slender cylinders can be used in storage tanks such as silos for solid fill or liquid filled tanks (R/t in the range of 1000 or more). Shells of intermediate slenderness may also be used for the construction of pressure vessels.

This chapter will concentrate on the area of cylindrical shells used in offshore construction but will contain material relevant to other areas. In general such shells will be fabricated as welded steel structures and the presence of residual stresses caused by welding and geometric imperfections caused by the fabrication process and general handling have a particular significance for shell structures with a high sensitivity of local buckling leads to imperfections for situations where membrane compression forms a significant component of the resisting internal forces.

Shell Buckling

The area of shell buckling has attracted much attention, particularly from analysts, in recent years. This relatively recent work followed much experimental speculation about the reasons for the wide variation of test results from seemingly similar specimens. The interest in this topic has been such that no fewer than one thousand six hundred references have been included in a review published in 1981 by Arbocz, a noted expert dealing with the field of cylindrical shell buckling[1].

While many of the advances in the understanding of shell buckling have occurred in the last thirty years, significant steps in this field were being made around the turn of the century, for example by such noted figures as Timoshenko[2], Southwell[3] and Von Mises[4]. This early work, however, related to small deflection theory and could not explain the inconsistencies in response obtained from experimental tests such as those carried out in the thirties by Wilson and Newmark[5], Donnell[6] and Fluegge[7].

For many years design was based around lower bound results from experiments[8,9] although an understanding of the non-linear behaviour of shells gradually developed. For example in 1934, Donnell included initial geometric imperfections in an approximate large deflection theory[10] although because of over simplification the results had shortcomings. Other work, for example by Esslinder and Geier[11] attempted to define a minimum post buckling load and provided reasonable correlations with the available experimental data.

Many of the studies carried out this century have related to the aircraft and aerospace industries. The importance to the UK of offshore oil recovery led to a range of related research in the early eighties. A summary of this work will be presented later.

Before dealing with the design of shell structures and the rationale for and design of stiffening it is important to understand the essence of the behaviour of cylindrical shells and in particular the importance and effect of initial imperfections and post-buckling response. It is primarily because the initial buckled form is in a state of unstable equilibrium for many shell problems that the response of a shell is very different to that, for example, of a column or a flat plate. A column essentially is in a state of neutral equilibrium after critical buckling while a plate is stable.

The practical implication of this is that elastic critical buckling loads, calculated by classical stability theories, can never be attained in practice with buckling loads generally substantially below the theoretical load. While the geometries are often such that the shell behaviour involves substantial material non-linearity, this basic influence still has a major effect on the system response in practice. This post-critical instability provides the shell with an enhanced imperfection sensitivity which occurs primarily when compressive components of membrane stress dominate the behaviour. The explanation of these characteristics and the resulting behaviour are described in the following sections.

Elastic Stability of the Perfect Shell
As with column and flat plate structures, a reference case for a consideration of shell buckling is often taken as a unstiffened cylindrical shell subjected to

a uniform axial compression at its ends acting axially around the circumference of the cylinder with no eccentricity relative to the shell plating. In this way no load eccentricities are introduced tending to bend the shell plating at its ends. This reference case illustrates the basic phenomena governing many aspects of shell buckling and provides a good example of the effects of imperfection sensitivity.

There are essentially three initial buckling modes for an axially compressed cylinder. One of these corresponds to column buckling which will occur when the unsupported length of the shell is large compared with its diameter. This is not relevant to the present discussion although its effect will be considered in more detail later. In brief, with all columns the important geometric parameter controlling the onset of critical buckling is the effective buckling length divided by its radius of gyration which, of course, in the case of a cylinder is equal in all directions. The Euler buckling load P_{cr}

$$= \frac{\Pi^2 EI}{l_E^2} = \Pi^2 EA \left(\frac{r}{l_E} \right)^2.$$ It can therefore be seen that the reduction in

column elastic critical buckling stress is directly proportional to the square of the effective slenderness l_E/r.

The other two critical buckling modes correspond to different types of small deflection local buckling of the shell wall. The onset of these latter modes is controlled by the value of the shell local slenderness (a combination of the radius to thickness ratio, R/t, and also the length).

In the first local buckling mode the shell has concentric radial displacements which have constant amplitude around the circumference. This mode involves significant stretching of the shell wall and is therefore generally limited to the small deflection regime although the mode can develop into the large deflection regime particularly if the cylinder is very short and the ends are restrained against radial expansion so that poisson expansion of the shell is a major influence. The presence of closely spaced ring stiffeners can also influence the critical elastic and large deflection behaviour although this is not the topic of this section. Figure 1 illustrates this mode of critical buckling. For a long shell the buckles ripple as shown in the figure.

The second local buckling mode takes the form of waves in both the longitudinal and circumferential directions giving a pattern of depressions and bulges over the cylinder surface. This is known as chessboard or checkerboard buckling and the mode is illustrated in Figure 2.

In the large deflection post-buckling regime the buckling mode changes in order to minimise energy and the elastic shell snaps into a depressed diamond pattern of the type shown in Figure 3 where the degree of membrane stretching is reduced.

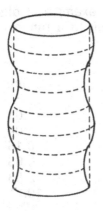 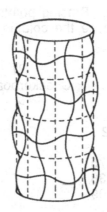 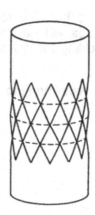

Figure 1.   Figure 2.   Figure 3.

Ring buckling.   Chessboard buckling.   Diamond buckling.

At this stage in the discussion it is relevant to introduce the Batdorf parameter Z which has been found to have a controlling influence on the type of critical buckling mode to actually occur. This parameter is a function of shell radius to thickness ratio and length, mentioned above as having a controlling influence on local buckling behaviour.

$$Z = \left(\frac{L}{R}\right)^2 \left(\frac{R}{t}\right) \sqrt{1 - v^2}$$

Axisymmetric critical buckling occurs for Z values less than 2.85 while for checkerboard buckling to occur the Z value needed is in the range between

2.85 and 4.8 $\dfrac{(R/t)2}{C}$ with C = $\dfrac{1}{\sqrt{3(1 - v^2)}}$ giving

$2.85 \leq Z < 8.314 \sqrt{1 - v^2} \, (R/t)^2$

For column buckling to occur as the critical elastic buckling mode the Z value needs to exceed that appropriate to checkerboard buckling

$Z \geq 8.314 \sqrt{1 - v^2} \, (R/t)^2$

To put these values into perspective, Table 1 shows the limiting R/t values for steel cylinders with a poisson's ratio of 0.3 for increments of L/R.

The general implications of the above figures are that axisymmetric buckling will not occur unless the cylinder length is less than about one half the radius and that column buckling will not occur unless the length is at least about eight times the radius. Checkerboard buckling therefore occurs for most slender cylinders (high R/t) unless closely spaced ring stiffeners cause

axisymmetric buckling to occur. Bracing points, for example in offshore
structure legs will normally reduce the column length to a range where
column buckling will not occur.

L/R	Axisymmetric	R/t Checkerboard	Column
	<	>	<
0.25	48		.01
0.5	12		.04
1.0	3		0.16
2.0	0.75		0.63
4.0	0.38		2.53
8.0	0.09		10.11

Table 1

The definition of the elastic critical buckling stress is different for the three
buckling modes.

For axisymmetric buckling the critical stress expression closely resembles
that of a flat plate.

$$\sigma_{cr} = K \frac{\Pi^2 E}{12(1 - v^2)(L/t)^2}$$

The buckling coefficient K depends on boundary conditions.

For simply supported edges $K = \dfrac{12 \, Z^2}{\Pi^4}$

For clamped edges $K = 4 + \dfrac{3 \, Z^2}{\Pi^4}$

For checkerboard (lobular) buckling the critical stress becomes independent
of the cylinder length.

The critical buckling stress is

$$\sigma_{cr} = \frac{C E}{R/t} \quad \text{where } C = \frac{1}{\sqrt{3(1 - v^2)}}$$

For a steel with poisson's ratio equal to 0.3 this gives

$$\sigma_{cr} = 0.605 \, \frac{E}{R/t}$$

For the column buckling mode the normal Euler critical buckling stress governs:-

$$\sigma_{cr} = \frac{\Pi^2 E}{(L/r)^2}$$ where r (the radius of gyration) for a cylinder = R/2.

$$\sigma_{cr} = \frac{\Pi^2 E}{(L/R)^2}$$

If the walls of the cylinder at the ends are free to move in the radial direction (i.e. expand freely) then the critical buckling stress for local buckling in the axisymmetric mode is halved for the simply supported case.

Having established the possible critical buckling modes under axial compression and the basis of calculating the appropriate critical buckling stress it becomes important to examine the post-buckling response of the cylindrical shell structure. An examination of the main differences between the post-buckling performance of columns, plates and shells provides the basis for an understanding of the imperfection sensitivity of shell structures.

Equilibrium paths for a perfectly straight column, a uniformly compressed perfect four sided simply supported plate and an axially loaded perfect cylindrical shell are shown by the solid lines in Figure 4. As long as the stress level is less than the elastic critical stress the primary equilibrium is stable for all cases. In this state, if a small disturbance is applied the structure will in theory return to its perfect state when the disturbance is removed. Any point on the pre-buckling line above the perfect elastic critical buckling load represents a state of unstable equilibrium and any notional disturbance will result in the onset of large instantaneous irreversible deformation in a plane normal to the loading. Because small disturbances will always occur in practice and indeed because the structure can never be perfect in its initial form in reality, it is virtually impossible to reach a load greater than the elastic critical load for the shell structure and in fact initial critical buckling can never realistically occur at load levels higher than the critical load for the column or plate.

For the perfect system which undergoes lateral deflection, once the critical load is reached, the equilibrium configuration changes and the post-buckling mode will contain deflections normal to the loading direction. It can be seen that the secondary equilibrium paths shown in Figure 4 are very different for the case of the column, plate and shell and it is this difference that results in the different behavioural characteristics of the real systems.

For the column the secondary equilibrium path is essentially horizontal although theoretical calculations will show small increases in load capacity

with large lateral deflections. In fact the strength only increases by about 1%
for deflections of the order of a tenth of the column length and for practical
purposes the post-buckling behaviour of a column can be regarded as
corresponding to a state of neutral equilibrium.

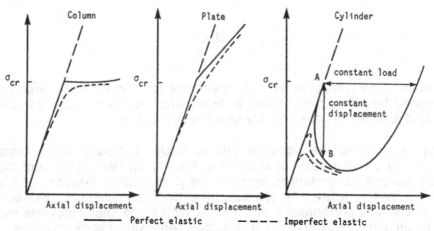

Figure 4.   Axial response of columns, plates and cylinder elements.

For the plate the secondary path rises at a significant although lower than
elastic stiffness. The lateral deflections (see Figure 5) are accompanied by a
gradually increasing load carrying capacity and the post-buckling response
is stable.   The enhanced load capacity is a function of the boundary
conditions but an increase of 30% over the perfect critical buckling load
could be expected for a lateral deflection of the order of the plate thickness.
The effect can therefore be seen to be significant.

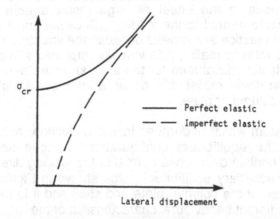

Figure 5.   Lateral deflection response of a plate element.

This stable post-buckling response can also be seen in the extensional stiffness of the plate where the average axial stress-strain response has a value as high as 75% of the initial stiffness if the edges are restrained or 41% if the edges are unrestrained for a square simply supported plate. This response is illustrated in Figure 4. The restrained and unrestrained conditions relate to the ability of the boundary to carry in-plane forces transverse to the edge of the plate and hence anchor the plate against in-plane edge movement. The restrained condition corresponds to an edge which is unable to move. The unrestrained condition corresponds to a boundary that can carry no transverse stress. These two conditions therefore represent extremes. It is important not to confuse the degree of in-plane restraint with the out-of-plane conditions such as simply supported or clamped edges. The in-plane and out-of-plane boundary conditions are in principle independent although they may be linked in practice. For example in a practical situation a clamped plate will often have restrained edges and a simply supported plate may well have unrestrained edges. The in-plane restraint is important because it represents the ability of the boundary to carry transverse membrane tensions, the development of which provides the post-critical reserve of the plate. Even with unrestrained boundaries, however, the membrane tensions can develop away from the boundaries in a self-equilibriating fashion and this is reflected in the significant post-buckling stiffness noted above for the unrestrained simply supported plate. The membrane tensions are caused primarily by the out-of-plane displacements producing stretching of the plate middle surface in the large deflection regime. The development of the membrane tension stresses tends to resist and slow down the increase in lateral deflections.

The behaviour of the compressed cylinder is significantly different to either that of the column or flat plate. The solid curve of Figure 4 represents the elastic response of the perfect cylinder and the key feature that can be seen is the dramatic unloading with reversal of displacement following the attainment of the elastic critical buckling load. The secondary equilibrium path is therefore highly unstable.

The reversal of displacement reflects the large drop in load and hence a partial recovery of elastic strain which can exceed the increase in end displacement caused by the lateral deflections.

It is interesting that two equilibrium positions exist for loads below the critical load and, because of the reversal of displacement, two equilibrium configurations also exist for displacements less than that corresponding to the critical load. This has implications on the response of the system that could be expected in different loading situations.

Considering a laboratory environment where the loading situation can be clearly defined and controlled, experimentalists will be well familiar with the concept of load control and displacement control. The former at its simplest level could be thought of as the application of load to the end of a cylinder using dead weight that remains in place as the cylinder buckles. For this form of loading the cylinder response would in theory jump from the critical point to the point on the rising curve at the same level of load (point A) with a resulting very dramatic increase in buckle deflections. For a loading system which is capable of instantaneous removal of the load to maintain a constant axial displacement the response would drop along the constant displacement line to point B. Whether the latter is practically achievable is of little concern, but the resulting buckle deflections would be more modest than for the former case. In experimental terms the only way in theory of obtaining the complete characteristic of the perfect response would be to employ an actuator with the ability to load in reverse so that the instantaneous relief of axial displacement could be followed and therefore permitted by the loading device.

While the above discussion is rather academic the key point is that in practice the surrounding structure would provide some rigidity and restraint that would almost certainly provide a response somewhere between the first two scenarios with significant reduction in load carrying capacity accompanied by some axial shortening and large buckling deflections. With a shell structure the interaction between the response of the element and the stiffness of the surrounding system will be very important. In most experimental tests where the test rig has significant internal straining and where the speed of response of the loading system is moderate the result would tend to be the same. In any situation the unstable response of the shell could lead to a catastrophic collapse situation. At this stage it is worth re-emphasising that the above relates to the perfect system and that the presence of imperfections of various types will affect the response.

The fundamental reasons behind the differences between column, plate and shell stability are not easy to explain but the following gives an indication of the forces involved.

Vandepitte[12], referring to Esslinger and Geier[13], explains the post-critical action in terms of the de-stabilising forces and internal reactions in the three systems.

The post-critical lateral equilibrium of an axially loaded strut is defined by the differential equation

$$F_{cr}\frac{d^2w}{dx^2} + EI\frac{d^4w}{dx^4} = 0$$

where w is the lateral deflection measured along the length x of the member. The first term represents the de-stabilising force due to the external load F_{cr}

while the second term represents the internal restoring force produced by bending action. The equality means that actions and internal forces balance each other and both sets are proportional to the lateral deflection level. As a consequence of this the equilibrium of the column is independent of the level of the deflection and equilibrium can be maintained with no change in F_{cr}. Distribution of the two force components are shown in Figure 6.

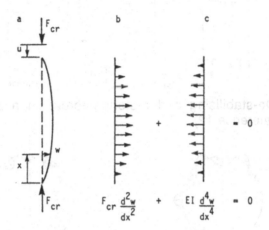

$$F_{cr} \frac{d^2w}{dx^2} \quad + \quad EI \frac{d^4w}{dx^4} \quad = 0$$

Figure 6.   De-stabilising and restoring forces in a column (Taken from reference 12).

For a buckling plate, the restoring forces are due to longitudinal and transverse bending moments but in addition, because the plate deflected surface is non-developable, lateral deflections cause in-plane membrane stretching which might anchor off the unloaded edges if these are capable of resisting the stresses or which are self equilibriating within the plate for other situations. Figure 7 provides an illustration of these forces. The contribution to stability provided by the membrane forces increases as loading progresses with the magnitude of these forces being proportional to the square of the lateral deflection. The increasing internal forces can therefore maintain an increase in the level of the applied force and the post-critical behaviour is stable.

The shell situation is different. Figure 8 shows the change in the resisting forces that occurs as the cylinder circumferential curvature increases. As the curvature of the inward displacements exceeds the curvature of the original cylinder the forces which initially resist buckling and which give the cylinder its enhanced resistance change to a system where the compressive membrane forces act in the direction of buckling. Once buckling has started the shell is unstable and can only attain equilibrium under a reducing applied axial load.

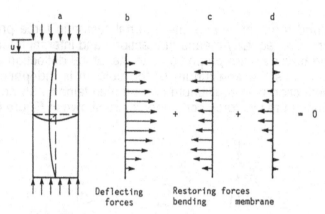

Figure 7.   De-stabilising and restoring forces in a plate (Taken from reference 12).

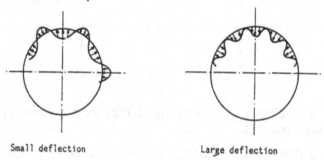

Figure 8.   Restoring forces for an axially compressed cylinder (Taken from reference 12).

Elastic Stability of the Imperfect Shell

In reality, imperfections of various types will always be present in a real system. As already mentioned, residual stresses will occur due to welding in

Civil Engineering structures and welding and general fabrication handling will cause various types of geometrical imperfections. For example, construction of a single sided weld, a stiffener fillet welded to a plate or shell, will provide a racking force across the weld that will rotate the plate or shell at that location and hence bend the plate in the direction of the stiffener outstand. The compressive stresses between welds in the plate will produce amplification of existing deformation and will tend to produce enhanced imperfections, a significant percentage of which is likely to be in the critical buckling mode. These residual stress and imperfection effects are illustrated in Figure 9.

In practice imperfection states will be multi-mode but certain modes, for example a single half wave for a panel or a mode-form in the critical buckling shape may tend to dominate the response.

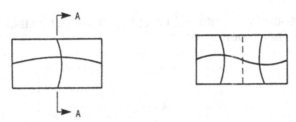

Possible key modes for a plate with an aspect ratio of 2:1

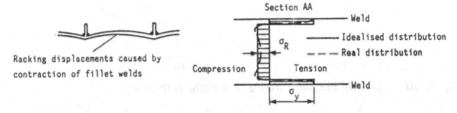

Figure 9.   Typical geometrical imperfections and residual stress patterns.

If an imperfection form generally sympathetic to the elastic critical buckling mode is considered in the context of the elastic buckling curves of Figure 4, the response of the column, plate and shell will tend to follow the dotted lines in the Figure with the amplitude of the imperfection tending to control the deviation of the imperfect curve from the original perfect response. It can be seen that in all cases the curves are smoother with the loss of the sudden change in response that occurs at the onset of critical buckling. However, there is a key difference that occurs between the behaviour of the column and plate, and that of the cylinder.

For the column and plate, which do not not suffer post-buckling elastic instability, the imperfect curves will eventually become asymptotic to the perfect path regardless of the level of imperfection. The larger the imperfections the larger the deflection will be at which the curves converge. For the shell it is evident from an examination of Figure 4 that with any imperfection which grows with applied load in the direction of the buckling form there will be a decrease in peak load capacity. For significant imperfections the peak load may be substantially below the perfect theoretical elastic critical load.

It is also worth contemplating the situation where imperfections are in a form which is not sympathetic to the critical buckling mode. This can most readily be appreciated in the case of a plate as shown in Figure 10 where the length in the direction of loading is twice the width (aspect ratio $\alpha = 2$). The critical buckling mode for this plate is with two longitudinal half waves as shown in Figure 10b and when an imperfection with this form is present the response of the plate will be 'soft'. A representative average stress-strain response for this case is shown in Figure 11. In contrast, if a single half wave imperfection is present, the plate will be stiffer in the early stages than with the 'sympathetic' initial imperfection.

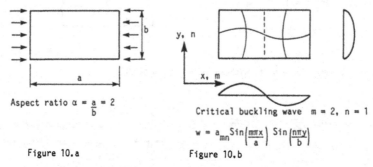

Aspect ratio α = $\frac{a}{b}$ = 2

Critical buckling wave m = 2, n = 1

$$w = a_{mn} Sin\left(\frac{m\pi x}{a}\right) Sin\left(\frac{n\pi y}{b}\right)$$

Figure 10.a   Figure 10.b

Figure 10.   Critical buckling mode of a plate with α = 2.

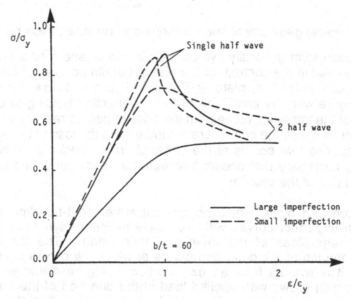

Figure 11.   Effect of imperfection mode on plate response.

At some point the non-critical mode would have to change in the large deflection regime and this change could be sudden in nature. In principle the curve could be of the form shown in figure 11 when the average stress actually falls as the mode changes. This form of behaviour is unlikely to occur in practice because the imperfection form will always be a mixture of modes but the precise combination of modes will affect the elastic stiffness. It must be remembered that this discussion still relates to elastic behaviour.

The same trend in behaviour will relate to shells but in this case both the imperfection forms and the critical modes tend to be of short wavelength.

There will therefore be an increased likelihood that some component of imperfection will correspond with the critical buckling mode and because of the unstable nature of the response, the imperfect system will always produce a peak load lower than the theoretical buckling load. Even when imperfections are very small, significant reductions in peak capacity will be obtained.

The basic sensitivity of peak elastic shell capacity to imperfections was a course of considerable uncertainty in early days of experimental work in shell buckling. Substantially varying experimental results were obtained from nominally identical structures, all of which were lower than the predicted theoretical buckling loads.

Fully Non-linear Response of Shells
In the range of shell geometries associated with offshore fabrication and indeed in many other environments, the onset of material yield or material non-linearity will have an appreciable effect on the response of the system. Perhaps fortunately, because the yield stress in steel cylinders puts in principal a limit to the possible strength, it reduces, for stockier geometries, the control of elastic buckling and in reality reduces to some degree the unstable nature of the system and the sensitivity to imperfections. For geometries where substantial plastic flow occurs, the response of the structure tends to be much softer.

The reader will be very familiar with the type of interaction that occurs between elastic critical buckling and yield for almost all types of instability phenomena associated with Civil Engineering structures. Figure 12a shows the average stress-strain response that could be expected for a cylinder of medium slenderness loaded by axial compression and Figure 12b the variation in peak capacity with variation in cylinder slenderness. The latter can be compared in Figure 12b with the responses of column and plate systems and the high slenderness relationship between imperfect collapse curves and critical buckling curves can readily be associated with the post-critical response of the system.

In all three cases the curve of Figure 12a will be similar although the degree of unloading will depend on the type of structure and its relative slenderness.

It is interesting to examine the effect of type of imperfection on the shell ultimate response curve of Figure 12b. Figure 13 shows the effect of varying the longitudinal wavelength of the imperfection form on the ultimate capacity of the cylinder (a ring stiffened cylinder was used in this computational example) and it can be seen that substantial variation in strength can occur even when the amplitude of the imperfection is kept constant. Variation in imperfection amplitude will itself, of course, produce a significant variation.

The difference in the curves is produced by 'sympathetic' and 'non-sympathetic' imperfection forms. In the latter case the non-sympathetic form is likely to change to the 'sympathetic' mode at or shortly after peak load, 'snap-through', with a resulting rapid loss in load capacity. This is illustrated in Figure 14.

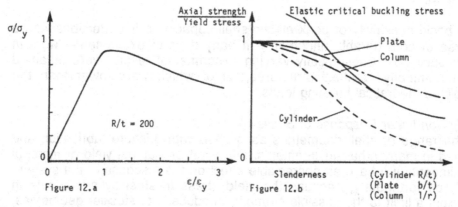

Figure 12. Response of a cylinder to axial loading.

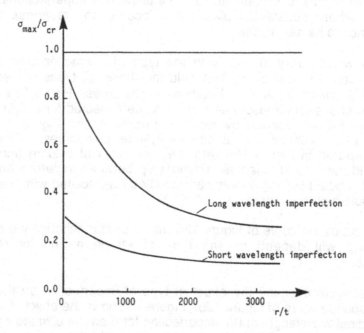

Figure 13. Effect of imperfection wavelength on ultimate capacity.

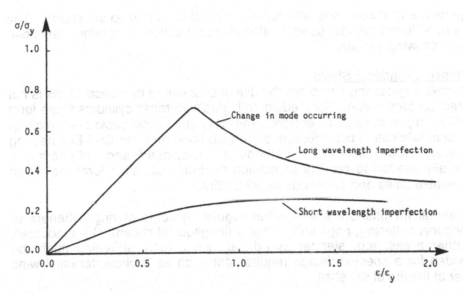

Figure 14.   Effect of imperfection mode on response.

Material non-linearity occurs in two ways. The division between these depends on the slenderness of the system and there is no clear sub-division between them. For a stocky system where the critical stress is significantly higher than the yield stress, growth in lateral deflections before peak load is small and yield occurs as a general phenomena due to predominantly membrane forces. This results in a stable response with the overall structure response reflecting the plateau characteristic of the steel material. Some gradual loss in load capacity may result as the axial shortening of the cylinder increases particularly as the geometry becomes more slender. Early yielding may occur in particular zones, for example the combination of compressive residual stresses combined with compressive loading will cause early membrane yield in this zone but if the geometry is stocky, this yield does not normally cause major problems and plastic redistribution can occur so that the cylinder still reaches its normal load capacity.

When the geometry is more slender, elastic deflections have most influence on the mode of failure although yield can still occur. In this case out-of-plane bending stresses combine with in-plane stresses to cause extreme fibre yield. This yielding causes a reduction in in-plane stiffness which combines with the elastic buckling response to define peak capacity and post peak behaviour. In this area there is a complex interaction between elastic buckling and material non-linearity. For very slender geometries the elastic behaviour dominates the response and yielding may only be relevant at a very advanced stage in the behaviour.

There is a need to look in more detail at the influences described above. This, however, will be undertaken in the context of the geometries appropriate to design and with regard to practical stiffened structures. The type of stiffened cylinder used in offshore construction is therefore described in the following section.

Stiffened Cylindrical Shells

For most Engineering structures the use of unstiffened cylindrical shells is not a practical proposition. For medium to high slenderness cylinders some form of stiffening is needed to enhance the strength of the basic shell plating. Practical fabrication considerations are also important. In Civil Engineering relatively crude fabrication methods are employed and stiffeners are generally needed to provide fabrication control, maintain tolerances within reasonable limits and to provide handleability.

In general for offshore shells either regular systems of ring stiffeners or orthogonal stiffening, rings and stringers (longitudinal stiffeners) are adopted. In other areas, e.g. slender cylindrical silos, local stiffening might be provided for a specific design requirement such as a circumferential wind girder at the top of the shell.

This chapter concentrates on typical offshore stiffening systems.

Ring stiffened shells are excellent for resisting external pressure because they help to maintain the overall circularity of the shell. They do not enhance strength substantially for axial compression or bending because the longitudinal buckle wavelength is very short and the presence of rings, unless very closely spaced, does little to influence the buckle waveform. If close enough, dependent on the R/t, they can serve to alter the lobular waveform to the axisymmetric form and in this case their beneficial effect can be substantial. Reference to the previous table showing the relationship between R/t and cylinder length for the different critical buckling modes can give general guidance in this area if the spacing of the rings is taken as the length criterion.

Rings, however, are often used as the sole stiffening system because they provide sufficient rigidity for controlling fabrication and they are relatively cheap to fabricate. Orthogonal systems of stiffeners have substantial cutting, shaping and hand welding at the cross-overs which make them relatively expensive.

Interactive loading arrangements also mean that the presence of ring stiffeners provides significant benefit in many instances.

Ring stiffened shells suffer from the basic disadvantage that their response is still significantly imperfection sensitive and they generally lack robustness against local damage because of their low stiffness to lateral patch loading against the shell plate.

In comparison orthogonally stiffened shells can show considerable structural advantage which has to be balanced against the higher fabrication cost. Longitudinal stiffeners will contribute in two ways to axial capacity. Firstly the area of the stiffeners contributes directly to axial capacity with the stiffeners and associated shell plating acting effectively as a column between support points. The latter is normally provided by the ring stiffeners. This column analogy is indicated in Figure 15. As long as the column slenderness of the effective stiffener/shell section is moderate the stiffener cross-section will be working at a significant percentage of the material yield stress. Secondly the stiffening system will provide some benefit in enhancing the buckling capacity of the shell plating by interrupting the natural buckling mode form. This benefit will be improved compared with rings alone because the buckle form is affected to some degree in both directions and in the circumferential direction is effectively constrained to be the same as the curved panel width between stringers. As the natural buckling wavelength in the circumferential direction tends to be longer than the corresponding axial wavelength, the effect of the stiffening can be somewhat greater. The other major benefit is that the imperfection sensitivity of the shell is generally reduced because of the relative stockiness of the individual elements including the shell plate. This results in a less severe post collapse unloading characteristic and an increased general robustness. The individual sub-panels are relatively stiff when subjected to local lateral load because of the close support provided by adjacent stringers and ring stiffeners. The shell plate can no longer deflect inwards under lateral load by bulging in adjacent regions to reduce resulting membrane straining unless the stiffeners themselves deflect to some degree.

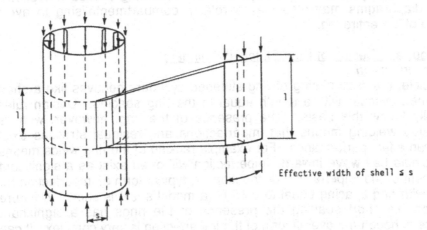

Figure 15.   Column analogy for stringer stiffener cylindrical panel.

It is always difficult to give 'normal' fabrication dimensions for structural systems but a ring stiffened shell would often have ring spacing of around the radius with values as low as 0.25 R or below in certain instances. For example floating platforms such as a semi-submissible or a tension leg platform might be designed in the latter range because the pressure loading combined with a need for robustness makes this viable.

Orthogonally stiffened shells will often have around 30-40 stiffeners around the circumference with S/t values (curved panel breadth/shell thickness) often around 40-60 giving a good strength capability compared with the yield stress. In this case the ring spacing is not likely to be lower than the cylinder radius because of the expense involved in too many cross-overs although the design parameter for the ring spacing would tend to be the column slenderness of the effective stiffener section.

The stiffeners themselves would often be flanged elements although flats are possible. Ring stiffeners would tend to be T-bar because of the larger inertia requirement related to their design and because of the requirement for the stringers to run continuously through them in the case of orthogonally stiffened shells. The latter is to avoid eccentricities in the longitudinal stress path. Accurate cutting, positioning and welding of longitudinal stiffeners at rings is difficult and expensive.

Longitudinal stiffeners would often be L section being smaller than the corresponding T-bar sections. Particularly in the case of floating structures, diaphragms or flats will be provided at intervals down the legs to maintain the overall circularity of the legs, important in enhancing pressure capacity. These diaphragms may also play a role in compartmentalising to avoid flooding of the entire leg.

Behaviour and Design of Ring Stiffened Cylinders
Axial compression
In principle, the shell plating of ring stiffened cylinders behaves like a short unstiffened cylinder with a length equal to the ring spacing. Design rules generally follow this basis. The presence of the ring, however, with its associated welding means that imperfections and residual stresses exist which can affect performance. For example racking of the ring welds means that a single half wave inwards imperfection will often exist as a significant component of the imperfection distribution. A typical form of imperfection for a shell with ring spacing equal to 0.15 R (a model shell) is shown in Figure 16. For this short spacing the presence of the rings has a significant influence although the overall form of the imperfection is very complex. It can be seen in the figure that both panel modes between rings and overall longitudinal modes between end supports exist as well as reasonably regular undulations in the circumferential direction. Failure in this model actually occurred at the weld line (the break line at which the cylinder has

been opened out), probably caused by some residual stresses remaining after stress relieving. The maximum imperfection noted of 0.47 mm from a 'best fit' cylindrical template can be compared with a shell thickness of 0.6 mm.

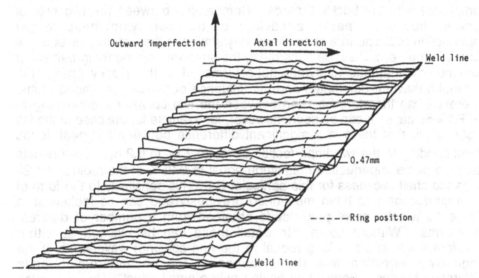

Figure 16.   Typical initial imperfections for a ring stiffened cylinder.

A better understanding of the behaviour of ring stiffened cylinders has been derived from non-linear numerical analysis using finite difference and finite element packages in combination with associated experimental work by the author of this chapter and others[14,23]. Finite difference work carried out in the early 80's[16] looked at the behaviour of ring stiffened panels with idealised boundary conditions in place of the rings. This work concentrated particularly on the effect of imperfection magnitude and mode on the behaviour and collapse capacity of the cylindrical panels under axial loading.

As an example of this work, Figure 17 shows the average axial stress-strain response of a cylinder with an R/t of 200 and a ring spacing of 0.75 R with an imperfection mode with a longitudinal half wavelength equal to one third of the ring spacing and with two half waves around the circumference. The exact nature of the imperfection form is not important but the results do demonstrate that even when the imperfection is not particularly 'sympathetic' to the natural mode, the resulting behaviour is still markedly sensitive to the actual magnitude of the imperfection with the strength varying from close to squash for a near perfect shell to around 30% of the capacity when the imperfection amplitude is of the order of the shell thickness. Figure 18 explores a different aspect of the response showing the different characteristic that can be obtained by varying the mode of the imperfection. In this case a shell with a shorter ring spacing and different R/t values has been analysed with two different imperfection forms corresponding to single and five half longitudinal waves between rings. As has already been mentioned the natural axial buckling length is very short and it can be seen from the 'soft' responses that the short wavelength imperfection is more compatible with the buckling mode. Comparison between the two sets of curves shows the peaky behaviour of the non-sympathetic longer imperfection reducing to a stress level vary similar to the eventual peak of the curves in the second set demonstrating the snap through buckling behaviour from non-sympathetic to sympathetic forms. For the stocky shells, the material behaviour dominates and the influence of buckling is limited. Little difference can therefore be seen between the two curves corresponding to the R/t = 50 case. It can be seen, however, for example for the case of the R/t = 500 shell, that there is a significant difference between the peak loads corresponding to the two imperfection forms (0.7 σ_y to 0.2 σ_y). Both results have the same imperfection magnitude equal to 0.25% of the radius or 1.25 times the shell thickness for this particular R/t. This sensitivity to the form of the imperfection has been mentioned previously and can be looked at in Figure 19 which shows peak loads obtained from a number of different waveforms. Without going into detail these can be classed as either waveforms with short or long (equal to the ring spacing) wavelength in the longitudinal direction with the influence of the circumferential mode (difference between modes 1-3) having only a small effect. The very large variation in peak capacity for the intermediate slenderness shells is readily apparent.

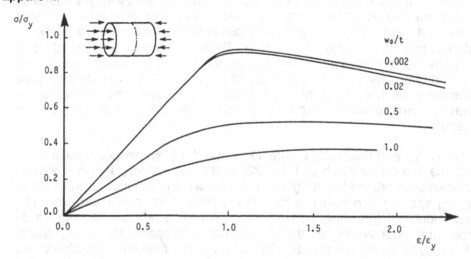

Figure 17.   Effect of imperfection level on an R/t = 200 cylinder.

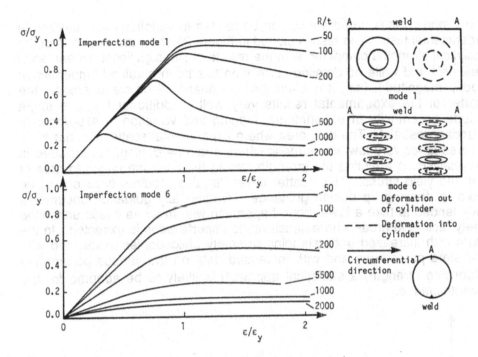

Figure 18.   Effect of imperfection mode on stress-strain response.

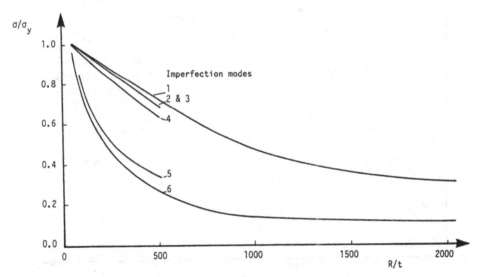

Figure 19.   Effect of imperfection mode on strength.

The various trends noted above can be related to variations encountered in experimental tests. The results plotted in Figure 20 show the extremes of the analytical strengths compared with the results of two significant experimental series[24] and while no detailed correlation has been obtained (imperfection mode, ,magnitude, etc.) it is clear that the analytical differences cover the scatter of the experimental results very well. Additional features might include eccentricity of experimental loading and variation in experimental boundary restraint. The only area where experimental scatter has not been covered is for very low slenderness where strain hardening, not included in the analysis, is affecting the strength raising this in some cases to greater than the yield stress. This scatter in results also requires decisions to be taken in drawing up design guidance. Traditionally guidance documents have tended to take a lower bound approach with in some cases enhanced safety factors in areas where sensitivity to imperfections is expected. In the future with increased understanding of safety philosophies associated with limit state procedures and with increased data on the various parameters influencing strength, a statistical approach is likely to be adopted for the strength criteria.

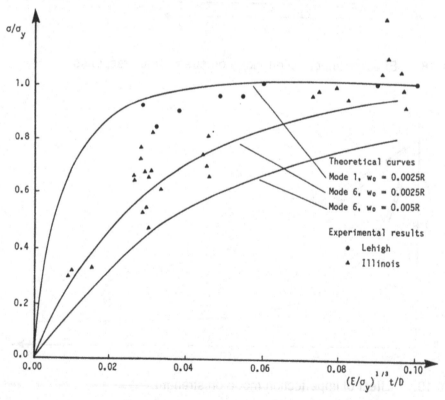

Figure 20.   Comparison of experimental and analytical results.

Three design curves are plotted on Figure 21 as an illustration of the lower bound concept.

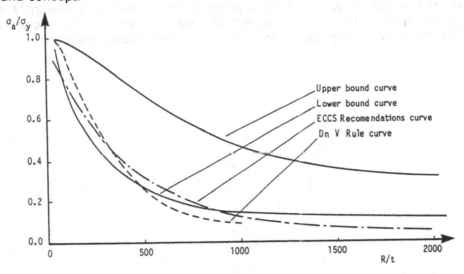

Figure 21.   Comparison between analytical strength predictions and design rules.

The effect of ring spacing is also an important consideration but this cannot be separated from a consideration of the imperfection form. Figure 22 shows the variation in compressive strength with ring spacing but with an imperfection half wavelength equal to the length of the panel. This clearly illustrates that minima can be obtained in the strength corresponding to particular l/R values. The dotted line in the figure corresponds to the longitudinal elastic critical buckling wavelengths and it can be seen that this accurately picks out the minima in the curves. This approach is very simplistic, however, because of the use of the single imperfection mode. Far less strength variation with ring spacing would be expected for a complex imperfection form and the minima would be far less evident if present at all. Nonetheless the results do highlight the fact that a decrease in ring spacing does not necessarily provide any benefit in terms of panel response under axial compression.

The strength of ring stiffened cylindrical panels subject to axial load with increasing eccentricity (combined axial load and bending) is effectively determined by the peak capacity of the most highly stressed compressive area of the cylinder circumference. While limited redistribution is possible because of the variation in stress around the cross-section, the short circumferential wavelength of the buckles tends to restrict the beneficial effect of the redistribution phenomenon. Figure 23 shows that the peak bending strength (M/M$_y$ where M$_y$ is the first yield moment) is only slightly greater than the axial capacity (σ/σ_y) with enhancement of perhaps 10-20% for critical imperfection forms. It can be seen that the straight line interaction commonly adopted in codes, while being in principal conservative, is generally a sensible representation of the analytical results as long as the reference strengths in axial compression and bending are accurately assessed.

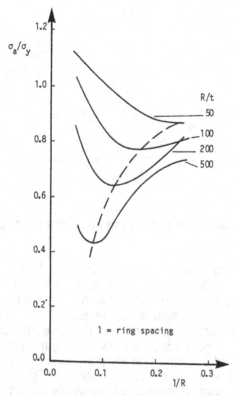

Figure 22.   Effect of ring spacing on strength.

This section has concentrated on the buckling behaviour and design of the shell panels under axial compression. There is a need to consider the role of the rings under this form of loading. In general terms the loading will be far less severe than that corresponding to external pressure with the rings generally subjected to circumferential tension and some local out-of-plane bending. It is this latter action due to oscillation of buckles in the circumferential direction which produces the local out-of-plane inertia requirement for the rings in order to maintain the effectively circular boundary restraint for the panel. Very limited analytical or experimental data is available in this area.

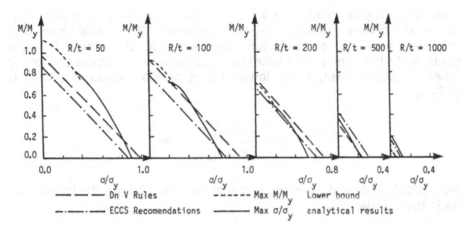

Figure 23.   Analytical curves and design rules - combined compression and bending.

External Pressure

The behaviour of ring stiffened shells subjected to external pressure is more complex in that the rings play a major role, in combination with the panels, in resisting the pressure and the structure can either fail by panel buckling between the rings or by overall failure involving the rings. In the former case the rings provide anchorage to the panels which can be substantial in the large deflection regime.

With external pressure loading both the panel and ring stiffeners are subject to membrane hoop compressive forces and the design of the rings therefore in principle should involve an area requirement to sustain the hoop compressive force as well as an inertia requirement to resist local out-of-plane bending resulting from lobular buckling of the panels. The area requirement, however, is not always apparent in design guidance.

At this stage it is important to differentiate between pure lateral pressure and hydrostatic pressure because these conditions have a significantly different effect on the cylindrical shell. The former represents the loading to be expected on a section of a structure leg which is continuous from the ground to above water level resulting in a pressure which is applied radially to the shell but with no end load on the cylinder inducing axial compressive stresses. The hydrostatic case represents an enclosed submerged cylinder in which an end load is present inducing a longitudinal compressive stress which can be directly related to the hoop compression, i.e a fixed ratio of the two induced stresses. In terms of a normal structural loading environment the hydrostatic case can be considered as a particular defined location on the pressure-axial compression interaction diagram and will be dealt with in the next section. Design rules, however, often have a section devoted to this specific load combination.

The lateral loading of cylinders induces a non-uniform hoop compression in the plate which varies between the ring stiffeners. This variation is often ignored with the hoop compression taken equal to pR/t where P is the pressure and t the shell plate thickness. The actual hoop stress at the ring stiffener location is likely to be lower and a possible value is quoted in DnV[25] as

$$\frac{pR}{t}\left[1 - \upsilon\sigma\frac{1.56\,t\sqrt{Rt}}{A_s + (1.56t\sqrt{Rt})} + \upsilon\sigma\right]$$

where υ is poissons ratio, A_s is the ring stiffener area and σ is the longitudinal stress in the cylinder.

The hoop stress σ_θ causes local buckling of the shell plate but definition of the elastic critical buckling stress is more complex then for the equivalent axial stress. It involves the minimisation of the governing equation with respect to the waveform and a closed form solution cannot be obtained. The elastic critical pressures are therefore normally presented graphically such as in the DnV[25] code. The background to the derivation can be found in Donnell[26].

Failure of the panel between the rings will be lobular in nature, generally with a significant number of internal buckles around the circumference.

When a ring stiffened cylinder fails by general collapse under pressure loading, the deflected shape along the axial direction will be affected by the restraints provided by the presence of intermediate non-deflecting heavy rings or diaphragms. In theory the elastic critical buckling form of a ring stiffened cylinder is two waves around the circumference if the cylinder is infinitely long. However, in practice there will almost certainly be some form of end restraint and in offshore structures there will generally be intermittent diaphragms which maintain circularity at particular locations. These diaphragms or flats alter the buckling mode substantially so that the ultimate failure will tend to occur around a limited arc of the circumference. Figure 24 shows a deflected profile of a ring stiffened cylinder failing under lateral pressure through the rings with end diaphragms five panels apart, i.e. with four ring stiffeners. The deflection is of a single half sine wave type form between diaphragms although the curvature is higher adjacent to the ends with reasonably uniform deformation along the central zone of the model. The failure deflections can be seen to occupy only about one eighth of the circumference. Appreciably plasticity is also involved with longitudinal yield lines at three locations being apparent. This particular model had on R/t of 267 and a ring spacing of 0.15 of the radius.

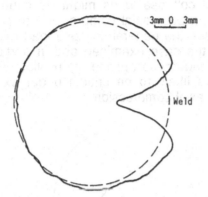

Figure 24.   Buckled shape of cylinder around centre of panel.

The restraint of the elastic buckling mode by the diaphragm has a substantial effect on the collapse resistance of the shell to pressure loading, a factor that is insufficiently recognised in many design codes. Figure 25 shows the results of sample analyses which demonstrate that the shell collapse pressure does decrease as the number of bays between rigid diaphragms increases but that even with sixteen stiffeners the difference with the strength of the infinite case (obtained from using a symmetry condition in a finite element analysis) is still substantial. The implications are that using a theoretical prediction may be unduly conservative but also that using the results of a model test with perhaps three or four bays may be non-conservative.

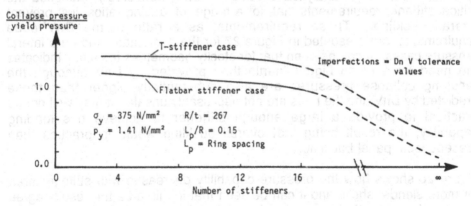

Figure 25.   Dependence of collapse pressure on stiffener number.

Interaction between axial compression and external pressure
This interactive loading environment is often found in offshore structures and is clearly important for floating rigs.

Interaction between stresses will have significant effects both on local and overall buckling modes. Figure 26 gives a schematic representation of how the panel and general collapse loads might be expected to vary as the loading ratio varies from pure axial compression to pure lateral pressure. The relationship between the general collapse load and the panel collapse mode will depend on the case examined and in particular the shell plate versus the stiffener slenderness but the figure illustrates that the general collapse mode is more likely to be critical under external pressure and unlikely to occur under axial compression.

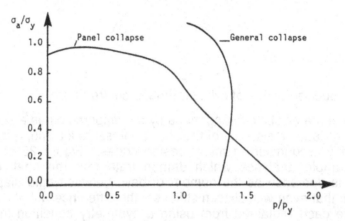

Figure 26.   Effect of loading ratio on collapse behaviour of cylinder.

Ignoring interaction between the mode forms, the point at which the curves cross indicates the load combination that would just cause general collapse to occur. A family of analytical studies[20] have, conversely, identified the critical stiffener requirements that, for a range, of loading ratios, just prevent overall buckling.   These requirements, as a ratio of the DnV' code requirements, are presented in Figure 27 for the hydrostatic and pure lateral pressure cases. It can be seen that for stocky geometries the study indicates the need for a much bigger inertia than provided by DnV although the resulting collapse pressures are also substantially bigger than those predicted by DnV and the rules are not necessarily unsafe. It may well not be practical to provide a large enough stiffener to achieve this loading capability, the result being that overall buckling would in practice then precede local panel buckling.

Figure 28 shows how the pressure capability decreases with stiffener area for more slender shells and it can be seen that in this area the results agree in general terms with the area requirement of the code and that only modest decreases in pressure capacity occur with significant reductions in stiffener size.

A general comment has been made above that pressure loading induces less imperfection sensitivity then axial compression. There is also little evidence in the analytical work carried out that the capacity is sensitive to mode interaction, for example when overall and local modes are predicted to occur at approximately the same load level. However, during analyses relating to panel buckling under interactive loading there were strong indications that a particular zone with an axial to hoop stress ratio of 1.5 produced a very high sensitivity to imperfections. It transpired that this zone corresponded to the load ratio where the effect of pressure counterbalanced the natural poisson expansion of the panel between rings resulting in a response resembling the perfect cylinder under axial compression. In this configuration the high sensitivity to imperfections can be readily understood.

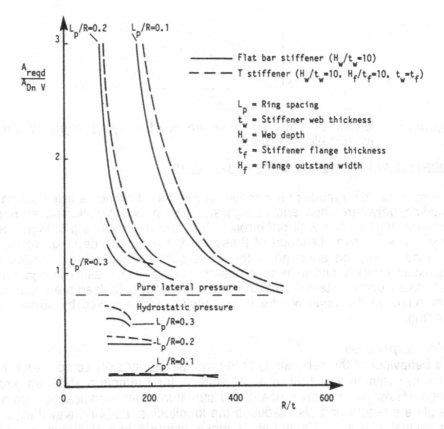

Figure 27.   Comparison of required stiffener areas with DnV predictions.

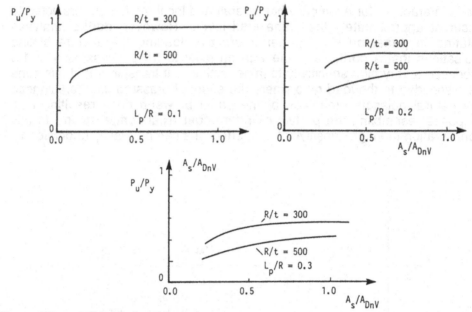

Figure 28.   Effect of stiffener area on load-carrying capacity under hydrostatic pressure.

Behaviour of Orthogonally Stiffened Cylinders

The main buckling modes for orthogonally stiffened cylinders are local panel buckling between rings and stringers, panel buckling including stringers between rings and overall orthotropic shell buckling including stringers and rings. The relative likelihood of these modes occurring depends again on the loading applied although in this case both of the first two modes are significant for axial and external pressure loading. Ring failure is again only really likely under external pressure because under axial load the major load component in the rings will be hoop tension accompanied by some local bending.

Axial compression
The behaviour of the sub-panels of the cylinder under axial compression has much in common with that of a flat plate in for example a stiffened girder flange. However, the presence of the curvature does significantly increase the plate strength and also reduces the longitudinal buckling wavelength of the critical mode. This tends to make inwards imperfections of short longitudinal wavelength more critical (inwards imperfections effectively reduce panel curvature and hence reduce strength) although Figure 29 shows that all modes but the single outward half wave for a square panel have a similar effect. The imperfection amplitude w_0 adopted in these finite difference analyses is typical of normal fabrication.

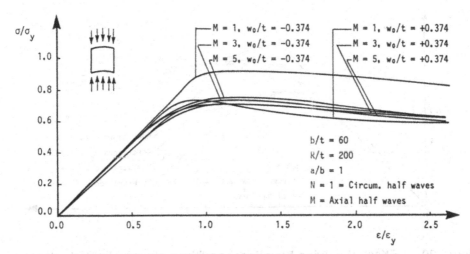

Figure 29.   Effect of varying shape of initial distortion.

The other important parameter in all panel behaviour is the degree of boundary restraint. Both in-plane and out-of-plane resistance to deflection are important but the normal assumption is that the surrounding stiffeners are sufficiently rigid to resist out-of-plane deflection of the panel edges while continuity across the stiffeners provides little moment restraint and the edges are therefore taken as simply supported.   The latter is a lower bound representation.   The in-plane restraint along the longitudinal edges is more difficult to determine but three commonly accepted mathematical boundary representations are examined in Figure 30.   The constrained condition provides an intermediate level of restraint in which the transverse in-plane stresses are self equilibriating so that no net force is transferred between panels but the edges are constrained to remain straight. This is felt to be the most appropriate simple idealisation of the cylindrical panel edges. Restrained and unrestrained conditions refer to non-deflecting and free edges (in the in-plane transverse direction) respectively.

Taking representative values of panel tolerance, parametric study was undertaken to examine existing design guidance and the results of Figure 31 show generally excellent correlation with the formulation in the DnV fixed offshore structure guidance document[25] over a wide range of panel and shell slenderness. For high slenderness panels (b/t = 60 and 80) the results also demonstrate the very substantial strength enhancement that can be gained through the effect of curvature but interestingly this effect is not as consistent within the range of higher R/t values as would be expected and indeed as represented by the code. As offshore structure main elements tend to fall in the R/t range of 200 to 400, however, the beneficial effect is well represented by the code.

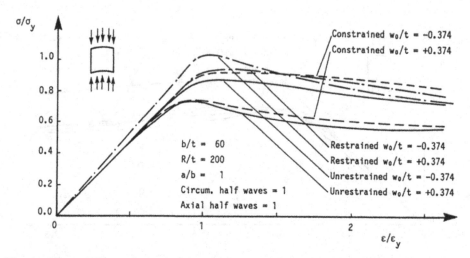

Figure 30.   Effect of varying boundary conditions along unloaded edges.

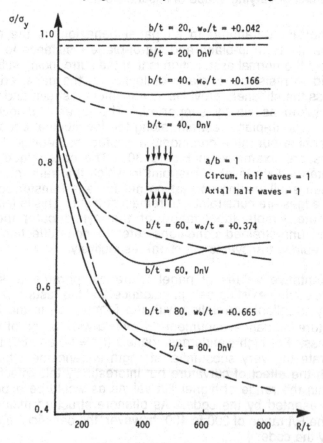

Figure 31.   Variation of strength with panel and shell slenderness.

The behaviour of the stringer stiffened panel between rings is more complex because, while the stringer, considered with an effective width of shell plate, tends to behave like an axially compressed strut between ring frames, overall buckling modes can develop involving several stringers that are not accurately modelled by the simple strut formulation. However, the strut formulation is commonly used in design and the strut slenderness provides a useful and generally relevant parameter for strength assessment. The work of Agelidis[27] examined the implication of the overall interaction between the deflection of adjacent stringers using finite element analyses.

The earlier parametric study[28] demonstrates clearly the effect of the column slenderness of the stiffener and associated shell plating on the collapse strength of the stiffened panel. A double stiffener model with idealised stiffened panel edge conditions was adopted. This does potentially restrict to some degree the deflection of the panel because longitudinal edges are prevented from displacing laterally but the model still demonstrates the effect of column displacement (column slenderness) on the collapse strength of the sub-panels.

Figure 32 shows the effect of column slenderness ($\lambda = l/r$ where r is the radius of gyration of the equivalent flat plate system and l is the ring spacing equal to the stiffened panel length) on the average stress strain response of the central panel comparing this with the idealised panel responses. γ is the depth to thickness ratio of the flat stiffener outstand and α is the stiffener to shell thickness ratio. b is the curved panel width.

The normal expected column response occurs and the λ value of 64 provides a support for the panel which does not have a deleterious effect on panel capacity. The panel in this case is relatively stocky and Figure 33 shows comparison results for a higher panel slenderness. In this case because the panel strength is lower, the effect of column deflections is greatly reduced for the entire practical range of λ values.

As panel strength increases with curvature, reduction in R/t, it would be expected that column strength would become important even with high panel slenderness.

The results presented, however, do not underline the important contribution of the area of the stringer stiffeners to the direct load carrying capacity of the cylinder. As with the panel these stiffeners will be contributing a proportion of their squash load that depends mainly on their column slenderness. The stiffening ratio, stiffener to shell plate area ratio, and the number of stiffeners for a given R/t are significant parameters in determining the collapse capacity of the cylinder as a whole.

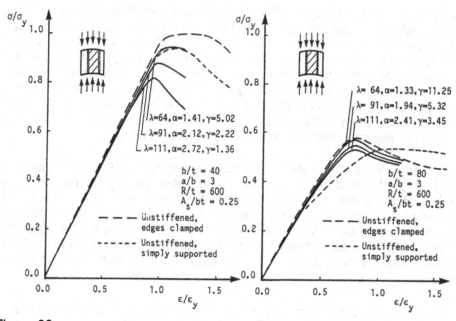

Figure 32.   Figure 33.

Variation in stress/strain response with stiffener column slenderness.

External Pressure

The failure of the sub-panels and the stringer stiffened panel between ring frames are relatively stable under external pressure loading. Unstiffened panels gain good membrane resistance to lateral pressure as out-of-plane deflections increase. The response of the stringers themselves resembles that of a laterally loaded beam with bending stresses generally providing the resistance limit. However, the latter has to be considered in terms of an overall buckling mode that is complicated by continuity around the cylinder and the influence of the general circumferential buckling mode of the cylinder, the waveform in the circumferential direction, can be significant. A cylinder with a low R/t ratio and a few stiffeners would for example be expected to offer a significantly different pressure resistance than that of an equivalent isolated beam. Detailed results will, however, will not be presented here.

While it is possible to obtain an orthotropic shell failure involving buckling of the rings under external pressure, the rings themselves are normally conservatively designed, the philosophy being that stringer stiffened shell buckling should precede overall buckling. While elastic orthotropic shell analysis is possible, virtually no information is available on the non-linear buckling of the rings in combination with the stringer stiffened panels.

Interaction between axial compression and external pressure
Figure 34, taken from Reference 27, shows in a general way how the resistance of the stringer stiffened panel varies depending on the ratio between axial and pressure loading and how it depends on the number of stiffeners and the spacing between the rings. An interesting conclusion is that the above geometrical parameters have a much more substantial effect on pressure capacity than on axial capacity but as the R/t increases to 400 (Figure 35) the variation in axial capacity is much more substantial. It seems likely that this is mainly a panel slenderness effect as for a given N value (the number of stringers) the panel slenderness doubles with the change from the R/t = 200 to the R/t = 400 case. The axial strength even in Figure 35 varies modestly with ring spacing indicating that the stringer column slenderness has a relatively small influence.

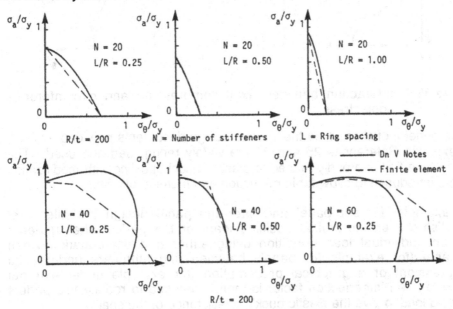

Figure 34.   Interaction between axial compression and circumferential hoop stress.

It can also be seen that the interaction is generally convex with poisson enhancement occurring between axial and hoop stresses for the stockiest geometries. This is associated with the biaxial compressive interaction of stresses on the yield surface.

Design Rules
It has already been stated that design rules tend to be of a lower bound nature and can be conservative. However, the studies that have recently been undertaken, some of which feature above, do not give significant cause for concern.

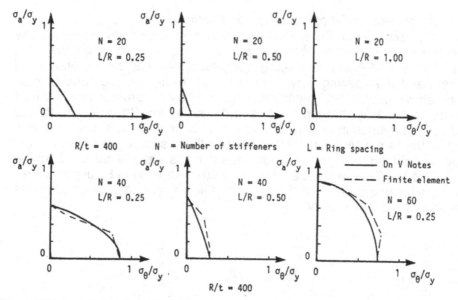

Figure 35.   Interaction between axial compression and circumferential
               hoop stress.

In the general offshore area the various design documents provided by DnV,
for example References 25 and 29 are widely recognised and used. The
basis of the first is broadly related to elastic critical buckling with modification
factors introduced to allow for imperfection and material effects.

For example, for sub-panel and inter-ring panel design the rules first
establish the elastic critical buckling load for the perfect shell under a
relevant individual load condition using either a direct equation when
available (for example for panels between ring stiffeners under axial
compression) or a graphical presentation (for example under external
pressure). An imperfection factor is then introduced to reduce this perfect
buckling load to give the elastic buckling resistance of the shell.

$$f_e = \rho f_i$$

An example of ρ is given in Figure 36 for the cylindrical panel between ring
stiffeners under axial compression and it can be seen that for this case ρ
values may be as low as 0.2. Z is the Batdorf parameter with the cylinder
length taken equal to the ring spacing.

A general slenderness is then defined in terms of the relationship between
the elastic resistance and the yield stress $\left(\lambda = \sqrt{\dfrac{f_y}{f_e}}\right)$ and the resistance as a

function of yield is given as

$$R_\kappa = \phi \; f_y \text{ where } \phi = \sqrt{\frac{1}{1+\lambda^4}} \; .$$

The result is a strength slenderness curve very similar in nature to the Perry type curve well known in the area of column buckling.

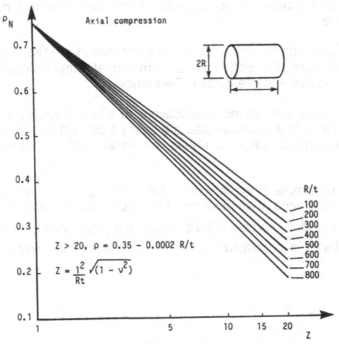

Figure 36.   Imperfection factor against cylinder slenderness.

Once the appropriate panel R_κ value has been determined for all relevant loading types, an interaction equation, dependent on the ratio of the applied stresses, is used to evaluate the ultimate collapse capacity under the combined load condition.

The stringer stiffeners for an orthogonally stiffened shell are designed as isolated struts although in this particular version of the DnV rules[25] benefit is not allowed in this particular design calculation for the curvature of the shell.

The design of the rings for both ring and orthogonally stiffened cylinders is based around an inertia requirement for each load type which is a function of the panel resistance to the appropriate loading and the geometry. For the orthogonally stiffened case the resistance of the stringer stiffener panel is adopted and an allowance made for the area of the stringer stiffeners by modifying the shell thickness to include the smeared effect of the longitudinal stiffeners. For combined loading, use is again made of an interaction equation where the inertia required is based on a sum of the individual inertia for each load case proportioned in terms of the stress resulting from each applied load relative to the shell resistance under that load. If for example a large axial compression is present together with a low external pressure the resulting inertia required will be slightly greater than that required for axial loading alone but substantially less than that needed for pressure alone.

Arguably the most up to date rules for cylinder design are those emanating from the European Convention for Constructional Steelwork (ECCS)[30]. These are described in some detail in Reference 12.

These are again based around modifications to elastic buckling predictions. One interesting feature of these rules is a safety factor which is increased for a higher slenderness range to reflect the wider scatter of experimental results.

Looking at an example of the format of the rules, Figure 37 shows the axial compressive resistance of an unstiffened cylinder or the cylindrical panel of a ring stiffened cylinder. A value of $\lambda\sqrt{2}$ divides the graph with different values of the safety factor incorporated in the curve for greater and lower λ values.

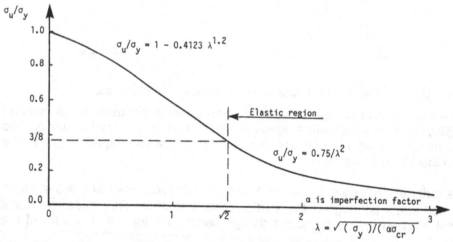

Figure 37.   Ultimate axial stress against cylinder slenderness.

The same limiting stress is applied to a cylinder under bending.

α is the imperfection factor which is again a function of the slenderness as in DnV.

The curve of Figure 38 shows that values of α for compression (α_0) are not dissimilar to DnV. Values for bending (α_b) are significantly higher providing an enhancement in strength for the bending case reflecting the fact that a bending stress has some redistribution capability in terms of its effect on buckling behaviour.

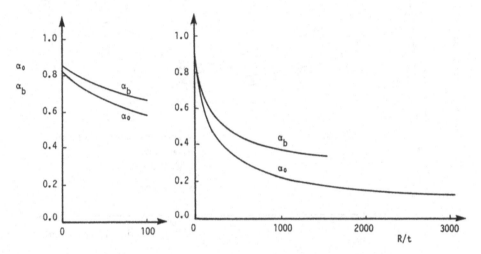

Figure 38.   Imperfection factors for axial and bending loading.

The pressure rules within the ECCS guidance are more complex and reference should be made to the description by Vandepitte[12].

Stiffener Outstand Behaviour
In most codes of practice there are restrictions on the local slenderness ratios of the various plate components of all stiffeners and these are included in offshore and other shell codes.

These limitations are present to prevent torsional buckling of the stiffener element or local buckling of part of the stiffener. Restrictions placed on stiffener outstands are normally conservative in nature, ensuring that the shell or stiffened shell buckling modes would occur prior to local outstand buckling. This is primarily because local failure of the stiffener outstand is a relatively unstable phenomenon which results in a high degree of stress shedding from the stiffener element to the remainder of the shell which could easily result in catastrophic collapse.

Figure 39 gives an example of the rapid unloading that can occur for a ring stiffener as the local outstand slenderness is increased. An external pressure load was applied to the shell. More severe characteristics can be encountered for axially loaded longitudinal stiffeners.

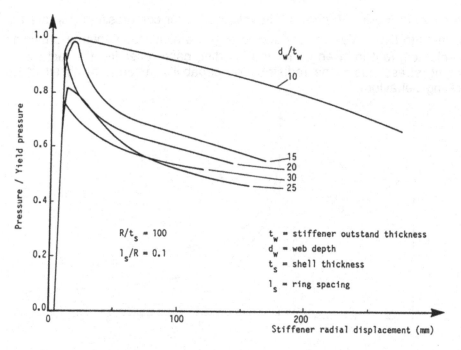

Figure 39.   Pressure against radial displacement for different stiffener web slendernesses.

In ship fabrication the possibility of a torsional mode is often precluded by the use of tripping brackets connecting the stiffener outstand at right angles to the plate and these are sometimes used in semi-submissible offshore rigs but they are not common in Civil Engineering fabrication. Because of the short buckling wavelength of tripping failures there may be some doubt about the effectiveness of these brackets.

A recent study carried out by Louca[31] examines the effect of various geometrical parameters on both flat stringer stiffeners in cylinders and flat and T-bar ring stiffeners

Figure 40 shows that stiffener curvature has little effect on the sensitivity of collapse load to stiffener outstand slenderness other then in globally enhancing the strength because of the curvature effect on panel capacity. The drop in the curves for low outstand slenderness and high plate slenderness is a secondary effect caused by reduction in column slenderness of the stiffener section as the stiffener depth reduces.

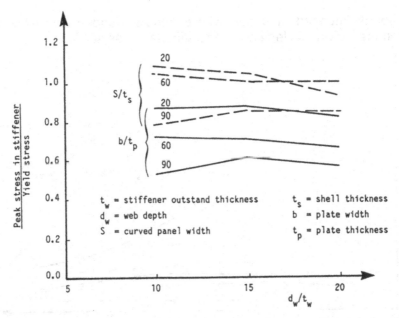

Figure 40.   Variation in stiffener stress with stiffener and panel slenderness.

Figure 41 shows the variation in elastic critical buckling pressure as the slenderness of the outstand of a flat ring stiffener increases and how this affects ultimate pressure. For R/t = 200 and 300 failure is dominated by a shell mode but for the R/t = 100 case the rapid decrease in the critical pressure with increasing outstand slenderness shows that the tripping mode is the critical consideration for the stockier shell geometry. This induces slight reductions in ultimate capacity as the elastic critical stress becomes assymptotic to the dotted curve. Higher stiffener slenderness values would be expected to have a more substantial effect on pressure capacity with the tripping mode dominating the collapse behaviour.

With a T-bar ring it is interesting to see, Figure 42, that the web slenderness has a more marked effect on local buckling capacity than the flange slenderness with the flange outstand slenderness exceeding twenty before a significant effect is noted. It can be seen, however, that the presence of a flange has a significant beneficial influence compared with the torsional behaviour of the flat ring stiffener. The reductions in capacity for low d/t values again reflect the decrease in web depth resulting in a lower radial capacity for the stiffened shell, a characteristic independent of stiffener tripping.

The results of the study illustrate the difficulty of isolating the tripping mode from the other buckling modes that can occur, with a strong inter-dependence between the various parameters. Generally the results support the philosophy imposed in codes where simple slenderness limits are imposed on the individual elements of the stiffener cross-section.

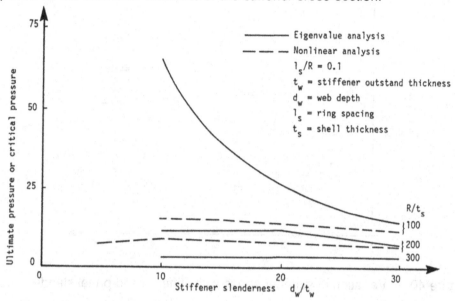

Figure 41.   Variation of ultimate and critical pressures with stiffener outstand slenderness.

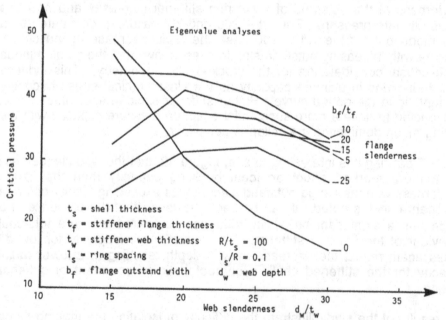

Figure 42.   Effect of stiffener flange and web slenderness on critical pressure.

The Effect of Damage on Cylindrical Shell Behaviour

Recent analytical and experimental work has been undertaken into the effect of damage on the load capacity of stiffened cylindrical shells.

This followed a concern about the possibility of damage occurring due to ship impact with a rig combined with an awareness of the imperfection sensitivity of such structures.

A study on the post damage axial capacity of ring stiffened cylinders[23] involved experimental and analytical work. In the former, local lateral load was applied to the cylinders to induce a local dent and axial load was then applied to obtain the residual strength. Detailed imperfection measurements were taken before and after testing so that non-linear finite element analyses could be performed on impact and damaged models with accurate imperfection information so that the effect of the damage could be predicted and the analyses calibrated against the experiments.

It was originally assumed that imperfection sensitivity would play a major role in the strength reduction process with the dent acting as a large imperfection. The detailed analyses and the forms of failure clearly indicated that the most decisive feature was the presence of a band of high stress in a line circumferentially adjacent to the dent corresponding to an essentially straight generator in the imperfection form. Stress has shed to this area from the less stiff dent zone and failure occurred when this stress level reached approximately the failure stress of the equivalent intact cylinder. When the form of the dent corresponded with the natural buckling mode there was some enhanced imperfection sensitivity. Strength reductions were around 30% on intact capacity for a dent depth of approximately ten times the wall thickness. An extended parametric study produced predictions of strength reduction versus dent depth such as those shown in Figure 43, compared in the figure with a single prediction obtained by ignoring the dent area, allowing for resulting load eccentricity, and taking the maximum bending stress to be limited to the intact axial stress at failure.

A corresponding study in the area of stringer stiffened shells[32] looked at the development of the damage deflection in the stringers of orthogonally stiffened cylinders and then its effect on axial capacity. The same effective section analysis was applied and found to be generally satisfactory although it was concluded that small radial imperfections, circumferentially remote from the dent, but produced during lateral loading, may have an additional effect on the compressive response of the most slender shells.

In a third study[33] a limited number of experimental tests were carried out on dented ring stiffened cylinders subjected to external pressure and subsequent axial compression. This work indicated possible severe problems occurring under external pressure where the rings suffered significant twisting during denting and where their torsional stiffness was insufficient to resist tripping under the deformed state. For one test in particular a sudden buckling response was obtained under incrementing pressure at a load level significantly lower than expected due to the plastic bending of one of the rings. The authors conclude that with a long shell, a cascade effect could occur with buckling of successive rings leading to a particularly catastrophic failure possibly restrained by the provision of rigid diaphragms. The same problem is not likely to occur with T-bar rings with their higher torsional and lateral inertia but it does reinforce doubts about the robustness of slender flat stiffeners. Conclusions can only be tentative because of the very limited data available.

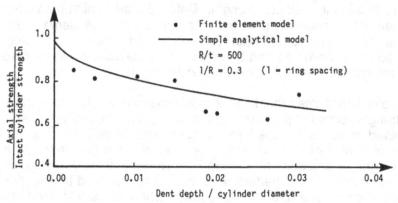

Figure 43.   Variation in strength with dent depth.

Summary
This chapter has looked very briefly at the types of buckling to occur in ring and orthogonally stiffened cylindrical shells and at the background fundamental influences to this behaviour. Reference has been made to recent research to illustrate the points being discussed. While this chapter may serve as an introduction to cylindrical shell buckling the reader is directed to the substantial references available in order to obtain a deeper knowledge of the subject and of the basis of the design of such systems.

References

1. Arbocz, J.: Past, present and future of shell stability analysis, Delft University of Tech, Department of Aerospace Eng Rept. LR-320, 1981

2. Timoshenko, S.: Einige Stabilitats - Probleme der Elastizitats - Theorie, Zeit Math. Phys., Vol 58, no 4, 1910.

3. Southwell, R.V.: On the general theory of elastic stability, Phil. Trans. Roy. Soc., London, Series A, no 213, 1914.

4. Von Mises, R.: Der kritische Ausserdruck zylinderischer Rohre, Z. Ver. Deutsch. Ing., vol 58, 1914, pp 750-755, English transl: David Taylor Model Basin Translation no. 5, Aug. 1931.

5. Wilson, W.M. and Newmark, N.M.: The strength of thin cylindrical shells as columns, Univ of Illinois, Eng. Station Bull. no. 255, 1933.

6. Donnell, L.H.: A new theory for the buckling of thin cylinders under axial compression and bending, Trans. ASME, vol. 56, no. 11, Nov. 1934, pp795-806.

7. Fluegge, N.: Statik und Dynamik der Schalen, Springer-Verlag, 1934.

8. Batdorf, S.B., Schildcrout, M. and Stein, M.: Critical stress of thin walled cylinders in axial compression, NACA report. 887, 1947.

9. Schilling, C.G.: Buckling strength of circular tubes, Proc. ASCE, J. Struct. Div., vol 91, ST5, Oct 1965, pp325.

10. Donnell, L.H.: Stability of thin-walled tubes under torsion, NACA Rept. 479, 1933.

11. Esslinger, M. and Geier, B.: Calculated post-buckling loads as lower limits for the buckling loads of thin-walled circular cylinders, Buckling of Structures, Proc IUTAM Symp. edit by Budiansky, B.: Harvard Univ. Springer-Verlag, 1976, pp274-290.

12. Vandepitte, D.: Buckling of Shell Structures, Constructional Steel Design, An International Guide, to be published by Elsevier Applied Science Publishers, 1992.

13. Eslinger, M.E. and Geier, B.M.: Buckling and post-bucking behaviour of thin-walled circular cylinders, International Colloquium on Progress of Shell Structures in the last 10 years and its future development, Madrid, 1969.

14. Dowling, P.J. and Harding, J.E.: Current research into the strength of cylindrical shells used in steel jacket construction. Proc. BOSS 79, Imperial College, London, August 1979.

15. Dowling, P.J. and Harding, J.E.: Experimental behaviour of ring and stringer stiffened shells. International conference on buckling of shells in offshore structures, Imperial College, April 1981.

16. Harding, J.E.: The elasto-plastic analysis of imperfect cylinders, Proc. Instn. Civ. Engrs. Part 2, Vol 65, December 1978.

17. Dowling, P.J. and Harding, J.E.: The strength of steel jacket leg components. Proc OTC, Houston, May 1979.

18. Dowling, P.J. and Harding, J.E.: Experimental behaviour of ring and stringer stiffened shells. International conference on buckling of shells in offshore structures, Imperial College, April 1981.

19. Tsang, S.K. and Harding, J.E.: Buckling behaviour under pressure of cylindrical shells reinforced by light ring stiffeners, Proc Instn. Civ. Engrs. Part 2, vol 79, 1985.

20. Tsang, S.K. and Harding, J.E.: Design of ring stiffened cylinders under external pressure, The Structural Engineer, Vol 63B, No 4, December 1985.

21. Harding, J.E. and Tsang, S.K.: The behaviour and design of ring - stiffened cylinders , 5th International symposium on offshore mechanics and arctic engineering, Tokyo, Japan, April 1986.

22. Tsang, S.K. and Harding, J.E.: Design of ring-stiffened cylinders under interactive loading, ASCE, Journal of Structural Engineering, Vol 113, No 9, September 1987.

23. Harding, J.E., Onoufriou, A and Tsang, S.K.: Behaviour and design of ring-stiffened cylinders including damage effects, Steel Construction Offshore/Onshore, Imperial College, April 1987.

24. Ostapenko, A.: Local buckling of welded tubular columns, Fritz Engineering Laboratory, Lehigh University, 1977, Report 406.11.

25. Det norske Veritas, Rules for the design, construction and inspection of fixed offshore structures, Det norske Veritas, Oslo, 1977.

26. Donnell, L.H.: Beams, Plates and Shells, McGraw Hill, 1976.

27. Agelidis, N.: Collapse of stringer stiffened cylinders, PhD Thesis, University of London, 1984.

28. Harding, J.E. and Dowling, P.J.: Analytical results for the behaviour of ring and stringer stiffened shells, International conference on buckling of shells in offshore structures, Imperial College, April 1987.

29. Det norske Veritas, Classification notes - buckling strength analysis of mobile offshore units, Note No 30.1, Hovik, Norway, 1984.

30. Buckling of steel shells; European Recommendations, Fourth Edition, European Convention for Constructional Steelwork, 1988.

31. Louca, L.A., Buckling behaviour of stiffener outstands, PhD Thesis, University of Surrey, 1991.

32. Ronalds, B.F. and Dowling, P.J.: Local damaged effects in cylinders stiffened by rings and stringers, Steel Construction Offshore/Onshore, Imperial College, April 1987.

33. Walker, A.C., McCall, S. and Kwok, R., Strength of damaged shells under pressure and axial loading, Steel Construction Offshore/Onshore, Imperial College, April 1987.

PART 3

STEEL COMPRESSED BARS AND BEAMS

DESIGN OF COLUMNS AS PART OF FRAMES
- SOME REMARKS

S. Vinnakota

Marquette University, Milwaukee, WI, USA

ABSTRACT

With the help of two simple rigid-jointed frames, this lecture shows that for frames with weak beam/strong column designs, use of effective length factors K less than one, given by bifurcation analysis, for design of such frames may be unconservative. Also, in the case of strong beam/weak column design, the beam has to support additional moments that are equal in magnitude to the restraining moments on the column.

1.0   INTRODUCTION

In general, columns do not occur as isolated members but as part of frames. Even for a simple case of loading of a simple frame, the loading history of a column forming part of this frame is rather complicated and depends upon the relative strength of different members of the frame. This interaction makes it necessary to consider the connecting members when designing columns of frames. The effective column length design procedure described in Chapter 15 of the SSRC Guide [1] is the most widely accepted procedure by which this interaction is taken into account in the USA. In this procedure, the axial force and end moments on a given column are obtained from a first order, elastic analysis of the given frame subjected to the given service loads (primary bending moments). The bifurcation solution for an equivalent or derived axially loaded frame provides a critical load for the column in the frame comparable to the critical load solution for an isolated axially loaded column. The framed column and the isolated pin-ended column are directly related to each other by the concept of effective column length. This effective length, KL, is used to determine the allowable axial load term in interaction formulas for design of the column subjected to both axial load and bending moments.

For braced, rigid frames, the use of the effective column length design procedure described above apparently leads to certain contradictions [2]. For example, the bifurcation analysis of a derived axially loaded braced frame results in effective length, KL, less than the actual length, L for the columns, indicating that the columns are being restrained by the beams. On the other hand, the end moments on the columns -- calculated by the elastic analysis of the frame -- are obtained on the assumption that the columns restrain the beams (note that the unbalanced fixed end moments at a joint are distributed between the beams and the columns at that joint in proportion to their elastic rigidities).

In this report, we consider the elastic and inelastic stability of rectangular, rigid jointed steel frames ABC and ABCD (Fig. 1) as influenced by primary bending moments. The results should give us an indication of the validity of the assumptions made in the effective column length design procedure.

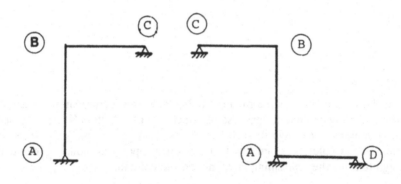

Fig. 1 - Braced Frames Under Study

2.0   HISTORY

Chwalla [3] was the first to investigate, in 1938, the effect of primary bending moment on stability of rectangular portal frames. He used the approach of integrating a system of differential equations that define the equilibrium of various members in the buckled state. The bending moments present in the members and their increments after buckling were taken into account in obtaining the basic equations. In 1961, Masur, Chang and Donnell [4] described a method based on slope-deflection equations to study the elastic stability of frames. Stability of pinned base portal frames with primary bending moments, in the elastic and partially plastic range, was studied by Lu [5]. Vinnakota [6,7] studied the effects of primary bending moments on the stability of rectangular and mill building frames with crane loads, using both slope-deflection and three moment equations. Because of the rapid development in the plastic methods of designing structures, increased attention has been given, since the 1960s, to the overall instability of elastic plastic frames as influenced by primary bending [5,7,8].

3.0 GENERAL SLOPE DEFLECTION EQUATIONS

The general slope deflection equations for a laterally loaded prismatic member 1-2 of length L subjected to an axial compressive load P (Fig. 2) are given by [9,10]:

$$M_{12} = M_{12(F)} + k \left[s\theta_1 + sc\theta_2 - s(1 + c) \frac{\Delta}{L} \right]$$

$$M_{12} = M_{21(F)} + k \left[sc\theta_1 + s\theta_2 - s(1 + c) \frac{\Delta}{L} \right]$$

(1)

where k (=EI/L) is the rigidity factor of the member; θ_1, θ_2 are the end slopes of the member; Δ is the relative deflection of the ends; s, c are the stability coefficients and $M_{12(F)}$, $M_{21(F)}$ are the fixed end moments induced in the member when both ends are kept fixed in position and direction. In the derivation of above relations, clockwise moments and rotations at the ends of the member are considered positive.

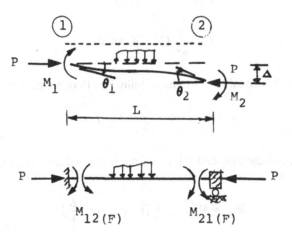

Fig. 2 - Member Forces and Deformations

The stiffness coefficient, s, and the carryover factor, c, are functions of the axial force, P, in the member. We have

$$s = \frac{(1 - 2\alpha \cot 2\alpha) \, \alpha}{(\tan \alpha - \alpha)}$$

$$c = \frac{(2\alpha - \sin 2\alpha)}{(\sin 2\alpha - 2\alpha \cos 2\alpha)}$$

(2)

wherein

$$\alpha = \frac{\pi}{2} \sqrt{\rho}$$

$$\rho = \frac{P}{P_E}$$

(3)

$$P_E = \frac{\pi^2 EI}{L^2}$$

Here, P_E is the pin-ended Euler load of the member 1-2 for buckling in the plane of the applied loads and bending moments.

For a member subjected to uniformly distributed lateral load, γq -- wherein γ is the load factor and q is the lateral load intensity at $\gamma = 1$ -- the fixed end moments are given by

where

$$- M_{12(F)} = M_{21(F)} = f \frac{\gamma q L^2}{12}$$

(4)

$$f = \frac{3}{\alpha^2} (1 - \alpha \cot \alpha)$$

(5)

When the axial force P is equal to zero, the stability functions simplify to:

$$s = 4; \qquad c = \frac{1}{2}; \qquad f = 1$$

(6)

When one end of the member, say end 2, is free from rotational restraint, the slope-deflection equation (1) reduces to:

$$M_{12} = M_{12(F)}'' + k [s'' \, \theta_1 - s'' \frac{\Delta}{L}]$$

(7)

where the modified stiffness coefficient s" is given by

$$s'' = s (1 - c^2) = \frac{4\alpha^2}{1 - 2\alpha \cot 2\alpha}$$

(8)

For uniformly distributed load, γq, the fixed moment is now given by

where
$$- M''_{12(F)} = [\tfrac{2}{3} f (1 + c)] \frac{\gamma q L^2}{8} = f'' \frac{\gamma q L^2}{8} \tag{9}$$

$$f'' = \frac{\tan \alpha - \alpha}{(1 - 2\alpha \cot 2\alpha) \alpha}$$

When the axial load is equal to zero, we have:

$$s'' = 3 \quad ; \quad f'' = 1 \tag{10}$$

For negative values of P, i.e. for axial tension, the trigonometric functions should be replaced by the corresponding hyperbolic functions. Thus if $\beta = \frac{\pi}{2} \sqrt{-\rho}$ then

$$s = \frac{(1 - 2\beta \coth 2\beta)\beta}{\tanh \beta - \beta} \tag{11}$$

$$c = \frac{2\beta - \sinh 2\beta}{(\sinh 2\beta - 2\beta \cosh 2\beta)}$$

$$f = \frac{3}{\beta^2} (\beta \coth \beta - 1)$$

Figure 3 shows the variation of the quantities s, c, s" and f as a function of P.

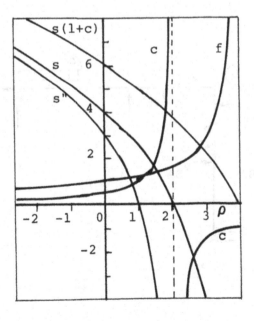

Fig. 3 - Stability Functions

4.0   ELASTIC STABILITY OF BRACED FRAME ABC

Consider a rigid jointed, braced frame ABC, hinged at A and C (Fig. 4). The load on the frame consists of a uniformly distributed load of intensity, γq, acting on the beam. The behavior of the frame is studied as the load factor γ is increased proportionally from 0 to its maximum value, γ_{cr}, at which the frame becomes unstable.

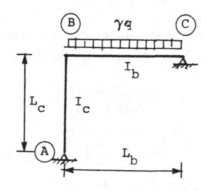

Fig. 4 - Braced Frame ABC and Loading

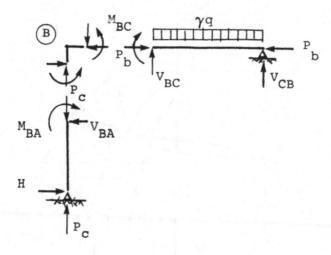

Fig. 5 - Forces on the Beam and Column

Moment equilibrium of joint B gives (Fig. 5):

$$M_{BA} + M_{BC} = 0 \tag{12}$$

We have

$$M_{BA} = k_c s_c'' \theta_B \tag{13}$$

$$M_{BC} = M_{BC(F)}'' + k_b s_b'' \theta_B$$

wherein the subscripts b and c stand for the beam and column, respectively.

Substitution of these values in Eq. (12) results in:

$$[k_c s_c'' + k_b s_b''] \, \theta_B = f_b'' \, \frac{\gamma q L_b^2}{8} \tag{14}$$

which can be written in the form

$$R \, \theta = F \tag{15}$$

where R represents the rigidity of the frame.
From Eq. (14), we get

$$\theta_B = \frac{\gamma q L_b^2}{8} \, \frac{f_b''}{[k_c s_c'' + k_b s_b'']} \tag{16}$$

From Eqs. (13) and (16) we get

$$M_{BA} = \frac{\gamma q L_b^2}{8} \, \frac{k_c s_c'' f_b''}{[k_c s_c'' + k_b s_b'']} \tag{17}$$

If V_{BC} and V_{BA} are the shear forces in the members BC and BA at joint B, and P_b, P_c the axial forces in the beam and the column, we have:

$$P_b = V_{BA} = \frac{M_{BA}}{L_c} = \frac{\gamma q L_b^2}{8 L_c} \, \frac{k_c s_c'' f_b''}{[k_c s_c'' + k_b s_b'']} \tag{18}$$

$$P_c = V_{BC} = \frac{\gamma q L_b}{2} - \frac{M_{BC}}{L_b} = \frac{\gamma q L_b}{2} + \frac{M_{BA}}{L_b} = \frac{\gamma q L_b}{2} + \frac{\gamma q L_b}{8} \, \frac{k_c s_c'' f_b''}{[k_c s_c'' + k_b s_b'']} \tag{19}$$

Iterative solution of Eq. (16), with the help of the relations 18 and 19 allows us to trace the load-deformation (γ - Θ) response of the frame. The maximum value, γ_{cr} for which the calculations converge is taken as the critical load as influenced by primary moments.

An equivalent axially loaded frame is obtained, say, by replacing the distributed load on the beam BC by a concentrated, vertical load, $Q(= qL_b/2)$ at joint B. For such a frame Eqs. 14-19 could be rewritten as

$$[k_c s_c'' + 3k_b]\theta_B = 0 \tag{20}$$

$$R^* \theta \qquad\qquad = 0$$

$$M_{BA} \qquad\qquad = 0$$

$$P_b \qquad\qquad = 0$$

$$P_c \qquad\qquad = \gamma^* Q$$

A bifurcation analysis of the frame gives the critical value, γ_{cr}^* , of the load factor. This value is obtained from the condition:

$$|R^*| = 0 \tag{21}$$

5.   NUMERICAL EXAMPLES

Using the relations given in the previous section, a small computer program was developed using an Apple II home computer. Three frames were then analyzed. The members, in all the three frames were of W8 x 31 section. Member lengths and frame loadings are given in Table 1. Here, $\gamma_p q$ represents the collapse load intensity as per a first order, plastic analysis.

The results are given in Tables 2 to 7 and Figures 6 to 8. Note that the Tables 2, 4, and 6 give the results of bifurcation analysis of frames 1, 2, and 3 with the distributed load on the beam replaced by a single concentrated load $\gamma^* Q$ acting at joint B along the column axis (with $Q = \dfrac{qL_b}{2} = 72$ kips).

TABLE 1: DATA FOR NUMERICAL ANALYSIS OF FRAMES 1-3

Material: A36 steel
             Yield stress = F_y = 36 ksi; Modulus of Elasticity = E = 29000 ksi

Section: W8 x 31; Major Axis Bending
             Moment of inertia = I_c = 110 in.^{4};
             Plastic moment = M_p = 1,094.4 kip. in.

Frame	L_c ft	L_b ft	qL_b kip	k_b kip-in.	P_{Ec} kip	$\gamma_p q$ kip/in.
1	12	12	144	22,150	1,518	0.616
2	24	12	144	22,150	759	0.616
3	12	24	144	11,075	1,518	0.154

TABLE 2: BIFURCATION BEHAVIOR - FRAME 1

NO	γ^*	$\dfrac{R^*}{k}$	$\dfrac{P_c}{P_{EC}}$
1	4.0	5.60	0.19
2	8.0	5.16	0.38
3	12.0	4.64	0.57
4	16.0	4.02	0.76
5	20.0	3.25	0.95
6	24.0	2.24	1.14
7	28.0	0.81	1.33
8	29.0	0.35	1.38
9	29.5	0.09	1.40
10	30.0	- 0.18	1.42
11	32.0	- 1.40	1.52

TABLE 3: STABILITY LIMIT LOAD - FRAME 1

NO	γ	θ_B rad	$\dfrac{M_{BA}}{M_P}$	$\dfrac{R}{k_b}$	$\dfrac{P_c}{P_{Ec}}$	$\dfrac{P_b}{P_{Eb}}$	$\dfrac{P_c}{\gamma q L_b}$
1	2.0	0.04	2.30	5.76	0.11	0.012	0.559
2	4.0	0.09	4.42	5.51	0.21	0.022	0.557
3	6.0	0.14	6.32	5.25	0.32	0.032	0.556
4	8.0	0.19	7.92	4.97	0.42	0.040	0.552
5	10.0	0.25	9.12	4.69	0.52	0.046	0.548
6	12.0	0.33	9.78	4.39	0.62	0.049	0.543
7	14.0	0.41	9.72	4.08	0.71	0.049	0.537
8	16.0	0.51	8.71	3.77	0.80	0.044	0.529
9	18.0	0.62	6.45	3.45	0.89	0.033	0.519
10	20.0	0.74	2.79	3.16	0.96	0.013	0.507
11	22.0	0.89	- 2.60	2.88	1.03	-0.015	0.493
12	24.0	1.05	- 9.76	2.64	1.09	-0.051	0.478

Limitations: 1. Elastic Behavior
2. Small Deformation Theory

Notations: γ = Load Factor
θ_B = Rotation of Joint B
M_{BA} = Internal Moment at B on Column BA
R = Frame Stiffness
P_c = Column Axial Force
P_b = Beam Axial Force
M_p = Plastic Moment of Section = 1095 kip-in.
P_{Ec} = Euler Load of Column = 1518 kip
P_{Eb} = Euler Load of Beam = 1518 kip

k_b = Rigidity Factor of Column = $\dfrac{EI_b}{L_b}$ = 22150

q = Intensity of Distributed Load (γ = 1) = 1 kip-in.
L_b = Beam Span = 144 in.

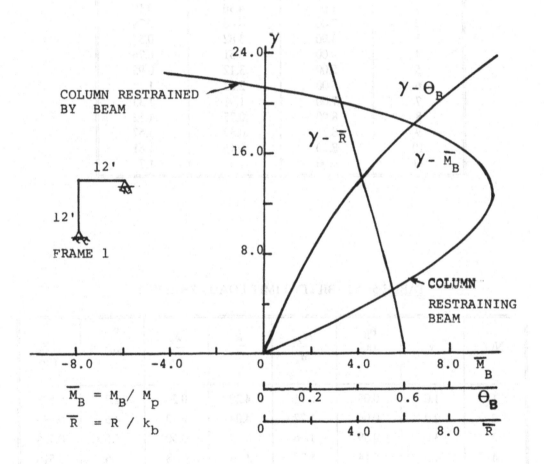

Fig. 6 - Results for Frame 1

TABLE 4: BIFURCATION BEHAVIOR - FRAME 2

NO	γ^*	$\dfrac{R^*}{k_b}$	$\dfrac{P_c}{P_{Bc}}$
1	1.00	4.30	0.19
2	2.00	4.08	0.38
3	3.00	3.82	0.57
4	4.00	3.51	0.76
5	5.00	3.12	0.95
6	6.00	2.62	1.14
7	7.00	1.91	1.33
8	8.00	0.75	1.52
9	8.25	0.35	1.57
10	8.50	- 0.16	1.61
11	9.00	- 1.55	1.71

TABLE 5: STABILITY LIMIT LOAD - FRAME 2

NO	γ	θ_B rad	$\dfrac{M_{BA}}{M_p}$	$\dfrac{R}{k_b}$	$\dfrac{P_c}{P_{Bc}}$	P_b kips	$\dfrac{P_c}{\gamma qL}$
1	1.0	0.03	0.71	4.29	0.20	2.7	0.538
2	2.0	0.06	1.22	4.04	0.40	4.6	0.533
3	3.0	0.09	1.46	3.77	0.60	5.5	0.525
4	4.0	0.14	1.27	3.46	0.78	4.6	0.516
5	5.0	0.19	0.39	3.10	0.96	1.6	0.505
6	6.0	0.26	- 1.49	2.72	1.11	-6.1	0.486
7	7.0	0.35	- 4.96	2.33	1.23	-18.2	0.464

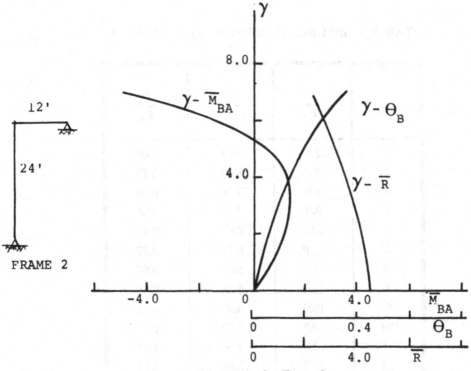

Fig. 7 - Results for Frame 2

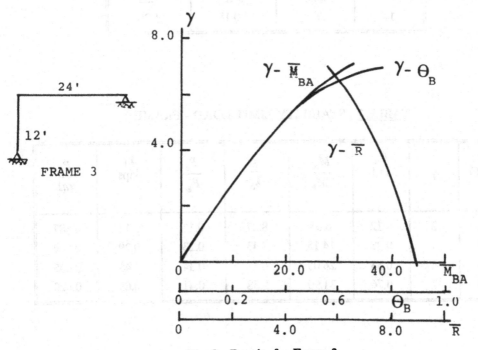

Fig. 8 - Results for Frame 3

TABLE 6: BIFURCATION BEHAVIOR - FRAME 3

NO	γ^*	$\dfrac{R^*}{k_b}$	$\dfrac{P_c}{P_{Ec}}$
1	2.0	8.61	0.09
2	4.0	8.20	0.19
3	6.0	7.76	0.28
4	8.0	7.30	0.38
5	10.0	6.81	0.47
6	12.0	6.27	0.57
7	14.0	5.68	0.66
8	16.0	5.01	0.76
9	18.0	4.30	0.85
10	20.0	3.50	0.95
11	22.0	2.56	1.04
12	24.0	2.48	1.14
13	26.0	0.18	1.23
14	26.5	- 0.18	1.26

TABLE 7: STABILITY LIMIT LOAD - FRAME 3

NO	γ	θ_B rad	$\dfrac{M_{BA}}{M_P}$	$\dfrac{R}{k_b}$	$\dfrac{P_c}{P_{Ec}}$	P_b kips	$\dfrac{P_c}{\gamma q L}$
1	2	0.12	6.65	8.27	0.11	0.13	0.587
2	4	0.28	14.15	7.43	0.23	0.29	0.594
3	6	0.53	24.05	6.38	0.34	0.48	0.605
4	7	0.76	31.90	5.58	0.41	0.63	0.620

From these results the following observations could be made:

1. As could be seen in Col. 4 of Tables 3 and 5, the beam which applies moments to the column at working load levels, generally restrains it while nearing the stability limit load of the frame (for column axial force values P_c greater than simple Euler load of the column, P_{Ec}). From Eq. (12), $M_{BC} = - M_{BA}$. From Figs. 6 and 7 we can observe that, at load levels for which the columns are being restrained by the beam, the beam itself has to be designed for moments greater than those in the corresponding simply supported beam (Fig. 9).

2. Primary bending slightly relieves the axial load in the column at high load factors.

3. As could be seen in Col. 7 of Tables 3 and 5, the axial force in the beam changes from compression, initially, to tension, towards the stability limit load ($P_c > P_{Ec}$).

These remarks are for a structure in which the column is relatively weak compared to the beam, as is the case with frames 1 and 2 considered. For a structure in which the beam is quite slender compared to the column the conclusions will be different.

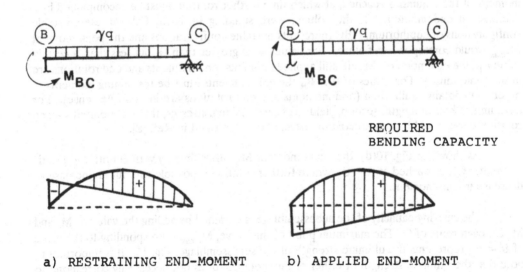

a) RESTRAINING END-MOMENT   b) APPLIED END-MOMENT

Fig. 9 - Beam-Design

When the primary bending is included and loading approaches the stability limit load, the frame deformations are too great to be considered small. Thus, the usual slope-deflection equations used (Eq. 1) are no longer valid. Also, the bending moments are quite high so that members will plastify well before the elastic stability limit load is attained.

6.0   INELASTIC STABILITY OF SUBASSEMBLAGE ABCD

Let us consider the subassemblage of Fig. 10(a) consisting of two identical beams rigidly jointed to a column. The beams, of length L_b, are pinned at their far ends. The column, of length L_c, is subjected to a compressive force P. An external moment M_e is applied at each joint so that the column is bent into a symmetric single curvature configuration. Due to symmetry, there is no axial force in the beams. In what follows we assume P is kept constant while M_e is increased monotonically. It is further assumed that out-of-plane deformations and lateral torsional buckling of the members are prevented by suitable lateral supports.

Because of symmetry, each joint is subjected to the same forces and so we need to consider the behavior of only one joint. Compatibility requires that the beam and the column ends both rotate through the same angle Θ. Equilibrium requires that $M_e = M_b + M_c$ (Fig. 10b), where M_b is the beam end moment and M_c is the column end moment. The proportion of M_e that is resisted by each member depends upon the end moment vs. end rotation relationship of each member. Such a relationship is shown in Fig. 10(c) for a column and in Fig. 10(d) for a beam.

It can be seen from Fig. 10(c) that at a certain value of $\Theta = \Theta_c$, the maximum end moment of the column is reached, at which time further rotation must be accompanied by a decrease in end moment. If the column were standing by itself, OA represents stable configurations of equilibrium, ABC represents unstable configurations and the peak moment M_{cmax} would correspond to failure. For values of Θ greater than Θ_c but less than a certain value Θ_R, the end moments M_c are still applied moments (end moments and end rotations are in the same sense). For values of $\Theta > \Theta_R$ the end moments must be restraining moments in order to maintain equilibrium (end moments and end rotations are in opposite sense). For a column of known length, section, yield stress and residual stresses, the end moment vs. end rotation curve could be constructed by the methods described in Ref. [8].

As shown in Fig. 10(d), the beam moment M_b varies linearly with Θ until the plastic moment, M_{bp}, is reached at $\Theta = \Theta_b$, where further rotation is possible without an increase or decrease in end moment.

The carrying capacity of the subassemblage is obtained by adding the value of M_c and M_b for each value of Θ. The maximum point of this curve, M_{emax}, corresponding to the value of $\Theta = \Theta_m$ represents the ultimate strength of the subassemblage, Fig. 11. It is important to note that the ultimate strength of either of the members does not necessarily correspond to the ultimate strength of the subassemblage.

7.0   ALTERNATIVE DESIGN METHODS

Depending on the relative strength of beam to column, various alternative design methods for the subassemblage can result, as shown in Fig. 11. The same column is chosen in all the alternatives.

When $\Theta_b < \Theta_c$ (Fig. 11a) a weak beam/strong column design is obtained. Near the

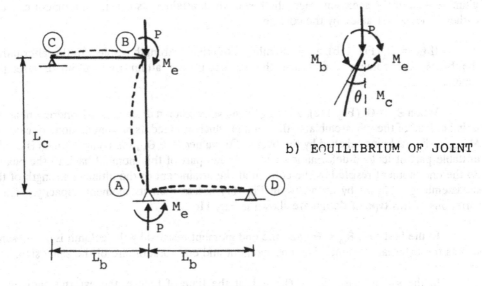

a) BEAM-COLUMN SUBASSEMBLAGE

b) EQUILIBRIUM OF JOINT

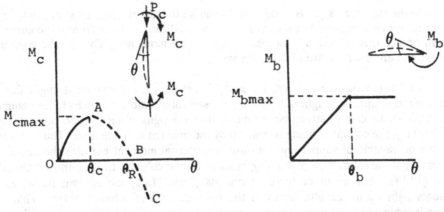

c) LOAD-DEFORMATION RESPONSE OF
    COLUMN ALONE, FOR A GIVEN P

b) LOAD-DEFORMATION
    RESPONSE OF BEAM

Fig. 10 - Braced Frame ABCD

ultimate load of the subassemblage, the beam which attained its maximum moment capacity earlier is being restrained by the column.

If $\Theta_b = \Theta_c$ (Fig. 11b), a subassemblage of critical strength results. In this alternative, the beam, the column and hence the subassemblage attain their ultimate strengths simultaneously.

When $\Theta_b > \Theta_c$ (Fig. 11c), a strong beam/weak column design results wherein near the ultimate load of the subassemblage, the column which attained its maximum moment capacity M_{cmax} earlier, is being relieved by the beam. For values $\Theta > \Theta_c$, the column which is on the unstable part of its load-deformation curve, throws part of this moment back to the beams. So the end moment resisted by the column at the attainment of the ultimate strength of the subassemblage (shown by an arrow in Fig. 11) is less than its moment capacity. Three variations of this type of design are shown in Fig. 11c.

In the first case $\Theta_m < \Theta_R$ and the end moment resisted by the column is of the same sign as the external moment. The end moment and end rotation are of the same sign.

In the second case, $\Theta_m = \Theta_R$ and, at the time of failure, the external moment is completely carried by the beam, the part resisted by the column being zero. In this case, the columns and beams could be designed independently. Columns are designed as members with an axial load P and an initial slope (imperfection) Θ_R. Beams are designed as simply supported beams or as continuous members resting on props [11].

In the last case, $\Theta_m > \Theta_R$ and near the ultimate load, the beam has not only to resist the entire external moment but should resist a destabilizing moment from the column. The column on the other hand is subjected to restraining moments. That is, the column end moment and end rotation are of opposite signs.

These remarks indicate that, for an optimum design, the column strength should be based on the ultimate strength of a suitably chosen subassemblage of which the column is a part. Only in the case of strong beam/weak column designs of such proportions that $\Theta_m > \Theta_R$ (Fig. 11c), the column is being restrained by the beam at ultimate load. The beam should however be capable of supporting not only the external moment but also the destabilizing moment from the column (= restraining moment on the column) to keep the subassemblage stable (12,13). For all other types of designs (Figs. 11a-c) the column resists external moments right up to the attainment of the ultimate load of subassemblage. That is, the column is subjected to symmetric single-curvature bending. Hence, it is doubtful that use of effective length factors < 1 is justifiable for columns in such braced frames.

8.0   FINAL REMARKS

Generally design of columns as part of frames is effected by using interaction equations (Eqs. 8-16 of Ref. 1 for example). The P_u term in the denominator is frequently based on an effective length other than unity. In braced frames with rigid framing, designers will frequently use K less than 1.0 because that is what the alignment chart gives them. It is, however, difficult to envisage that the rigidly framed beam which is inducing the moment can

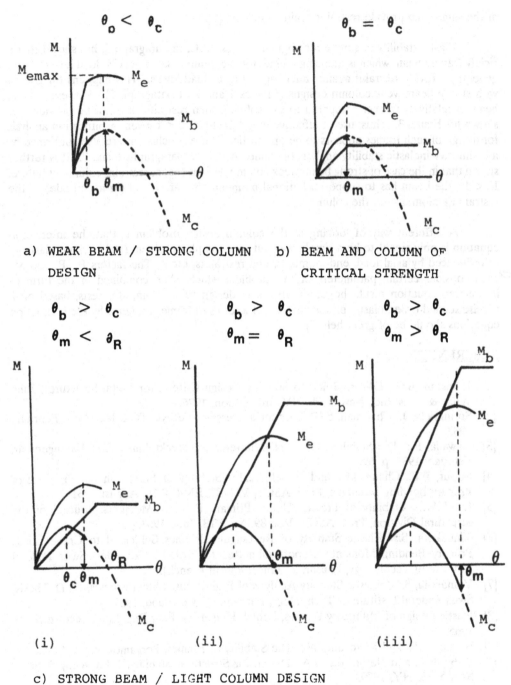

a) WEAK BEAM / STRONG COLUMN   b) BEAM AND COLUMN OF
   DESIGN   CRITICAL STRENGTH

(i)   (ii)   (iii)

c) STRONG BEAM / LIGHT COLUMN DESIGN

Fig. 11 - Classification of Beam-Column Subassemblage

at the same time provide restraint against buckling [2].

Elastic stability of simple rectangular frames ABC, in paragraph 5, has shown that a rigidly framed beam which is inducing moment in a column -- say at service load levels -- can, generally, provide restraint against buckling at higher load levels. This happens for frames with strong beam/weak column designs (Frames 1 and 2 of paragraph 5). However, if the beam is relatively weaker compared to the column, such restraint may not be provided, as shown for Frame 3. Thus, use of effective length factors, K < 1 given by bifucation analysis for design of such frames appears to be optimistic. These conclusions are also confirmed by a qualitative inelastic stability analysis of frames ABCD in paragraphs 6 and 7. It is further shown that in the case of strong beam/weak column designs (beam restrains column at failure, K < 1), the beam has to support additional moments that are equal (in magnitude) to the restraining moments on the column.

A different way of looking at the column design problem is that the interaction equation is only meant to be a simple way out to a complex problem (stability and plasticity as influenced by axial load, end moments, end restraints, etc.). The factors P_u, P_e and M_u each consider certain parameters of this problem which when combined in the form of interaction equation result, hopefully, in a safe design rule. Computer tests, based on a precise second-order, elastic-plastic analysis of a series of frames designed by the interaction equations should be of great help[7].

REFERENCES

[1]   Johnston, B.G., Editor, Guide to Stability Design Criteria for Metal Structures, John Wiley & Sons, Inc., New York, NY, 3rd Edition, 1976.
[2]   Springfield, J., Chairman, SSRC, Letter addressed to SSRC TG-3 Members, February 1983.
[3]   Chwalla, E.: Die Stabilitat Lotrecht Belasteter Rechteckrahmen, Der Bauingenieur, February 1938, p. 69.
[4]   Masur, E.F., Chang, I.C., and Donell, L.H.: Stability of Frames in the Presence of Primary Bending Moments, Proc. ASCE, Vol. 87, EM4, P 19, August 1961.
[5]   Lu, L.W.: Stability of Frames Under Primary Bending Moments, Journal of the Structural Division, Proc. ASCE, Vol. 89, No. ST3, June 1963.
[6]   Vinnakota, S.: Elastic Stability of Rectangular Frames Subjected to Crane Loads Primary Bending Moments, Report submitted to Prof. M. Cosandey, Swiss Federal Institute of Technology, Lausanne, June 1964 (in French).
[7]   Vinnakota, S.: Inelastic Stability Analysis of Rigid Jointed Steel Frames, Ph.D. Thesis, Swiss Federal Institute of Technology, Lausanne, Switzerland, 1967.
[8]   Plastic Design of Multistory Frames, Lehigh University, Fritz Eng. Lab., Lecture Notes 1965.
[9]   Horne, M.R. and Merchant, W.: The Stability of Frames, Pergamon Press, 1965.
[10] Kirby, P.A. and Nethercot, D.A., Design for Structural Stability, John Wiley & Sons, New York, NY, 1979.
[11] Gent, A.R., Elastic-Plastic Column Stability and the Design of NoSway Frames, Proceedings Institute of Civil Engineers, Vol. 34, June 1966, pp. 129-151.

[12] Kanchanalai, T., and Lu, L.W.: Analysis and Design of Framed Columns under Minor Axis Bending, Engineering Journal, AISC, April 1979.

[13] Vinnakota, S.: Design of Columns in Planar Frames - A Few Comments, Proceedings of the National Conference on "Tall Buildings", held at New Delhi, January 1973.

INELASTIC EFFECTIVE LENGTH FACTORS FOR COLUMNS IN UNBRACED FRAMES

S. Vinnakota

Marquette University, Milwaukee, WI, USA

ABSTRACT

The use of effective length factors in the design of columns in frames is quite widespread in the USA. Thus, the latest AISC Allowable Stress Design Specification and the recently adopted AISC Load and Resistance Factor Design Specification include effective length factors in the expressions for the strength of columns and beam-columns. This lecture reviews the basis of elastic and inelastic effective length factors for columns in unbraced frames.

1.0 INTRODUCTION

The effective length factor is still widely used by practioners in the USA, for design of columns in unbraced frames. However, the use of inelasic effective length factors as suggested by Yura in 1971 [1], appeared unreasonable to some designers. AISC (American Institute of Steel Construction) endorsed the use of inelastic effective length factors in its 7th edition of the Allowable Stress Design (ASD) Steel Manual [2] and, more recently, in its 1st edition of the Load and Resistance Factor Design (LRFD) Steel Manual [3]. The LRFD Specifications are based on the limit state design of structures. In view of this change in philosophy, this report re-evaluates the use of the inelastic effective length factors for the design of columns in unbraced frames.

2.0 ELASTIC AND INELASTIC STRENGTHS OF PINNED COLUMNS

For a pin-ended, axially loaded, elastic column (Fig. 1) the critical load at which the column buckles is given by the **Euler load**, namely:

$$P_{cr} = P_E = \frac{\pi^2 EI}{L^2} \tag{1}$$

where EI is the bending stiffness of the column section about the buckling axis. This equation can also be expressed in the form of a critical stress as:

$$\sigma_{cr} = \sigma_E = \frac{P}{A} = \frac{\pi^2 E}{(L/r)^2} \tag{2}$$

where (L/r) is the slenderness ratio of the column, in the plane of buckling. For a pin ended, axially compressed column made of a homogeneous material having a non-linear stress-strain curve (Fig. 2), the critical load at which the column may start to bend is given by the **Shanely load** or **tangent modulus load** [4], namely,

$$P_{cr} = P_S = \frac{\pi^2 E_t I}{L^2} \tag{3}$$

The corresponding critical stress is (Fig. 2):

$$\sigma_{cr} = \sigma_S = \frac{\pi^2 E_t}{(L/r)^2} \tag{4}$$

In the above relations E_t is the tangent modulus (slope of stress-strain curve) of the material at the stress level σ_S.

Mild structural steels have stress-strain curves which are linearly elastic-perfectly plastic before strain hardening sets in at large strains (Fig. 3). By the strict application of the tangent modulus concept the critical stress below σ_y is then governed by the elastic formula, and the column curve would take the form shown in Fig. 3. A great variety of carefully performed tests on well centered and relatively straight specimens have shown that the column strength predicted by Fig. 3b is usually higher than the actual strength for hot rolled steel columns of intermediate length. Osgood [5] attributed this reduction in strength to the presence of residual or locked-in stresses in hot rolled steel sections.

It can be shown that for hot rolled steel sections, the part of the member cooling most slowly will be left in residual tension. Thus, in the case of a wide flange section, the flange tips (which cool the most rapidly) are left in compression while the junctions of the flange and web are left in tension. Maximum residual stresses at flange tips of rolled shapes are of the order of 10 to 15 ksi, although values higher than 20 ksi have been measured. The yield stress of the steel has little effect on the magnitude of the residual stress in rolled shapes.

The presence of residual stresses influences the shape of the averge stress-strain curve

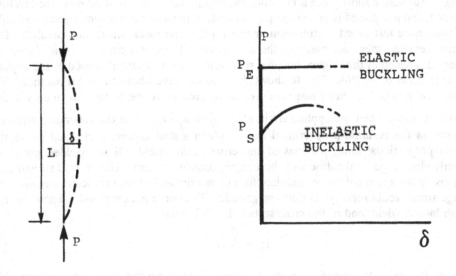

Fig. 1 - Buckling Load of Axially Loaded Column

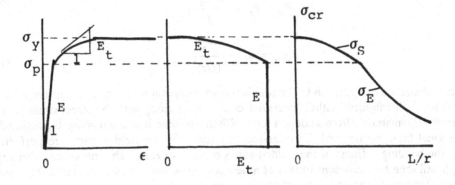

Fig. 2 - Tangent Modulus Column Curve

of the section as a whole, as distinct from the curves obtained on coupons cut from the same member. Suppose a short piece is cut from the column shape of Fig. 4, its ends are carefully machined, and it is placed between the plates of a testing machine and compressed gradually. In a concentric test called a **stub-column test** [6] the end faces will remain parallel. The distance between them decreases as the compression force is increased. This change of distance Δ, divided by the original length s, represents the unit strain ϵ^* caused by the applied (average) stress $\sigma^* = P/A$. Fig. 4c shows the type of curve obtained in this manner. If a compressive residual strain of magnitude ϵ_{rc} were present at the flange tips, then yielding would commence when the applied strain $\epsilon^* = \epsilon_p^* = \epsilon_y - \epsilon_{rc}$. The corresponding stress σ_p^* is known as the **residual-proportional limit**. When a stub column is strained above the residual proportional limit, portions of the cross-section yield. If the yielded parts are perfectly plastic, the axial stiffness of those zones reduces to zero. The over-all stress-strain relationship for a stub column would therefore be nonlinear; but with sufficient straining the average stress would reach yield point magnitude. The corresponding load is known as the **squash load** or **yield load** of the cross section P_y. We have

$$P_y = A \cdot \sigma_y \tag{5}$$

Thus the influence of residual stresses in columns is to make the stress-strain relationship nonlinear above a residual proportional limit.

There are two general methods by which strength of pin-ended steel columns loaded into the plastic domain could be obtained.

The first is to experimentally determine an average stress-strain (σ^*-ϵ^*) diagram from a stub column test. Column strength can then be determined using the tangent modulus E_t^* of this average stress-strain diagram along with the slenderness ratio that is proper for strong or weak axis bending (Fig. 5). This approach does not require knowledge about residual stresses.

$$P_{cr} = P_t^* = \frac{\pi^2 E_t^* I}{L^2} \tag{6}$$

$$\sigma_{cr} = \sigma_t^* = \frac{\pi^2 E_t^*}{(L/r)^2} \tag{7}$$

The second method of approach to the problem is an analytical solution. It makes use of the residual stress distribution (either measured or assumed) along with the stress-strain (σ-ϵ) diagram for the material (from a coupon test). When a column is strained above the residual-proportional limit, portions of the cross-section yield. As the yielded parts are perfectly plastic, the bending stiffness of the yielded zones reduces to zero. The theoretical buckling strength will then be equivalent to that of a new column whose moment of inertia, I_e, is the moment of inertia of the remaining elastic cross-section, or

$$P_{cr} = P_H = \frac{\pi^2 E I_e}{L^2} \tag{8}$$

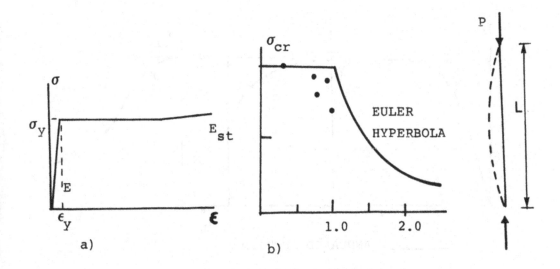

Fig. 3 - Buckling Stress for an Annealed Steel Column

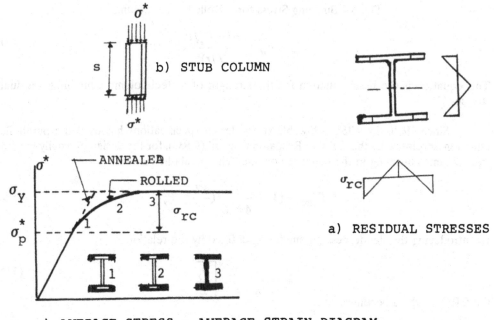

c) AVERAGE STRESS - AVERAGE STRAIN DIAGRAM

Fig. 4 - Stub Column Test

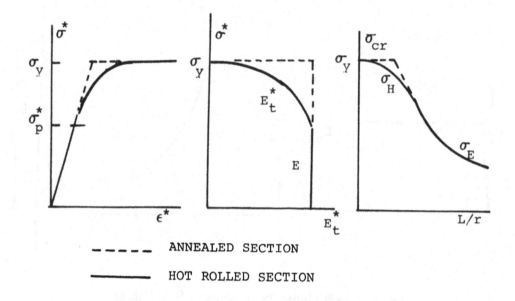

ANNEALED SECTION

HOT ROLLED SECTION

Fig. 5 - Buckling Stress for a Rolled Steel Column

$$\sigma_{cr} = \sigma_H = \frac{\pi^2 E \ (I_e/I)}{(L/r)^2} \qquad (9)$$

This equation is the basic equation for the strength of a steel column containing residual stresses [7].

Since 1960, the AISC allowable stress design specifications has used the parabolic equation developed by the Column Research Council (CRC), for the design of axially loaded steel columns buckling in the inelastic domain. The parabolic curve is:

$$P_{CRC} = [1 - \frac{\sigma_y}{4\pi^2 E} (\frac{KL}{r})^2] P_y \qquad (10)$$

By introducing the slenderness parameter λ_c defined by the relation:

$$\lambda_c = \frac{1}{\pi} \frac{KL}{r} \sqrt{\frac{\sigma_y}{E}} \qquad (11)$$

the CRC parabola becomes:

$$\sigma_{CRC} = [1 - \frac{\lambda_c^2}{4}] \sigma_y \qquad for \ \lambda_c \leq \sqrt{2} \qquad (12)$$

In Eqs. 10 and 11, K is the effective length factor, to be studied in detail in the next section. Note that $\lambda_c = \sqrt{2}$, when the parabola and the Euler hyperbola become tangent to each other. Thus, Eq. 10 applies for $\lambda_c \leq \sqrt{2}$ and for greater values of λ_c, the generalized Euler equation applies, namely

$$\sigma_{EK} = \frac{\pi^2 E}{(KL/r)^2} = \frac{1}{\lambda_c^2} \sigma_y \qquad for \ \lambda_c > \sqrt{2} \qquad (13)$$

Equation 12 provides a reasonable approximation for the strength of initially straight columns reflecting essentially the effect of residual stresses. Traditionally, accidental eccentricity and initial crookedness were accounted for in ASD Specifications, by using a safety factor that increased with slenderness from 1.67 to 1.92.

The AISC LRFD Specification uses on the other hand, an equation developed to fit closely the SSRC Curve 2 [6,8] that includes the effects of an initial out-of-straightness of about L/1500 in addition to residual stresses. The nominal strength for rolled shape compression member is given by:

$$P_{LRFD} = [0.658^{\lambda_c^2}] P_y \qquad for \ \lambda_c \leq 1.5 \qquad (14)$$

$$\qquad (15)$$

$$= [\frac{0.877}{\lambda_c^2}] \ P_y \qquad for \ \lambda_c > 1.5$$

The design strength is obtained by multiplying the nominal strength by a resistance factor of 0.85.

3.0 ELASTIC EFFECTIVE LENGTH FACTOR

The effective length concept is one approach for estimating the interaction effects of the total frame on the stability of a compression element being considered. This concept uses K-factors to equate the buckling strength of a framed compression element of length L to an equivalent pin-ended member of length KL subject to axial load only. Thus:

$$P_{cr} = P_{EK} = \frac{\pi^2 EI}{(KL)^2} \qquad (16)$$

or

$$\sigma_{cr} = \sigma_{EK} = \frac{\pi E}{(KL/r)^2} \qquad (17)$$

The AISC Specifications [2,3] require the determination of K-factors by some rational analysis. Most engineers use the alignment charts which furnish approximate elastic solutions (K-factors) in lieu of a detailed stability analysis.

Figure 6 shows the alignment chart for determining the effective length factor K of columns in multistory frames with sidesway uninhibited.

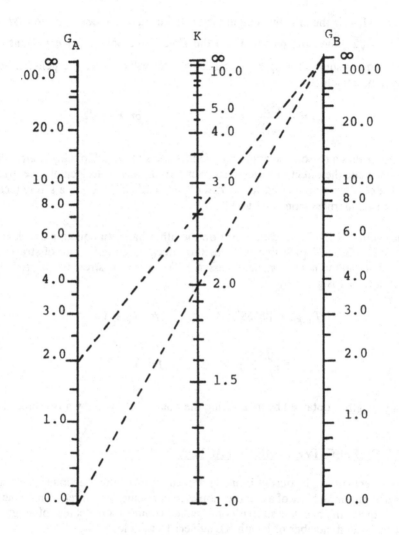

Fig. 6 - Alignment Chart for Columns in Unbraced Frames

To enter the chart, evaluation of the relative stiffness of the members (columns and beams) of the frame at the ends of the column under consideration must be made. The relative stiffness G is equal to the sum of column stiffnesses at a joint (destabilizing members), divided by the sum of the beam stiffnesses at that joint (restraining members). This computation is performed at each end of the column. The alignment chart is entered with these two parameters (G at each end of the column), and a straight line drawn between these two known values of G gives the corresponding value of K. For example, for column AB in Fig. 7, we have:

$$G_A = \frac{\sum (I/L) \; column}{\sum (I/L) \; beam} = \frac{\dfrac{I_c}{L_c} + \dfrac{I_{c1}}{L_{c1}}}{\dfrac{I_{b1}}{L_{b1}} + \dfrac{I_{b2}}{L_{b2}}} \tag{18}$$

$$G_B = \frac{\sum (I/L) \; column}{\sum (I/L) \; beam} = \frac{\dfrac{I_c}{L_c} + \dfrac{I_{c2}}{L_{c2}}}{\dfrac{I_{b3}}{L_{b1}} + \dfrac{I_{b4}}{L_{b2}}}$$

I_c and I_b are taken about the axes perpendicular to the plane of buckling. With moduli included the relation for G becomes:

$$G = \frac{\sum (EI/L) \; column}{\sum (EI/L) \; beam} \tag{19}$$

Note that the alignment chart is the graphical representation of the solutions for the characteristic equation [9]:

$$\frac{G_A G_B (\pi/K)^2}{6(G_A + G_B)} = \frac{(\pi/K)}{\tan(\pi/K)} \tag{20}$$

derived for the sway buckling of the subassemblage shown in Fig. 7.a. The buckled configuration of the subassemblage is as shown in Fig. 7.b. Equation 20 was derived on the basis of the following assumptions.

1. The structure consists of symmetrical rectangular frames.
2. All members have constant cross-section.
3. All joints are rigid.
4. At a joint, the restraining moment provided by the beams is distributed to the column above and below the joint in proportion to their stiffnesses of the columns.
5. The beams carry no axial loads.
6. Behavior is purely elastic.
7. The stiffness parameters $L\sqrt{P/(EI)}$ of all columns are equal.
8. All columns reach their buckling loads simultaneously.
9. For unbraced frames under consideration, at the onset of buckling, the rotations at the opposite ends of the beams are equal in magnitude and opposite in sign thus producing reverse curvature bending.

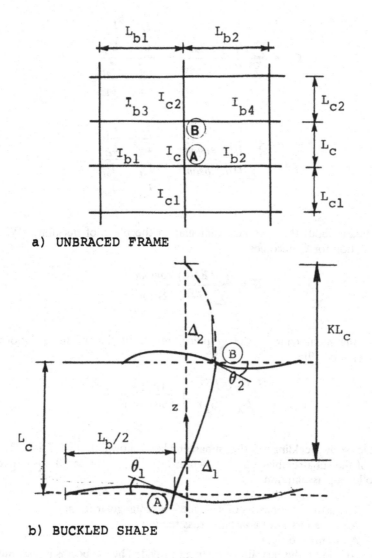

a) UNBRACED FRAME

b) BUCKLED SHAPE

Fig. 7 - Effective Length of a Column

Where the actual conditions differ from these assumptions unrealistic designs may result. Several modifications in the calculation of G for use in the alignment chart are suggested to give results more truly representative of conditions in real structures. Thus:

1. For the case of the far ends of the beam hinged and fixed (Fig. 8), the beam stiffnesses are $3\ EI_b/L_b$ and $4\ EI_b/L_b$, respectively, instead of the value $6\ EI_b/L_b$ used in the derivation of Eq. 20. Hence, to adjust for far ends of beams hinged, multiply I_b/L_b by 0.5, and for the far ends of beams fixed, multiply I_b/L_b by 2/3.

2.   For a column end supported by but not rigidly connected to a footing (pinned base) G_A may theoretically be taken as infinity. However, unless actually provided with a true friction-free pin, G_A may be taken as 10 for practical designs. If the end A were rigidly attached to a properly designed footing, a value of $G_A = 1.0$ is used. (Theoretically, if the base were truly fixed, $G_A = 0.0$.)

Notice that the alignment chart for the case of columns with uninhibited sidesway yields values of K from 1 to infinity. The alignment chart is shown to give conservative results by Yura [1] for short and intermediate length columns, buckling in the inelastic range.

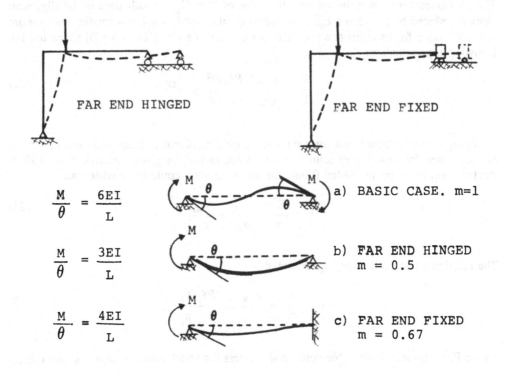

$$\frac{M}{\theta} = \frac{6EI}{L}$$

$$\frac{M}{\theta} = \frac{3EI}{L}$$

$$\frac{M}{\theta} = \frac{4EI}{L}$$

a) BASIC CASE. m=1

b) FAR END HINGED
   m = 0.5

c) FAR END FIXED
   m = 0.67

FAR END HINGED

FAR END FIXED

Fig. 8 - Corrections for G

4.0 INELASTIC EFFECTIVE LENGTH FACTOR, K_{iY}

The discussion in the previous section was restricted to buckling of perfectly elastic frames. However, in reality, frame instability is more likely to take place after the stresses at some parts of the frame have exceeded the yield limit.

The stiffness of a column section in the elastic range is proportional to EI; however, its stiffness in the inelastic range can be more accurately taken as $E_t I$, where E_t is the tangent modulus.

The relative stiffness coefficients G_A and G_B used in the development of the alignment chart expressions (Eq. 19) involve the modulus of elasticity E for both columns and beams. For elastic behavior, E cancels out from the equation, resulting in the familiar expression for G given in Eq. 18. If the elastic E applies for the beam members but the inelastic E_t applies for the columns, this can be accounted for by an adjustment of G values; thus

$$G_i = \frac{\sum (E_t I/L)_c}{\sum (EI/L)_b} = \frac{E_t}{E} G_e = \tau \, G_e \tag{21}$$

Thus, for frames with inelastic columns the value of $G = G_e$ normally used in the alignment chart is reduced by the factor E_t/E. A reduced value for G implies a smaller and a more favorable value for the effective length factor K. For a given KL/r, Lu [10] suggested the following approximation:

$$\tau = \frac{E_t}{E} = \frac{\pi^2 E_t/(KL/r)^2}{\pi^2 E/(KL/r)^2} \approx \frac{\sigma_{SK}}{\sigma_{EK}} \tag{22}$$

where σ_{EK} is the critical stress for a pin-ended, elastic column of length KL and σ_{SK} is the Shanely stress for a pin ended column of length KL and of the given material. For a rolled-steel column in an axially loaded frame, the above relation could be rewritten as:

$$\tau = \frac{E_t^*}{E} = \frac{\sigma_{tK}^*}{\sigma_{EK}} = \frac{\sigma_{HK}}{\sigma_{EK}} \tag{23}$$

The relation could be further approximated by:

$$\tau = \frac{E_t^*}{E} \approx \frac{\sigma_{CRC}}{\sigma_{EK}} = \frac{FS_1.F_a}{FS_2.F_e'} \tag{24}$$

where F_a is the AISC allowable compressive stress for short columns, equal to the critical stress σ divided by a factor of safety FS_1. F_e' is the Euler stress σ_{EK} divided by a factor of safety FS_2. Yura [1] suggested that

$$\tau = \frac{E_t}{E} \approx \frac{F_a}{F_e'} \tag{25}$$

This will be conservative since FS_1 used for short columns ranges from 1.67 to 1.92, whereas the long column equation uses $FS_2 = 1.92$. Both F_a and F_e' are tabulated in AISC Specification for different values of KL/r. In the elastic range, $F_a = F_e'$, so the nomograph procedures would be unchanged for elastic frames. Disque [11] proposed using $0.6 \, F_y$ for the numerator along with F'_e in the denominator. That approach is conservative and practical. Further, Disque [11] suggested that for members carrying little bending moment, the nominal stress $f_a = P/A$ could be used for the numerator. Thus, the Disque approach is:

$$\tau = \frac{E_t}{E} \approx \frac{0.6 F_y}{F'_e} \quad or \quad \frac{E_t}{E} \approx \frac{f_a}{F'_e} \tag{26}$$

Smith [12] suggested using the basic equation, Eq. 24, while Matz [13] has provided a table of values of E_t/E for various KL/r values for this approach. Disque has compiled a table (for $F_y = 36$ ksi and $F_y = 50$ ksi steels) from which f_a/F'_e values can be obtained for various values of f_a under the title **stiffness reduction factor** [2,3].

The design procedure may be itemized as follows:
1.   Choose a column section.
2.   Compute axial stress, $f_a = P/A$.
3.   With f_a, enter the column stress tables and find the KL/r corresponding to f_a.
4.   With the KL/r from (3), enter the F'_e tables and find the elastic stress, F'_e. [Steps 3 and 4 may be eliminated by entering the stiffness reduction factor tables found in Part 3 of the AISC-ASD Manual entering with f_a and reading off $\dfrac{f_a}{F'_e}$.]

5.   Compute $G_i = \dfrac{f_a}{F'_e} \cdot G_e$.

6.   Obtain $K = K_{iy}$ from the alignment chart using G_i.
7.   With $K_{iY} \, L/r$ enter the column stress tables to obtain F_a and compute $F_a \cdot A = P_{all}$ for the chosen column section. If $P_{all} \geq P_{actual}$, the choice of section is valid.

The approach is similar in the LRFD format. In his papers [1,14] Yura notes: "The inelastic approach can produce significant reductions in the effective length factor. When the elastic KL/r is reasonably low (about 50 or less)l the actual K will usually converge to values close to 1.0. This observation indicates that columns in multistory frames can often be designed on the basis of K = 1.0; that is the actual story height." Over the last two decades, practicing engineers have taken advantage of the Yura-Disque approach for the design of columns. However, several engineers [15,16,17,18] were apprehensive about the design of columns in unbraced frames with values of K quite close to 1, as there was no formal explanation of the Yura approach. The present lecture will attempt to alleviate this deficiency.

5.0 TANGENT MODULUS LOAD OF A FRAMED COLUMN, P_{tV}

Let us consider now the simple frame shown in Fig. 9 in which the compressed vertical member AB is free to move laterally at the top. The column is hinged at the lower end A and elastically restrained at the upper end B by two beams BC and BD, rigidly connected to the column at B and roller supported at the far ends C and D. The members are assumed to be initially straight. The column is subjected to a monotonically increasing axial load P. The primary moments in the frame are zero.

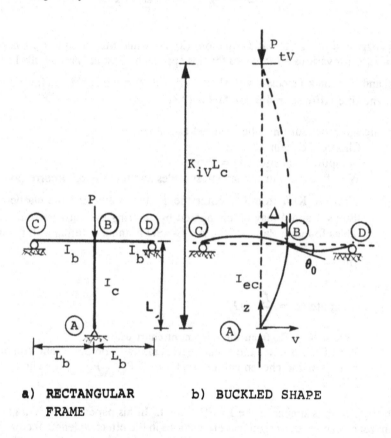

a) RECTANGULAR FRAME

b) BUCKLED SHAPE

Fig. 9 - Inelastic Buckling Load, P_{tV}

It is assumed that suitable lateral bracing is provided to prevent lateral buckling of members. Also, the width/thickness ratios of the elements of the cross-section are such that no local buckling occurs before the overall buckling.

When the steel column, underload P is strained above the residual-proportional limit, portions of the cross-section yield. The strain ϵ will increase uniformly until bending starts, corresponding to the tangent modulus load of the Shanley theory. This buckling load of the framed column is indicated by P_{tV} [19].

Moment equilibrium of a length z of the column, for the infinitesimal bent position about the x-axis results in:

$$\sum (M) = M_{ext} + M_{int} \tag{27}$$

$$= P_{tV}v + \int_A E(y) \; \delta\epsilon \; y \; dA = 0$$

Using the relationship

$$\frac{\delta\epsilon}{y} = curvature \approx \frac{d^2v}{dz^2} \tag{28}$$

the above equation becomes:

$$P_{tV}v + \frac{d^2v}{dz^2} \int_A E(y) \; y^2 \; dA = 0 \tag{29}$$

If the yielded parts of the section are perfectly plastic, the bending stiffness of those parts reduces to zero. For the ideal stress-strain curve of structural steel, the integral in Eq. 29 can therefore be evaluated as:

$$\int_A E(y) \; y^2 \; dA = E\int_{Ae} y^2 \; dA = EI_{ec} \tag{30}$$

where I_{ec} is the moment of inertia of the elastic part of the column cross-section. The differential equation 29 therefore becomes:

$$EI_{ec} \frac{d^2v}{dz^2} + P_{tV}v = 0 \tag{31}$$

or

$$\frac{d^2v}{dz^2} + \frac{P_{tV}}{EI_{ec}} v = 0 \tag{32}$$

with the abbreviations

$$\phi^2 = \frac{P_{tV}L_c^2}{EI_{ec}} \; ; \; \bar{z} = \frac{z}{L_c} \tag{33}$$

the general solution of the homogeneous, second-order differential equation (32) can be written as:

$$v = A \; \sin(\phi\bar{z}) + B \; \cos(\phi\bar{z}) \tag{34}$$

where A and B are the integration constants. The boundary condition at A, namely $v = 0$ at $z = 0$ results in $B = 0$ and then the boundary condition at B, namely $v = \Delta$ at $z = L$, results in a value for A

$$\Delta = A \sin\phi \rightarrow A = \frac{\Delta}{\sin\phi} \tag{35}$$

The general solution for the deflection is

$$v = A \sin(\phi\bar{z}) \tag{36}$$

Thus the column forms a portion of a half-sine wave which terminates on the vertical through its lower support. The stabilizing moment in the column at the joint B is $P_{tV}\Delta$. If M_{BC} and M_{BD} are the restraining moments developed in the beams, we have from the moment equilibrium of joint B:

$$M_{BC} + M_{BD} = M_{BA} = P_{tV}\Delta \tag{37}$$

or, in view of the symmetry of the structure and the loading:

$$M_{BD} = \frac{1}{2} P_{tV}\Delta \tag{38}$$

By conjugate beam principles, the slope at the end B of the beam BD is:

$$\theta_o = \frac{1}{L_b} \left(\frac{1}{2} \frac{M_{BD} L_b}{EI_b}\right) \frac{2}{3} L_b = \frac{P_{tV} \Delta L_b}{6 EI_b} \tag{39}$$

Also, in the column

$$\theta_o = \frac{dv}{dz} \Big|_{at\ \bar{z}\ =\ 1} = A\phi \cos\phi \tag{40}$$

Substituting the value of A from Eq. 35 in Eq. 40 and eliminating θ_o from Eqs. 39 and 40, we obtain:

$$\Delta\phi \cot\phi = \frac{P_{tV}\Delta L_b}{6 EI_b}$$

or

$$\phi\left[\cot\phi - \frac{\phi}{6} \frac{EI_{ec}/L_c}{EI_b/L_b}\right] \Delta = 0 \tag{41}$$

This is satisfied by putting $\Delta = 0$ (i.e., zero deformation); $\phi = 0$ (i.e., $P = 0$) both of which are trivial conditions, or by:

$$\cot\phi = \frac{\phi}{6} \frac{EI_{ec}/L_c}{EI_b/L_b} \tag{42}$$

Defining the inelastic effective length factor K_{iV} given by the formula

$$P_{cr} = P_{tV} = \frac{\pi^2 EI_{ec}}{(K_{iV}L_c)^2} \tag{43}$$

we obtain, with the help of Eq. 33, the relation

$$\phi = \frac{\pi}{K_{iV}} \tag{44}$$

The corresponding critical stress is given by:

$$\sigma_{cr} = \sigma_{tV} = \frac{\pi^2 E\,(I_{ec}/I_c)}{(\dfrac{K_{iV}L_c}{r})^2} \tag{45}$$

The inelastic buckling condition for the inelastically loaded frame (Eq. 42) can now be written as:

$$(\frac{\pi}{K_{iV}})^2 \frac{G_{iB}}{6} = \frac{(\pi/K_{iV})}{\tan(\pi/K_{iV})} \tag{46}$$

where

$$G_{iB} = \frac{\dfrac{EI_{ec}}{L_c}}{2 \times \dfrac{1}{2} \times \dfrac{EI_b}{L_b}} \tag{47}$$

is the relative stiffness coefficient at the joint B of the frame shown in Fig. 9, loaded into the inelastic range. Solution of Eq. 46 results in the effective length factor K_{iV} and hence the inelastic buckling load P_{tV} given by Eq. 43.

Let us consider Eq. 20 which is the stability condition for the sidesway buckling of the subassemblage shown in Fig. 7 which formed the basis for the alignment chart for effective length factors given in Fig. 6. This stability condition, applied to the hinged bases frame (theoretical value of $G_A = \infty$) shown in Fig. 9 reduces to:

$$(\frac{\pi}{K})^2 \frac{G_B}{6} = \frac{(\pi/K)}{\tan(\pi/K)} \tag{48}$$

where

$$G_B = \frac{(EI_c/L_c)}{2 \times \dfrac{1}{2} (EI_b/L_b)} \tag{49}$$

$$P_{cre} = \frac{\pi^2 E I_c}{(K L_c)^2} \tag{50}$$

These results indicate that K_{iV} could be obtained from the alignment chart (Fig. 6) using G_{iB} defined by Eq. 47. Relations 43 and 47 could be rewritten as:

$$G_{iB} = \frac{I_{ec}}{I_e} G_B \tag{51}$$

$$\frac{P_{tV}}{P_y} = \frac{\pi^2 E}{\sigma_y} \cdot \frac{1}{(\frac{K L_c}{r})^2} \cdot \frac{I_{ec}}{I_c} \cdot (\frac{K}{K_{iV}})^2 \tag{52}$$

$$= \frac{1}{\lambda_c^2} \cdot \frac{I_{ec}}{I} \cdot (\frac{K}{K_{iV}})^2$$

This relation shows that the inelastic buckling load of a column in an axially loaded, unbraced, rectangular frame can be obtained from its elastic buckling load by applying two corrections. The first factor, I_{ec}/I, takes care of the decrease in strength due to the decrease in stiffness of the cross section. The second factor, $(K/K_{iV})^2$, takes care of the increase in strength due to the increase in the relative stiffness of the rest of the frame (beams) as the column under consideration is being plastified. As both I_{ec} and K_{iV} in Eq. 52 are dependent on the load P_{tV}, Eq. 52 can only be solved by iteration.

6.0 NUMERICAL EXAMPLE

To clarify the relations derived in the previous section, we will consider the rectangular frame shown in Fig. 10. Here, the compressed vertical member AB is a W12x65, 96 inches long. Two W8x31 beams, 122 inches long each, are rigidly connected to the column web at B and roller supported at the far ends C and D. The material, A36 steel with an yield stress, $\sigma_y = 36$ ksi, is assumed to have an elastic-perfectly plastic stress-strain diagram. We study the buckling of this frame, in the plane of the paper, as shown in Fig. 9b.

To further simplify the presentation, without loss of generality, we assume that the column consists of a 12 x 1.2 in. rectangular plate bent about its major axis. A step-wise residual stress distribution with a maximum compression residual stress of 10.8 ksi ($= 0.3\ \sigma_y$) is assumed in the calculations. The stress distribution will be assumed constant over the thickness of the plate.

The column will be subjected to various increases in axial strain (axial load) termed stages in loading. At the end of each stage, additional areas plastify, as the total strain, ϵ_t (equal to the residual strain, ϵ_r plus the applied strain ϵ_a) reaches the yield strain, ϵ_y. The elastic core dimensions decrease with the load, as shown in Fig. 10c. The resulting moment of inertia of the elastic core, I_{ec} lets us determine G_{iB} from Eq. 47 and hence the inelastic

TABLE 1: DATA FOR NUMERICAL EXAMPLE

W12x65: $b_f = 12.00$ in.; $d = 12.12$ in.

$t_f = 0.605$ in.; $t_w = 0.39$ in.

$I_x = 533$ in.^{4}; $I_y = 174$ in.^{4}

$r_x = 5.28$ in.; $r_y = 3.02$ in.

Plate: 12 in.x1.2 in.: $A = 14.4$ in.^{2}
$I = 173$ in.^{4}; $r_y = 3.42$ in.

W8x31: $I_x = 110$ in.^{4}

Column length, $L_c = 96$ in.
Moment of inertia, $I_c = 173$ in.^{4}

Beam length, $L_b = 122$ in.
Moment of inertia, $I_b = 110$ in.^{4}
Modulus of elasticity, $E = 29{,}000$ ksi

Euler load of column, $P_E = \dfrac{\pi^2 \times 29000 \times 173}{96^2} = 5373 \; kips$

$G_B = \dfrac{(173/96)}{2 \times \dfrac{1}{2} (110/122)} = 2;$ $G_A = \infty$

Elastic effective length factor, $K = 2.635$

Elastic critical load of the column, $P_{EK} = \dfrac{5373}{(2.635)^2} = 774$ kips

Squash load of the column, $P_y = 14.4 \times 36 = 518.4$ kips

$\dfrac{P_{EK}}{P_y} = 1.493$

Maximum compressive residual stress, $\sigma_{rc} = 10.8$ ksi

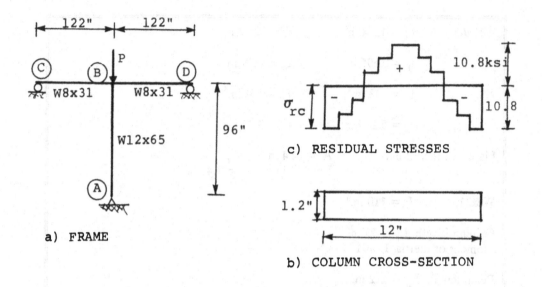

Fig. 10 - Numerical Example

TABLE 2:   INFLUENCE OF PLASTICITY ON STRENGTH OF
FRAMED COLUMNS

	ϵ_a	$\dfrac{P}{P_y}$	Stage	b_e in.	$\dfrac{I_{ec}}{I_c}$	K_{iV}	$\dfrac{I_{ec}}{I} \cdot \dfrac{1}{\lambda_c^2}$	$\dfrac{P_{tV}}{P_y} =$
1	0.0	0.0	I	12	1.000	2.635	1.490	1.490
2	0.7	0.700	II	10	0.579	2.378	0.863	1.055
3	0.8	0.783	III	8	0.296	2.196	0.441	0.635
4	0.9	0.850	IV	6	0.125	2.083	0.186	0.298
5	1.1	0.950	V	4	0.037	2.035	0.055	0.093
6	1.2	0.983	VI	2	0.005	2.003	0.007	0.012
7	1.3	1.000	VII	0	0.000	2.000	0.0	0.0

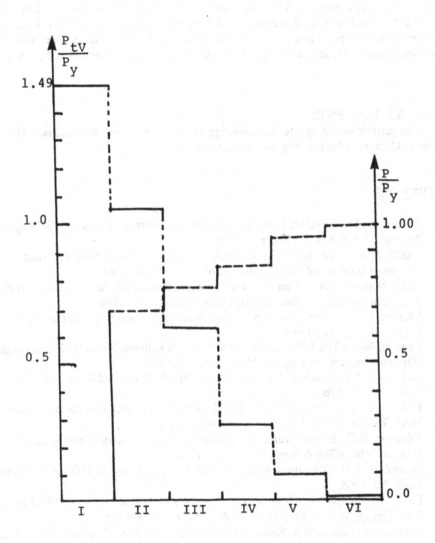

Fig. 11 - Variation of P_{tv} and P as a Function of Plasticity

effective length factor K_{iv} from the alignment chart (Fig. 6) or from Eq. 46. Next a value for P_{tv}/P_y is evaluated using Eq. 52.

In the initial stages of the loading process, the load required to buckle the frame will be higher than the existing load on the frame, at that stage, indicating that the system is stable. As the load is increased and more and more areas plastify, the inelastic buckling load of the system decreases. When these two values become equal, the frame becomes unstable and failure occurs. Table 2 and Fig. 11 also show the effect of assuming $K_{iv} = K$ at all stages.

ACKNOWLEDGEMENTS

The author would like to acknowledge the help of C. Foley, Graduate Student at Marquette University for carrying out the calculations.

REFERENCES

[1]   Yura, J.A.: The Effective Length of Columns in Unbraced Frames, AISC Engineering Journal, Vol. 8, No. 2, 1971, pp. 37-42.
[2]   AISC: Manual of Steel Construction: Allowable Stress Design, Ninth Edition, American Institute of Steel Construction, Chicago, IL, 1989.
[3]   AISC: Manual of Steel Construction: Load and Resistance Factor Design, 1st Edition, American Institute of Steel Construction, Chicago, IL, 1986.
[4]   Shanley, F.R.: Inelastic Column Theory, Journal Aeronautical Sciences, Vol. 14, No. 5, pp. 261-167, May 1947.
[5]   Osgood, W.R.: The Effect of Residual Stress of Column Strength, Proceedings First National Congress of Applied Mechanics, June 1951.
[6]   Galambos, T.V.: Guide to Stability Design Criteria For Metal Structures, 4th Edition, Wiley & Sons, 1988.
[7]   Huber, A.W., and Beedle, L.S.: Residual Stress and the Compressive Strength of Steel, Welding Journal, Vol. 33, Dec. 1954, p. 5598.
[8]   Johnston, R.G., Editor.: Guide to Stability Design Criteria for Metal Structures, Third Edition, John Wiley & Sons, New York, 1976.
[9]   Galambos, T.V.: Structural Members and Frames, Prentice Hall, Inc., Englewood Cliffs, NJ, 1968.
[10] Lu, L.W.: Compression Members in Frames and Trusses, Chapter 10 of Structural Steel Design, L. Tall, Ed., The Ronald Press Co., NY 1964.
[11] Disque, R.O.: Inelastic K-Factor for Column Design, AISC Engineering Journal Vol. 10, No. 2, 1973.
[12] Smith, Jr., C.V.: On Inelastic Column Buckling, AISC Engineering Journal, Vol. 13, No. 3, 1976, pp. 86-88.
[13] Matz, C.A.: Discussion of On Inelastic Column Buckling, by C.V. Smith Jr., AISC Engineering Journal, Vol. 14, No. 1, 1977, pp. 47-48.
[14] Yura, J.A.: Response to Discussions to The Effective Length of Columns in Unbraced Frames, AISC Engineering Journal, January 1972.

[15]   Adams, P.E.: Discussion of the Effective Length of Columns in Unbraced Frames, by
        Yura, J., AISC Engineering Journal, January, 1972.

[16]   Chiang-Siat-Moy.: K-Factor Paradox, Journal of the Structural Division, ASCE, Vol.
        112, No. 8, August 1986, pp. 1747-60.

[17]   Foley, C.: The Use of K-Factors in Unbraced Frames, A Critical Study, M.S. Thesis,
        Marquette University, Milwaukee, WI, July 1989.

[18]   Johnston, R.G.: Discussion of The Effective Length of Columns in Unbraced Frames,
        by Yura, AISC Engineering Journal, Chicago, IL, January 1972, p. 46.13.

[19]   Vinnakota, S.: LRFD of Steel Members, Lecture Notes, Dept. of Civil Engineering,
        Marquette University, Fall 1986.

MOMENT-ROTATION RESPONSE OF DOUBLE-ANGLE BOLTED CONNECTIONS

S. Vinnakota

Marquette University, Milwaukee, WI, USA

ABSTRACT:

Moment-rotation curves for double-angle, bolted, gusset plate connections are obtained by eleven full scale tests, using single connection cantilever arrangement. The specimen consisted of two 5 x 3½ x ⅜ in., A36 steel angles connected to a ⅜ in. thick gusset plate with 1, 2, 3, or 4 numbers of ¾ in. diameter, A325-N type high strength bolts. Also given are predicted moment rotation curves obtained by regression analysis using a polynomial fit. The influence of connection restraint on elastic stability of double angle columns is then briefly described.

1.0 INTRODUCTION

Bracings in large building structures (Fig. 1) are generally achieved by using double angles back-to-back and connected to the main framing via gusset plates. The gusset plates are normally bolted to the bracing member and connected to the column and beam by bolts or welds.

At present, the design of such bracing members is done in the USA by the following specification clauses and/or practical rules [1,2,3]: 1) To provide adequate lateral restraint for a compression member, the brace must have sufficient stiffness and strength. One simple procedure used in practice is to provide a bracing system to resist a force of 2% of the compression force it restrains. 2) The size of bracing is often dictated by slenderness limitation, such as KL/r <200, which will necessitate the use of a section larger than that

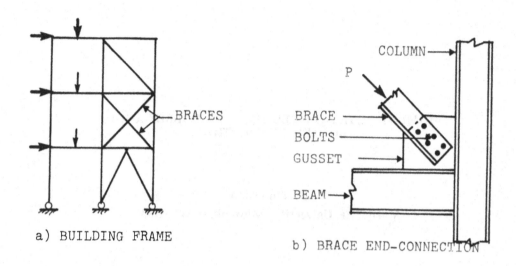

a) BUILDING FRAME

b) BRACE END-CONNECTION

Fig. 1 - Double Angle Bracing in Building Structures

which strength alone would require. 3) A depth of upstanding leg of not less than 1/90 to the unsupported length is frequently used in proportioning angle bracing. 4) The end connections are proportioned for the given force in the member or in the absence of a given force by the provision of LRFD Specification [3] Section J1.5, which stipulates a minimum factored design load of 10 kips. 5) When the connection is by bolts the usual practice is to provide a minimum of two bolts at each end of such members.

As the structure is loaded progressively, the compressive force in the member increases, the member begins to buckle, and the end connections (the bolt system and gusset plate) offer resistance to this tendency to buckle and therefore increase the critical load beyond the Euler load. The end restraint to rotation offered generally lies between that of a pinned connection (zero stiffness) and a fixed connection (infinite stiffness). the beneficial effect of this restraint is generally neglected in most specifications, as the connections are assumed to be pinned. The purpose of the present study is to experimentally determine the moment-rotating characteristics of bolted, double angle steel connections [4].

2.0 TEST SETUP

The single-connection cantilever arrangement [5] shown in Fig. 2 is used. The test setup was designed to duplicate the connection that exists between the gusset plate and the beam and column of the structure. A W10 x 112 column stub, welded to a 1½ inch thick base plate and bolted to the test frame beam with 1 inch diameter A354 Grade BD high strength bolts, had bolt holes prepunched in the flanges in a specific pattern. Each specimen that was to be tested was attached to this column stub using ⅞ inch diameter A354 Grade BD high strength bolts that were tightened using a standard shop wrench, providing the required rigidity of the actual field conditions.

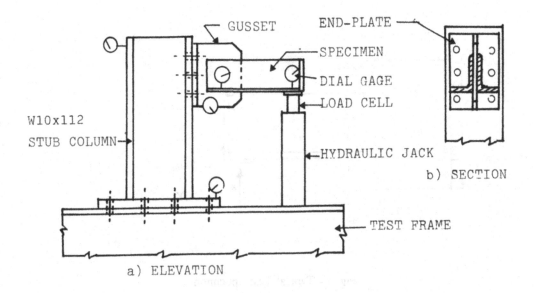

Fig. 2 - Test Set-up

The moment applied to the test specimen was introduced using an applied vertical force at a horizontal eccentricity of 12 in. measured from the center of gravity of the bolt group (Fig. 3). The vertical force was applied using an ENERPAC hydraulic jack powered by an ENERPAC electric powered hydraulic pump; and the load applied was monitored using ENERPAC load cells, models LH-102 of 10 kip capacity and LH-2506 of 50 kip capacity. The rotations of each connection were measured via four Starrett dial gages, two on each side of the double-angle. Dial gages were placed on both sides of the specimen to allow for the monitoring of any twisting of the specimen while under load. In order to verify that the test specimen remained fixed to the column and that the column stub itself did not deform under test loading, Starrett dial gages were attached in such a way as to monitor any movement of either member. The vertical force was generally applied in increments of 500 pounds-force. All dial gage measurements were taken at each load increment.

3.0 TEST SPECIMEN

A schematic drawing of a test specimen is shown in Fig. 3. A total of eleven tests were performed utilizing five different connection configurations (Fig. 4). The configurations ranged from a single bolt approximating a pin, through various bolt configurations that reveal the various levels of restraint, to a welded connection that approximates a rigid connection.

Each test specimen had been fabricated by a local steel fabricator. All of the test specimens were coated with whitewash, a simple solution of lime and water, prior to testing. Each test specimen consists of two 5 x 3½ x ⅝ inch rolled steel angles 16 inches long, one ⅝ inch thick end-plate, and ⅝ inch thick gusset plate and fasteners. A vertical end-stiffener

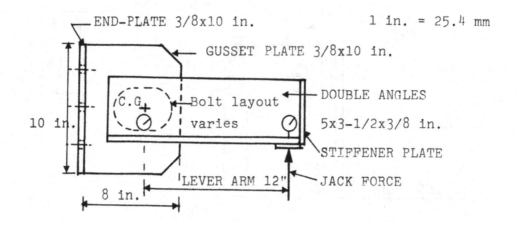

Fig. 3 - Typical Test Specimen

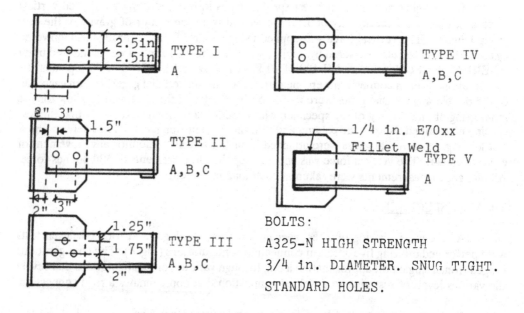

Fig. 4 - Connection Types Tested

plate was used to ensure that the two angles act as a unit and to provide stability against local buckling. The angles and the gusset are of mild structural steel with a nominal (specified minimum) yield stress of 36 ksi and a measured static yield stress of approximately 42 ksi as determined by tension tests of material taken from the double-angles.

The connections were bearing type connections using ¾ inch diameter A325-N high strength bolts in standard punched holes and E70XX fillet welds. The bolts were fastened to a snug tight condition, which requires only the full effort of a worker with a spud wrench. Specimen Type I utilizes a single A325-N bolt; specimen Types II, III and IV utilize 2, 3, and 4 bolts respectively, to yield various levels of restraint; and specimen Type V uses ¼ inch E70XX fillet welds all around the connection to approximate the rigid end condition.

4.0 TEST RESULTS

4.1 Connection Type I

Connection Type I consists of a single A325 high strength bolt (Fig. 4a). When the specimen was subjected to loading, the only restraint offered was due to the friction of the bolt due to tightening and that restraint vanished as a small load was applied. Thus, the connection indeed acted as a pin.

4.2 Connection Type II

Connection Type II consists of two bolts in line (Fig. 4b). Three specimens of this type were tested. Specimen IIA exhibited an approximately linear moment-rotation curve from zero moment to about 65 kip-inches of moment (Fig. 5a). The moment-rotation curve beyond 98 kip-inches of moment begins to level off. The maximum value of moment obtained was approximately 107 kip-inches. Specimen IIB's moment-rotation curve (Fig. 5b) is approximately linear from zero moment to approximately 90 kip-inches of moment. The curve becomes horizontal from approximately .0050 to .0400 radians of rotation at a moment of 90 kip-inches, indicating bolt slip. The specimen then exhibits a slight increase in restraint from approximately 95 kip-inches of moment to 110 kip-inches of moment where the remaining restraint diminishes greatly. It is evident that the bolt group begins to undergo large rotations beyond the point where the curve is no longer linear. Specimen IIC (Fig. 5c) is almost similar to IIB in every way. The bolt slip occurs at approximately 80 kip-inches of moment. Upon inspection of the specimens after testing, it was revealed that the failure or loss of restraint of the individual specimen occurred by the elongation of the bolt holes in the gusset plate.

4.3 Connection Type III

Connection Type III consists of three bolts in a row (see Fig. 4c). Three specimens of this type were tested. Specimen IIIA (Fig. 6a) was loaded until failure, in 500 pound increments, starting at 3,500 pounds of force. It exhibited an approximately linear moment-rotation curve from zero moment to approximately 114 kip-inches of moment (from zero to 0.0020 radians of rotation). The specimen failed at approximately 114 kip-inches of moment, at which point the specimen commenced to undergo unrestrained rotation. Specimen IIIB

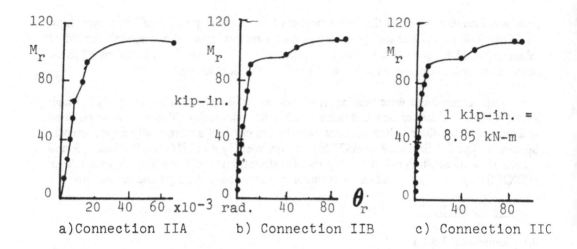

a)Connection IIA   b) Connection IIB   c) Connection IIC

Fig. 5 - Moment - Rotation Curves: Type II Connections

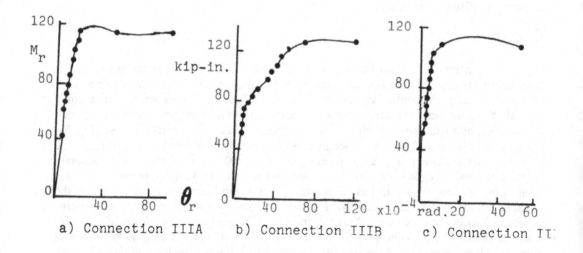

a) Connection IIIA   b) Connection IIIB   c) Connection II

Fig. 6 - Moment - Rotation Curves: Type III Connections

was loaded similar to specimen IIIA, but yielded different results. Specimen IIIB exhibited
an approximately linear moment-rotation curve from zero to 80 kip-inches of moment (zero
to 0.0020 radians of rotation), at which point the specimen underwent a loss of restraint.
From 80 to 120 kip-inches of moment (from 0.0020 to 0.0050 radians of rotation) the
specimen exhibited additional restraint but at a lower level than that of specimen IIIA.
Specimen IIIB was considered to have failed at 120 kip-inches of moment and 0.0050 radians
of rotation (Fig. 6b). Specimen IIIC exhibited a liner moment-rotation curve from zero to

approximately 100 kip-inches of moment (zero to 0.0010 radians of rotation), at which point the curve levels off indicating the loss of restraint. Thus, beyond 100 kip-inches of moment (approximately 0.0010 radians of rotation) the connection can be considered to offer no restraint (Fig. 6c). The failure of Type III connections was evident through a loud "popping" sound followed immediately by the loss of restraint in the gusset plate due to its deformation around the bolts.

4.4 Connection Type IV

Connection Type IV consists of four bolts (Fig. 4d). Difficulties were encountered in the course of testing Type IV connections. Specimen IVA began to pry away from the stub column of the testing apparatus at approximately 54 kip-inches of moment (Fig. 7a). The prying action was first made evident by the cracking and flaking of the whitewash around the area of the connection of the gusset plate to the ⅜ inch end plate. Therefore, since the specimen was not rigidly attached to the test setup, any rotations measured beyond the point that the specimen began to pry away from the column stub cannot be considered to be valid data. Specimens IVB and IVC were returned to the steel fabricator and were refitted with a ¾ inch end plate, replacing the original ⅜ inch end plate and thereby providing the required strength to counter any prying action and achieve reliable results. Specimen IVB had fabrication errors in that the specimen's bolt spacing on the end plate did not match the standard spacings on the stub column within tolerance limits. Consequently, in the process of leveling and forcing the test specimen in place, some loss of restraint occurred which was observed by the "popping" sound that usually denotes some loss of restraint in the connection. Because of this turn of events the data acquired from testing specimen IVB cannot be considered accurate. Specimen IVC was tested without any difficulty and is considered to yield the only reliable results of the Type IV connections. Specimen IVC exhibited a linear moment-rotation curve from zero to approximately 170 kip-inches of moment (from zero to 0.0006 radians of rotation) as shown in Fig. 7c. It failed at approximately 180 kip-inches of moment (0.0012 radians of rotation). Again, the failure of the connection was characterized by a loud "popping" sound and a significant loss of restraint. Failure of the connection occurred in the gusset plate and is characterized by the deformation of the gusset plate around the bolts. The ¾ inch end plate provided sufficient strength and did not undergo prying action.

4.5 Connection Type V

Type V connection consists of double-angles connected to a gusset plate via ¼ inch E70XX fillet welds all around (Fig. 4e). Prying action did occur during the testing of specimen VA at approximately 55 kip-inches of moment (Fig. 8) and thus confirmed that prying actin does manifest itself if the connecting member (in this case the ⅜ inch end plate) is not strong enough to counteract this action.

5.0 PREDICTION EQUATIONS

The data obtained from the tests was used to determine the prediction equations and to plot the regression curves. Sommer [6] determined that the moment-rotation characteristics of various connections can be approximated by using a higher order polynomial

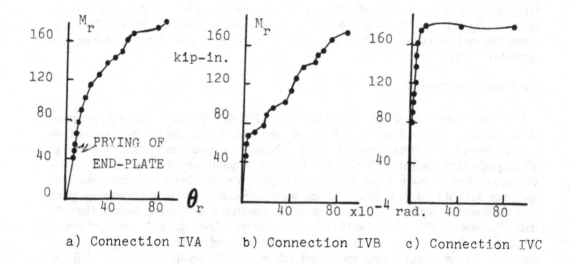

a) Connection IVA   b) Connection IVB   c) Connection IVC

Fig. 7 - Moment - Rotation Curves - Type IV Connections

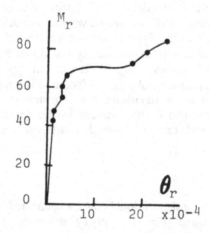

Fig. 8 - Moment - Rotation Curves - Connection VA

equation to model the moment-rotation curve. The general form of the equation is given by the following:

$$\theta_r = \sum_{i=0}^{n} c_i (B \, M_r^i) \tag{1}$$

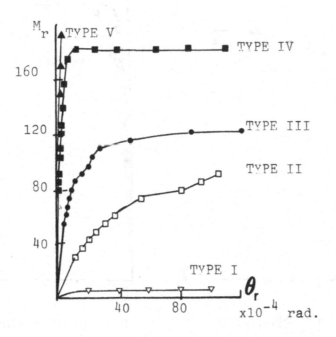

Fig. 9 - Experimental Moment - (Average) Rotation Curves

where: θ_r, is the rotation; i, order of polynomial; c, coefficient of regression; B, constant based on parameters; and M_r = moment.

In the present study, the only parameter that varies is the number of bolts and so the constant based on the parameters will be left out. A computer program (BMDP-P5R) that performs non-linear regression analysis was used to determine prediction equations for Connection Types II, III and IV [7]. They are

$$\theta_r = 1.52 \times 10^{-3} + 4.55 \times 10^{-4} M_r - 3.47 \times 10^{-5} M_{r2} + 1.17 \times 1011^{-6} M_r^3$$
$$- 1.65 \times 10^{-8} M_r^4 + 8.27 \times 10^{-11} M_r^5 \qquad (2)$$

$$\theta_r = 4.44 \times 10^{-2} - 2.93 \times 10^{-3} M_r + 7.46 \times 10^{-5} M_r^2 - 8.89 \times 10^{-7} M_r^3$$
$$+ 5.12 \times 10^{-9} M_r^4 - 1.07 \times 10^{-11} M_r^5 \qquad (3)$$

$$\theta_r = 4.97 \times 10^{-1} + 2.18 \times 10^{-2} M_r - 3.75 \times 10^{-4} M_r^2 + 3.17 \times 10^{-6} M_r^3$$
$$- 1.31 \times 10^{-8} M_r^4 + 2.14 \times 10^{-11} M_r^5 \qquad (4)$$

Figures 10a and 10b show plots of the moment-rotation curves of the typical experimental data vs. the prediction equation derived for Connection types III and IV respectively. By definition, 'typical experimental data' means the average of the results obtained for the particular connection type under consideration. The prediction curves approximate the actual experimental results reasonably well, though in order to have a higher degree of confidence in the experimental results, and thus the above equations, more tests are needed.

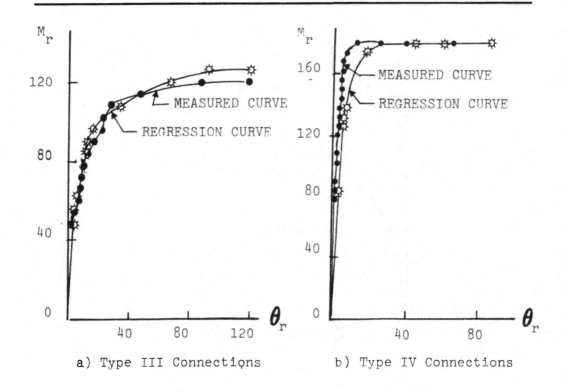

a) Type III Connections   b) Type IV Connections

Fig. 10 - Regression Curves

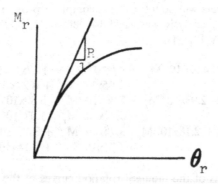

Fig. 11 - Connection Stiffness

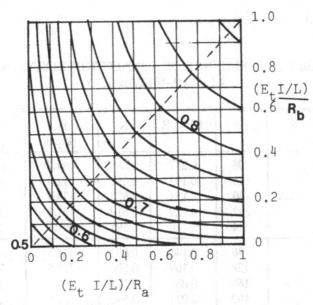

Fig. 12 - Nomograph for Effective Length Factor K of Restrained Column

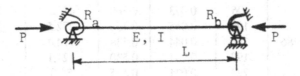

Fig. 13 - Rotationally Restrained Column

6.0 CRITICAL LOAD OF RESTRAINED COLUMNS

Using the moment-rotation regression curves determined in the previous section, the initial slope of the curve is determined. The initial rotational stiffness, denoted by R in Fig. 11, is then used in conjunction with the modulus of elasticity of the material, E; the moment of inertia of the structural shape about the strong bending axis, I; and the length of the member under consideration, L; to determine the effective length factor K from the nomograph shown in Fig. 12, assuming perfectly elastic behavior [8]. With the effective length factor determined, one can calculate the elastic critical load of the member.

The critical load was determined for various bolt configurations and member lengths and is compared to the Euler Critical Load in Table 1. The critical load has increased substantially when the rotational restraint of the end connection is taken into consideration. The compression member in Fig. 13 utilizing Connection Type II, III and IV yielded, on the average, an elastic critical load of 2½, 3 and 3½ times greater than the Euler critical load, when the effect of the end connection was taken into account. These remarks are based on the assumption that the material is elastic. However, material nonlinearity, residual stresses in rolled sections, and initial crookedness of columns, reduce this increase in strength beyond the Euler load and have to be considered [9, 10, 11].

TABLE 1: INFLUENCE OF CONNECTION RESTRAINT
ON COLUMN STABILITY

Conn. Type	R	L (in.)	(EI/L)/R	K	P_E (kips)	P_{EK} (kips)
II	18333	96	0.257	0.688	484.5	1114.5
		128	0.193	0.660	272.5	625.6
		160	0.154	0.625	174.4	446.5
		192	0.129	0.610	121.1	325.5
		224	0.110	0.591	89.0	254.4
III	32308	96	0.146	0.624	484.5	1244.3
		128	0.109	0.590	272.5	782.9
		160	0.088	0.575	174.4	527.5
		192	0.073	0.560	121.1	368.2
		224	0.063	0.552	89.0	292.0
IV	64286	96	0.073	0.560	484.5	1545.0
		128	0.055	0.550	272.5	900.9
		160	0.044	0.538	174.4	602.6
		192	0.037	0.530	121.1	431.2
		224	0.031	0.525	89.0	332.9

7.0 FINAL REMARKS

Additional tests on double angle bolted gusset plate connections using cruciform, pure moment loading arrangement [5] have been completed recently and will be presented elsewhere. Numerical studies to simulate the moment-rotation response of such connections using ANSYS finite element program [12] with 3-D isoparametric elements and 3-D nonlinear interface elements are under progress. The moment-rotation responses from all these studies will be used to determine the ultimate strength of double angle bolted columns, as influenced by connection restraint.

ACKNOWLEDGEMENTS

The present report is based on a Master's Thesis by Scott Romenesko [4] at Marquette University. The author would like to thank Robert Lorenz of AISC, Chicago and David Matthews of ACE Iron and Steel Co., Milwaukee for providing the steel specimen used in this study.

REFERENCES

[1] AISC, Manual of Steel Construction, 8th Edition, American Institute of Steel Construction, Chicago, IL, 1980.

[2] AISC, Engineering for Steel Construction, American Institute of Steel Construction, Chicago, IL, 1980.

[3] AISC, Manual of Steel Construction: Load and Resistance Factor Design, 1st Edition, Chicago, Il, 1986.

[4] Romenesko, S.: Experimental Results of Moment-Rotation Characteristics of Double-Angle Gusset Plate Connections, M.S. Thesis, Marquette University, Milwaukee, WI, December 1985.

[5] SSRC, Testing of Beam-to-Column Connections, Technical Memorandum (Draft), Tank Group 25, Structural Stability Research Council, USA.

[6] Sommer, W.H.: Behavior of Header Plate Connections, Master's Thesis, University of Toronto, Ontario, 1969.

[7] Dixon, W.J., ed., BMDP Statistical Software, University of California, Berkeley, CA, 1983.

[8] Kavanagh, T.C., Effective Length of Framed Columns, Transactions ASCE, Vol. 127, Part II, 1962.

[9] Vinnakota, S.: Planar Strength of Restrained Beam-Columns, Journal of the Structural Division, ASCE, Vol. 109, No. ST11, November 1982.

[10] Vinnakota, S.: Strength of Initially Crooked Double Angle and T-Section Steel Columns, Paper presented at the ASCE Structures Congress III, San Francisco, CA, October 1984.

[11] Vinnakota, S.: Planar Strength of Restrained Mono-Symmetric Steel Beam-Columns, Proceedings of the Structural Stability Research Council, Cleveland, OH, April 1985.

[12] ANSYS, Users Manual, Version 4.2b, Swanson Analysis Systems Inc., PA, 1986.

PART 4

FRAMES

STEEL FRAME STABILITY

M. Ivanyi
Technical University of Budapest, Budapest, Hungary

ABSTRACT

Collapse of frames is related not to a bifurcational phenomenon , but rather to divergence of equilibrium. The up–to–date stability analysis specifications are based on the analysis of so–called imperfect models.

Limit design of steel structures is based on the model of the plastic hinge. In its traditional form it is usually combined with rigid–plastic consitutive law. The model of "interactive plastic hinge" is suggested, which can take into account several phenomena (as residual stresses, strain–hardening, plate buckling, lateral buckling).

Chapter 1 gives definitions of load–deformation response of steel frames. Chapter 2 gives the concepts of the imperfect steel frames.

Chapter 3 is dealing with the model of interactive plastic hinge and its analytical interpretation. Chapter 4 gives the prediction of ultimate load of steel frames. Finally, Chapter 5 is devoted to the evaluation of load bearing capacity of steel frames.

1. INTRODUCTION

1.1. LOAD–DEFORMATION RESPONSE OF FRAMES

The beams, columns, and beam-columns do not occur in isolation, but many of them joined together make up a structural *frame*. This frame is the skeleton which supports the loads which the structure is called upon to support.

The purpose of frame analysis is to determine the limits of structural usefulness of a given frame and to compare the predicted performance with the required one. Such an analysis is part of the design process, wherein adjustments and new analyses are made until the predicted performance matches as closely as possible the design requirements.

In this article the methods of determining the maximum load capacity and the deformation response of frames will be examined. This topic is a vast one and we shall only be able to cover a small portion of it. Emphasis will be placed on basic behaviour and the discussion will center around very simple examples.

Frame behaviour is characterized by the relationship between the loads, as they vary during the loading history, and the resulting deformations. A typical load–deflection curve is shown in Figure 1.1. The relationship is non–linear from the beginning because of second-order geometric effects (that is, the forces produce deformations which in turn influence the forces). After the elastic stage is reached, the slope of the curve is further reduced, and finally the slope becomes zero at the maximum load P_M.

A curve, such as in Figure 1.1. gives the value of the maximum load which can be carried by the frame, as well as the magnitude of the deformation corresponding to any load intensity. Furthermore, at least the ascending branch of the curve is in stable equilibrium. In design the curve can be used to check if (1) the ratio P_M/P_W (where P_W is the actual or working load) is sufficiently near a specified *load–factor* (as determined by judgment or prescribed by a code or a specification), and (2) the deflection at working load v_W is less than or equal to a specified maximum value. In the design operation we try to match these requirements with the structural behaviour, each time adjusting the structure until the requirements are met.

Ideally it would be desirable to construct a load–deflection curve for each structure. We then could obtain the various items of information which we are interested in. Unfortunately we are only able to construct load–deflection curves for very simple structures. For more complex frames we need to introduce assumptions which will permit us eventually to obtain bounds for the value of P_M.

The purpose of this article is to describe (1) how the load–deflection curve of frames may be constructed in as exact a manner as possible and (2) to describe approximate methods whereby the load–deflection curve, and particularly P_M, can be estimed.

1.2. A BRIEF HISTORICAL SKETCH

In several periods of developing engineering practice different importance was attributed to the experimental and computational tools of structural design: (MCGUIRE 1984)

First, consider the simple column. Figure 1.2 contains two column design curves in use in the early 1960's. One is the curve introduced into the AISC Specification in 1963.

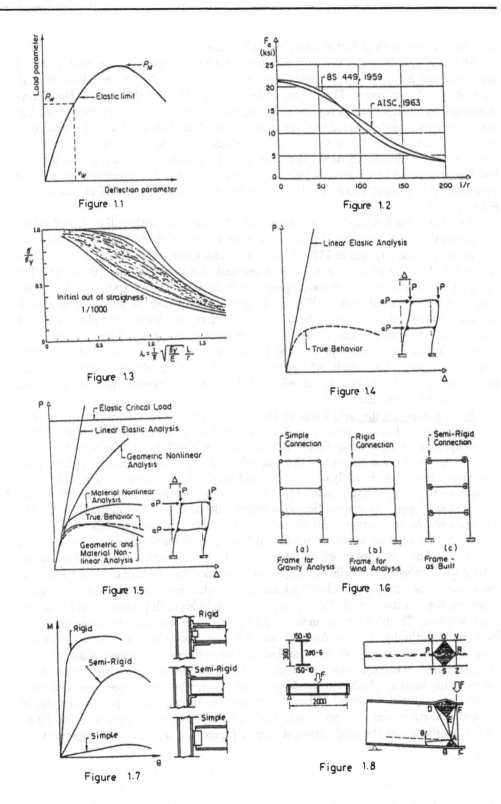

Figure 1.1

Figure 1.2

Figure 1.3

Figure 1.4

Figure 1.5

Figure 1.6

Figure 1.7

Figure 1.8

The other is the curve of British Standard 449 at the time.

Practically, it did not make much difference which curve one used, the result required the same amount of steel for a given load condition. But they were rationalized in completely different ways. The British curve was explained on the basis of the Perry-Robertson formula, a formula resulting the analysis of a strut that is initially bowed, but free of residual stresses. The American curve was explained on the basis of the tangent modulus formula, a formula resulting from the analysis of a perfectly straight strut, but one encumbered by residual stress. Fortunately, on both sides of the Atlantic the theories had been tempered by reasonable judgment of test data, and the agreement shown was achieved. This is as it should be, but it can hardly be said to be perfect analysis on either side of the ocean.

The computer has made it easy to include the effects of both initial imperfections and residual stresses in simple columns. Figure 1.3 contains the results of computer calculations of the strength of 112 column shapes and types, each having an initial bow of 1/1000 of the length and a prescribed residual stress distribution (BJORHOVDE 1972). The calculations are in good agreement with tests. This is gratifying, but what should be done with this power? Figure 1.3 represents only a small sample of an almost unlimited variety of shape, imperfection, and initial stress parameters. No analysis of isolated columns can furnish a clue to the interaction among members of a frame, a factor of fundamental importance. The figure shattered any illusions that may have remained regarding the sanctity of simple curves in Figure 1.2. But the older curves are still serviceable. Should they be abandoned? There are, at present, no clearly correct answers to these questions.

Second, consider the rigid frame in Figure 1.4. It has been known for a long time that, if loaded to failure, the response of frame will be as described by the non–linear load–displacement curve in the figure. However, before World War II, the best that could be done in practice was to analyze this frame by a linear elastic method. As the figure indicates, a linear elastic analysis may represent service load behaviour very well, but it reveals nothing regarding ultimate resistance. To make elastic analysis generally useful, engineers employed it in conjunction with design rules that reflected in numerous, often hidden, ways the actual non-linear behaviour that all real structures exhibit.

With the computer, a number of different types of frame analysis become practicable. The results of five ones are shown in Figure 1.5: (1) a linear elastic analysis; (2) an elastic critical load analysis; (3) an elastic analysis that includes the effects of displacements (geometric non-linearity); (4) an elastic-plastic analysis that includes the effects of yielding (material non-linearity); and (5) an analysis that includes both geometric and material non-linearities. The higher order methods may predict the total structural response of a frame quite faithfully. But the frame shown is a simple one. Is it economically feasible to apply the more exact methods to realistically large three-dimensional frames of complex geometry? Is enough known about the real ultimate resistance of three-dimensional frames to use these methods with confidence and, if not, what experiments are needed to provide the required evidence? What sort of member proportioning procedures should be used in conjunction with non-linear analysis? The non-linear results of the figure are far in advance of the single straight line of Figure 1.4, but they still represent the analysis of a frame alone, and not of a building with all of the components that may provide effective load resistance. Again, there are questions to answer and work to be done before the full potential of higher order methods can be realized.

Third, consider the treatment of joints in pre–computer times. A practice that had come down from the 19th century was to analyze a building frame as simply connected under gravity load and moment-resistant under wind load. It is only a mild exaggeration to say that in design this way the engineer deals with three buildings: the first is the building he analyzes for vertical loads (Figure 1.6a), the second is the building he analyzes for horizontal loads (Figure 1.6b), and the third is the building he builds (Figure 1.6c).

For several reasons, this design procedure is still used and defensible. One reason is the difficulty of reproducing in analysis the real semi–rigid behaviour of most connections. Methods for including connection deformation in the slope deflection method were developed in the 1930's, but did not find favour. They complicated an already laborious analysis, and were still only a linear approximation of decidedly non–linear phenomenon.

Fourth, consider the target models deduced from tests. Recent decades changes in the purport of stability analyses — in particular, due to computer facilities — are surveyed. The classic problem of stability analysis — to find the equilibrium bifurcation and calculation of the relevant critical load — is linked to the application of mathematical models reflecting earlier, restricted computing possibilities, simulating the behaviour of the structure by assuming small deformations and usually elastic material characteristics, at an often significant simplification of the geometry of the structure, limiting the degrees of freedom of displacements, making use of symmetry. More refined mathematical models resulting from omission of these restrictions — able to accomodate geometrical and residual stress data reflecting real, manufacturing features — may eliminate separate stability analysis, considering instability as one form of the loss of load capacity without fractures. Thus, criticism to the applied mathematical model for suitability — completeness — becomes a fundamental problem of stability analysis methods. Since actual computer facilities may fall short of these demands, so–called *target models* deduced from tests may be applied.

Two illustrative examples are presented:

(A) Semi — Rigid Connection

The type of beam–to–column connection used is a primary determinant of the behaviour of the frame. The construction types are defined in terms of connection rotational stiffness and moment resistance as represented by a moment–rotation ($M - \theta$) diagram (Figure 1.7). Generally, accurate $M - \theta$ diagrams can only be obtained experimentally by tests. Figure 1.7 contains $M - \theta$ diagrams for typical connections and indicates how they would normally be classified. The non–linearities in behaviour which are shown are the result of yielding of the connection components or local regions of the connected members, or slip of the fasteners.

Corrections are so diverse and so complex that large amounts of experimental and analytical data on connection deformation must be collected and systematized before reliable semi–rigid frame analysis can become common practice (NETHERCOT 1985).

(B) Interactive Plastic Hinge

The traditional concept of plastic design of steel structures is based on the assumption that under gradually increasing static loads plastic zones develop and grow in size and number, and eventually cause unstricted, increasing deflections; thus loading to the onset of ultimate limit state of the structure. The concept was first introduced by KAZINCZY (1914) by establishing concept of the "plastic hinge". Some basic questions are still discussed. Among them are the effects of the difference between ideal–plastic constitutive law and actual behaviour of steel material and the consequense of local instability (plate buckling; lateral buckling). The element of the bar is considered to be built up of plate elements (following the pattern of steel structures) instead of a compact section. Then the behaviour of the "plastic hinge" can be characterized by tests with simple supported beam (Figure 1.8). Based on these tests a yield–mechanism for the bar–element can be introduced, giving basis for a mechanism curve: defining thus the descending branch of the moment–rotation diagram. We introduce the concept of "interactive plastic hinge" which can substitute the classical concept of plastic hinge in the traditional methods of limit design, but can reflect the effect of phenomena like strain–hardening, residual stresses, plate buckling and lateral buckling (IVÁNYI 1985).

Local and lateral bucklings are so diverse and so complex that large amounts of experimental and analytical data must be collected and systematized thus reliable frame analysis can become common practice (BAKSAI, IVÁNYI, PAPP 1985).

1.3. RECENT DEVELOPMENT

The special scope of design theory is the theory of stability. Concerning the ultimate load capacity of structures a traditionally distinct (strength and stability) analytical system has been customary. Separation was made possible by the fact that the applied computation models were ideal, and it was necessitated by a non–conscious optimization. This modelling process involves several elements that lead to useless, excessive simplifications.

International efforts have contributed to the feasibility to "demolish" formulation of the given design theory problem to its foundations, and to reconstruct it by making use of new findings, new methods. The two factors of the new findings and methods are the international accumulation of experimental results in recent decades, and the development of computing techniques lending themselves to multiparameter problems to several degrees of freedom, and of simulation.

Present tasks of research on the theory of stability are:
- the introduction of imperfect models lending themselves to multiparameter system adapted to several degrees of freedom;
- the analysis of structures involving elements in postcritical condition, suitable failure simulation problems for a more comprehensive formulation of the safety concept.

2. THE CONCEPTS AND THE APPLICATIONS OF THE IMPERFECT STEEL FRAMES

2.1. SOME ASPECTS OF THE STABILITY CRITERION

In case of steel structures the designed ideal structure differs from the one completed according to the plans, and this difference can be characterized by numerous larger and smaller defects, irregularities, imperfections (Figure 2.1).

The equilibrium of the designed structure is stable if small imperfections and defects cause sufficiently small differencies with respect to the ideal operational conditions (Figure 2.2).

If small imperfections cause disproportionately big differences then equilibrium is unstable (Figure 2.3).

The designer should choose the material and dimensions of the structure such as at all possible load combinations, considering all possible disturbances, equilibrium of the structure be stable, furthermore stability must have on the safety side.

The term stability (instability) represents the relation between the disturbing factors and the requirement due to them.

Stability condition in case of structures can be expressed as follows: the equilibrium condition of ideal structure — during operation — must be stable against all these disturbances that differentiate the ideal structure from the real one. This way stability has a close relation with the choise of the calculation method, the calculation model.

(A) The traditional stability analysis methods assume ideal models and determine bifurcation (equivalent effective length).

Let's assume that the calculation model is a homogeneous, perfectly straight, concentrically loaded bar (Figure 2.4a).

The ideal model differs from the real one in the initial out–of–straightness and the eccentricity of the compression force. The statement, that straightness of bar (ideal model) is stable in regard with the above differences (disturbances), is equivalent to the statement, that by limiting the initial out–of–straightness and eccentricities under adequately small value of displacements perpendicular to the axis of bar be less than a given value. If compression force N is smaller than the Euler force N_E the straight bar is stable. If $N > N_E$ the least disturbances might cause infinite displacements.

(B) Trials with the goals to demonstrate the loss of stability with bifurcation (buckling, lateral buckling, plate buckling) have the information that theoretically assumed behaviour (bifurcation) can be achieved only with specimen (perfectly straight bar, perfect cylider or sphere, etc.) made with extreme care; behaviour of civil engineering structures might be significantly different. Inverse of this basic statement hints that in cases where analysis of model for such exaggerated abstractions (simplifications in structural geometry, conditions referring to the symmetry of arragement and loading), from which the least deviation (small geometric defects of structure, small asymmetry, etc.) — at least in certain load spectrum — can lead to a significant deviation in the assumed and real behaviour of the structure to be modelled. Aim of the up–to–date

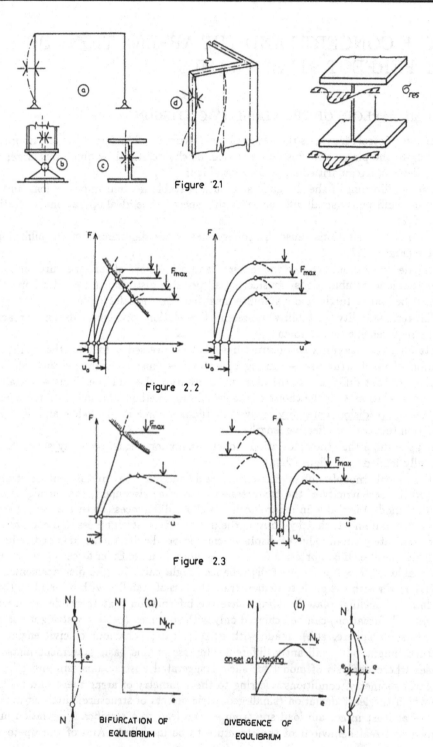

Figure 2.1

Figure 2.2

Figure 2.3

Figure 2.4

stability analysis — using the before mentioned term for stability of equilibrium — is the limitation of application of the above mentioned, too much simplified ideal models and the searching for such — real models (so–called *imperfect models*) that can reflect the generally negligible, but in given cases (so–called disturbing) effects of significant consequences.

Therefore up–to–date stability analysis specifications are based on the analysis of real models. In case of buckling and lateral buckling analysis slightly crooked and eccentrically loaded bars with residual stresses, while at buckling of plates and shells surfaces with geometrical imperfection are to be taken into consideration.

Let's assume that initial crookedness and eccentricity are given and calculation is carried out on the model of crooked and eccentrically loaded bar (Figure 2.4b). If the problem is expressed this way stability problems of the straight bar in respect with the given disturbances are eliminated and are replaced by determination of stress and deformation states by a more accurate calculation model. (However, in this case such a new problem may occur that analyzes the stability of this state in respect with certain new classes of disturbances.)

2.2. TAKING THE INITIAL (REAL AND FICTIOUS) IMPERFECTIONS

In case of elastic material, and applying plane model — primarily for computer aid — real and fictious initial imperfections can be taken in an arrangement that internal forces computed from the conjugate effect of loads and initial imperfections be the least favorable in respect with load carrying capacity.

Initial imperfections include the differences between ideal and real models, that is the effects of initial crookedness, inaccurate location of load, geometric defects of cross section, residual stresses, etc.

During investigations the above initial imperfections can be:
- of geometric or
- of load characteristic.

In respect with their physical meaning substituting geometric imperfections or substituting loads express their role more accurately, namely to replace all those phenomena between ideal and real model (IVÁNYI 1987).

2.21. Initial Imperfections of the Geometric Characteristic

(A) Element of non–sway plane framework: By assuming an initial crookedness of the half–sine wave or parabola shape on a single bar, initial crookedness should be chosen such as the modified strength analysis approach the buckled shape of the compression bar determined by tests with a specified accuracy.

Deflection (e_0) specified for compression members in the MSZ 15024-85 specification in the function of shape of cross section is given in Table I.

Initial crookedness of each eccentrically loaded compression member must be taken in the most unfavorable combination for the structure itself. No initial crookedness should be taken into consideration for tension, eccentrically tensioned and bent members. Initial crookedness of eccentrically loaded compression members can be neglected if parameter characteristic of bar stiffness is

	Buckling curve (MSZ 15024/1–85)	Ihitial imperfection e_o
$\varepsilon_i > 1{,}0$	a	$\dfrac{l}{500}$
	b	$\dfrac{l}{250}$
	c	$\dfrac{l}{200}$

Table I.

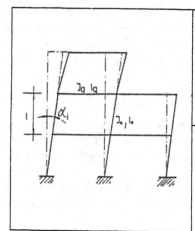

Range of application:

1.) $\quad \varepsilon_i = l_i \sqrt{\dfrac{N_i}{E \cdot l_i}} \;\leqq\; 1{,}6$

2.) $\quad \nu_j = \dfrac{\Sigma^{J_g}/l_q}{\Sigma^{l_o}/l_o} \;\geqq\; 1{,}0$

$$\alpha_i = \frac{1}{150} \cdot \gamma \qquad \gamma = \frac{1}{2}\left(1 \cdot \frac{1}{n-1}\right)$$

n = number of columns in one storey in the plane

Table II.

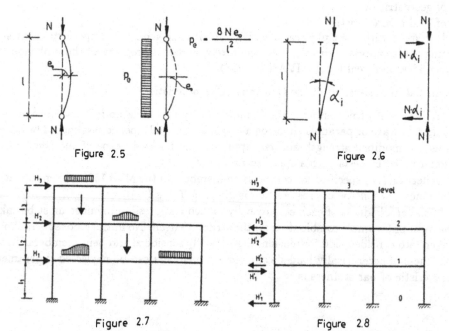

$$P_e = \frac{8\,N\,e_o}{l^2}$$

Figure 2.5   Figure 2.6

Figure 2.7   Figure 2.8

$$\varepsilon_i = \ell_i \sqrt{\frac{N_i}{E \cdot I_i}} \leq 1.0$$

where N_i is the compression force of bar
  ℓ_i is the length of bar
  I_i is the moment of inertia about the axis perpendicular to the plane
    of frame

(B) Element of sway framework: In case of applying plane model an initial out–of–straightness of $\alpha_i = \alpha_0 \cdot r$ must be assumed (Table II),
where $\alpha_0 =$ is the basic value of crookedness,
  $r =$ depends on number of columns on the levels.
    If $n = 1$, then $r = 1$ should be taken into consideration.

2.22. Initial Imperfections of the Load Characteristic

(A) Element of non–sway plane framework: A parabola shape with the deflection of ℓ_0 in a compression bar be caused by a uniformly distributed load perpendicular to the axis of bar (Figure 2.5).
  The non–sway plane framework is a geometrically ideal model and can be analyzed by taking initial imperfections of the load type characteristic into consideration.
  (B) Element of sway plane framework: The initial out–of–straightness of α_i of the compression member under investigation can be inverted to an initial imperfection of the load characteristic, taking the so–called principle of small deflections into consideration (Figure 2.6).

2.3. SIMPLIFICATION POSSIBILITIES OF SECOND–ORDER THEORY IN COMPUTER CALCULATION OF SWAY FRAMES

Second–order analysis of sway frames or vertical columns and horizontal beams according to Figure 2.7 can be simplified in different rates if ε_i (Section 2.2.1. (A)) for all eccentrically loaded compression bars meet the $\varepsilon_i \leq 1.6$ requirement.

2.31. First the internal forces of the structure should be determined, along with the horizontal displacements, δ_j's, of the levels with first–order analysis, but taking the initial crookedness for columns according to Section 2.2.1. (B) into consideration. The latter can be taken into consideration for the levels according to Figure 2.8 by H'_j complex of forces as external forces

$$H'_j = \alpha_i \cdot V_j \quad \text{and} \quad V_j = \sum N_{j,i}$$

where α_j is the assumed out–of–straightness of columns between the levels
  $j - 1 \cong j$ according to Section 2.2.1. (B)
  V_j is the sum of compression forces $N_{j,i}$, in the same columns.
    From horizontal displacements, δ_j, — calculated by first–order theory — the increment of out–of–straightness

$$\Delta \alpha_j = \frac{\delta_j}{\ell_j}$$

comes out.

Afterward the increment of "shifting" force

$$\Delta H_j = V_j \cdot \Delta \alpha_j$$

can be determined and the value of $\Delta H_j / Q_j$, where Q_j is the original shifting force

$$Q_j = \sum_{j}^{m} H_i + H_j'$$

In the formula of Q_j summation includes the horizontal forces above level "j", m is the number of levels.

2.32. If condition

$$\frac{\Delta H_j}{Q_j} \le 0.1$$

is met on each level of the frame then internal forces calculated by first–order theory, but taken the initial out–of–straightness of columns into consideration are considerably accurate.

If condition

$$0.1 < \frac{\Delta H_j}{Q_j} \le 0.25$$

is met on each level of the frame and the distribution of internal forces due second-order effect is similar to the one of the first-order theory, then internal forces can be calculated by first-order analysis, increased by

$$Q_j' = \frac{1}{1 - \frac{\Delta H_j}{Q_j}} \cdot Q_j$$

shifting force on each level.

2.4. SECOND-ORDER THEORY IN MANUAL CALCULATIONS; STABILITY FUNCTIONS

First-order solution of statically underterminant structures, based on the slope-deflection method can be transformed into a process based on second-order analysis with the so-called stability functions (HORNE, MERCHANT 1965).

2.41. Rotations φ_A and φ_B of nodes A and B, and the nodal moments (Figure 2.9) due to $\gamma_{AB} = e_B / \ell_i$ rotation of axis of straight-axis bar can be calculated by the following system of equations:

$$M_{AB} = \frac{E \cdot I_i}{\ell_i} \left[s_i \varphi_A + s_i c_i \varphi_B - s_i (1 + c_i) \gamma_{AB} \right]$$

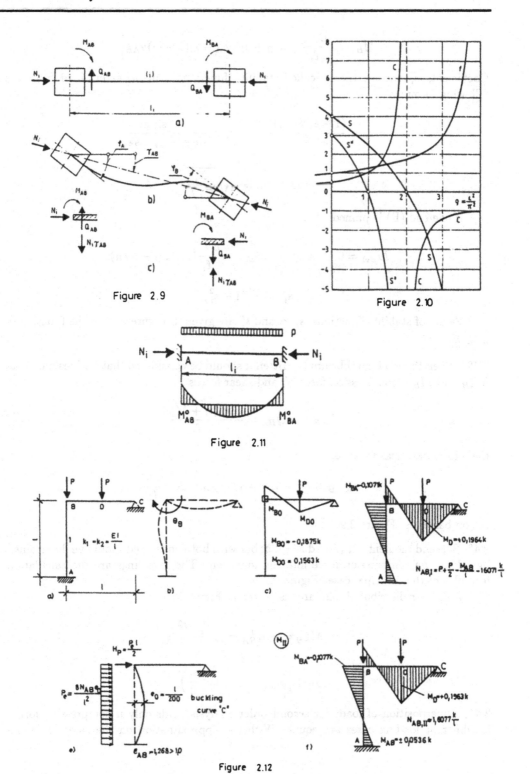

Figure 2.9

Figure 2.10

Figure 2.11

Figure 2.12

$$MBA = \frac{E \cdot I_i}{\ell_i} \left[s_i \varphi_B + s_i c_i \varphi_A - s_i (1 + c_i) \gamma_{AB} \right]$$

Coefficients s_i and c_i, the so-called stability functions, can be determined from the following formulas

$$s_i = \frac{(1 - \varepsilon_i \cot \varepsilon_i) \frac{\varepsilon_i}{2}}{\tan \frac{\varepsilon_i}{2} - \frac{\varepsilon_i}{2}} \qquad c_i = \frac{\varepsilon_i - \sin \varepsilon_i}{\sin \varepsilon_i - \varepsilon_i \cos \varepsilon_i}$$

$$\text{and} \qquad \varepsilon_i = \ell_i \sqrt{\frac{N_i}{E \cdot I_i}}$$

If one end (B) is pinned, then

$$M_{BA} = 0 \qquad \text{and} \qquad M_{BA} = \frac{E \cdot I_i}{\ell_i} s_i'' (\varphi_A - \gamma_{AB});$$

$$s_i'' = s_i \cdot (1 - c_i^2)$$

Values of stability functions, s_i, c_i and s_i'' are given in Figure 2.10 in the function of $\rho_i = \frac{\varepsilon_i^2}{\pi^2}$.

2.42. When the joint equilibrium is written it should be considered that besides moments M_{AB} and M_{BA}, compression force N_i and shear forces

$$Q_{AB} = -Q_{BA} = -\frac{M_{AB} + M_{BA}}{\ell_i}$$

the additional shear forces of

$$Q_{AB} = -Q_{BA} = -N_i \cdot \gamma_{AB} = N_i \cdot \frac{e_B}{\ell_i}$$

act on bar ends (Figure 2.9c).

2.43. The end moments M_{AB} and M_{BA}, of bar with both ends fixed should be determined by taking the compression force into consideration. The followings are the calculation formulae of the principal case (Figure 2.11).
    Values for distributed load are also given in Figure 2.10.

$$M_{AB}^0 = -M_{BA}^0 = f_i \cdot \frac{p \cdot \ell_i^2}{12}$$

$$f_i = \frac{12}{\varepsilon_i^2} \cdot \left(1 - \frac{\varepsilon_i}{2} \cdot \cot \frac{\varepsilon_i}{2} \right)$$

2.44. Superposition of loads for second–order analysis holds only if compression forces for different loading cases are equal. With the approximation on the safety side the compression forces belonging to the most unfavorable cases can be assumed for each loading cases.

2.5. EXAMPLES

EXAMPLE 1. Analysis of Non-sway Steel Structure (fixed nodes)

Let's determine the load-bearing capacity of the plane steel structure of Figure 2.12a.

a) Calculation of framework by disassembling to elements (beams, columns). The structure is analyzed by slope–deflection method. Degree of freedom of the plane structure is equal to 1 (one internal joint). Figure 2.12b shows the deflection, Figure 2.12c shows the moment diagram of the primary system.

The equilibrium equation:

$$(s_1 k_1 + s_2'' k_2) \cdot \Theta_B = M_B$$

During first–order analysis we neglect the effect of axial forces, therefore the equilibrium equation and its solutions are:

$$(4+3) \cdot k \cdot \Theta_B = 0.1875 \cdot k$$
$$\Theta_B = 0.0268$$

Figure 2.12d shows the distribution of the first–order moments.

The bending moment calculated by first–order analysis shall be modified. In order to do this, the effective length of the column must be determined. The factor showing the growth of eccentricity is (MSZ 15024-85)

$$\nu = 0.5 + \frac{1}{6\mu + 4} = 0.618 \qquad \mu = \frac{0.75 \cdot I_2 \cdot \ell_i}{I_1 \cdot \ell_g} = 0.75$$

$$\psi = \frac{1}{1 - \frac{N}{N_E}} = \frac{1}{1 - \frac{N}{A\sigma_H} \left[\frac{\lambda}{\lambda_E}\right]^2} = \frac{1}{1 - \frac{N\nu^2}{\pi^2}} = 1.066$$

Load–bearing capacity of column AB by taking the maximum internal forces into consideration:

$$\frac{N_I}{N_H} + \frac{\psi M_I}{M_H} = 1 \qquad \psi \cdot M_I = 1.066 \cdot 0.1071\,k = 0.1142 \cdot k$$

b) Calculation of plane framework (statical skeleton). The initial inaccuracies taken into consideration at planar statical skeletons make possible that, beside strength analysis carried out by second–order analysis, only plate element shall be checked for plate buckling and the beam shall be checked in the plane perpendicular to it.

Following are (Figure 2.12e) the initial inaccuracies according to Section 2.22.

$$N_{AB} = 1.607 \frac{k}{\ell} \qquad \varepsilon_{AB} = 1.268$$

$$p_e = \frac{8 N_{AB} e_0}{\ell^2} = \frac{8 \cdot 1.607\, k}{200\, \ell^2} = 0.0643 \frac{k}{\ell^2}$$

$$H_P = \frac{p_e \ell}{2} = 0.0321 \frac{k}{\ell}$$

$$\sum H = H_P = 0.0321 \frac{k}{\ell} \approx 0.$$

Having determined the axial force of bar AB by first–order analysis, we can find the values of the stability functions and may put down the equilibrium equation with their solutions:

$$\rho = \frac{\epsilon^2}{\pi^2} = 0.163 \qquad s = 3.7890 \qquad f = 1.0279 \qquad c = 0.5439$$

$$M_{AB,P} = f \cdot \frac{p_e \, \ell^2}{12} = 1.0279 \frac{0.0643}{12} \cdot k = 0.00551 k$$

$$(3.7890 + 3) \cdot k \cdot \Theta_B = 0.193 k$$

$$\Theta_B = 0.0248 \qquad M_{BA} = 0.1077 k$$

The second order moment distribution is shown in Figure 2.12f:
Load–bearing capacity analysis of column AB:

$$\frac{N_{AB,I}}{N_{AB,II}} = \frac{1.6971}{1.6077} \approx 1.00$$

$$\frac{\psi M_{AB,I}}{M_{AB,II}} = \frac{0.1142}{0.1077} = 1.06$$

Therefore analyses a) and b) give almost the same forces. Comparison of the bending moment analysis a) gives 6% higher internal forces than analysis b).

EXAMPLE 2. Analysis of Sway Steel Structure

Let's determine the load–bearing capacity of the planar steel bar construction sketched in Figure 2.13a.

a) Calculation of the structure by disassembling to elements (beams, columns)

The analysis is carried out using the slope–deflection method. The degree of freedom of the planar structure is equal to 2 (one internal joint, one swaying floor). Figure 2.13b shows the deflection, Figure 2.13c shows the bending moment diagram of the primary system.

The equilibrium equations are:

$$\begin{bmatrix} (s_1 + s_2)k & \vdots & -s_1(1 + c_1)\frac{k}{\ell} \\ \cdots\cdots\cdots\cdots\cdots & & \cdots\cdots\cdots\cdots\cdots \\ -s_1(1 + c_1)\frac{k}{\ell} & \vdots & \frac{2 s_1(1 + c_1)}{m_1} \frac{k}{\ell} \end{bmatrix} \begin{bmatrix} \Theta_B \\ \cdots \\ e_B \end{bmatrix} = \begin{bmatrix} M_B \\ \cdots \\ H_B \end{bmatrix}$$

During the first–order analysis we neglect the effect of axial forces, therefore the equilibrium equations and their solution are:

$$\begin{bmatrix} (4+3)k & \vdots & -6\frac{k}{\ell} \\ \cdots\cdots\cdots\cdots \\ -6\frac{k}{\ell} & \vdots & 12\frac{k}{\ell} \end{bmatrix} \begin{bmatrix} \Theta_B \\ \cdots \\ e_B \end{bmatrix} = \begin{bmatrix} 0.18575k \\ \cdots \\ 0.5\frac{k}{\ell} \end{bmatrix}$$

$$\Theta_B = 0.1094 \qquad e_B = 0.0964\ell$$

Figure 2.13d shows the first–order bending moment distribution.

The bending moment calculated by first–order analysis, shall be modified, therefore the effective length of the column shall be determined. The ψ factor showing the increment of the eccentricity is:

$$\nu = \sqrt{\frac{1+m}{2}} \cdot \sqrt{1 + 0.35(c + 6\alpha) - 0.017(c + 6\alpha)^2}$$

$$\text{if} \qquad m = 1 \qquad c = 2 \qquad \alpha \approx 0 \qquad \nu = 1.29$$

$$\psi = \frac{1}{1 - \dfrac{N}{N_E}} = \frac{1}{\dfrac{N_{AB}}{\frac{\pi^2 EI}{(\nu\ell)^2}}} = \frac{1}{1 - \dfrac{1.36 N^2}{\pi^2}} = 1.298.$$

Load–bearing capacity of column AB by taking the maximum internal forces into consideration

$$\frac{N_I}{N_H} + \frac{\psi M_I}{M_H} = 1; \qquad \psi \cdot M_I = 1.298 \cdot 0.359k = 0.466k$$

b) Calculation of plane framework (statical skeleton). The initial inaccuracies taken into consideration at planar statical skeletons, make possible that, beside strength analysis carried out by second–order analysis, only plate elements shall be checked in the plane perpendicular to it.

Following are (Figure 2.13e) the initial inaccuracies according to Section 2.22:

$$N_{AB} = 1.36 \frac{k}{\ell} \qquad \varepsilon_{AB} = 1.36$$

$$H_\alpha = N_{AB} \cdot \alpha_0 = 1.36 \frac{k}{\ell} \frac{1}{150} = 0.00905 \frac{k}{\ell}$$

$$\sum H = H + H_\alpha = 0.5091 \frac{k}{\ell}$$

Having determined the axial force of bar by first–order analysis, we can find the values of the stability functions and the equilibrium equations can be written with their solutions:

$$\rho = \frac{\varepsilon^2}{\pi^2} = 0.138; \qquad \begin{array}{ll} s = 3.515; & c = 0.537; \\ m = 1.1314; & s(1 + c) = 5.863; \end{array}$$

$$\left[\begin{array}{ccc} 6.815k & \vdots & -5.863\frac{k}{\ell} \\ \cdots\cdots\cdots\cdots & \\ -5.863\frac{k}{\ell} & \vdots & 10.363\frac{k}{\ell} \end{array}\right] \left[\begin{array}{c} \Theta_B \\ \cdots \\ e_B \end{array}\right] = \left[\begin{array}{c} 0.1875k \\ \cdots \\ .5091\frac{k}{\ell} \end{array}\right]$$

$$\Theta_B = 0.1395 \qquad e_B = 0.1251\ell$$

The second–order moment distribution is shown in Figure 2.13f.
Load–bearing capacity analysis of column AB:

$$\frac{N_{AB,I}}{N_{AB,II}} = \frac{1.36}{1.27} = 1.07$$

$$\frac{\psi M_{AB,I}}{M_{AB,II}} = \frac{0.466}{0.483} = 0.96$$

Therefore analysis a) gives 7% higher axial force than analysis b). Looking at the bending moments analysis a) gives an about 4% lower initial force than analysis b).

Analyzing the effect of simplification possibilities according to Section 2.3

$$H_j = 0.50906\frac{k}{\ell}$$

$\delta_j = 0.0977\ell$ calculated by first–order analysis is horizontal deflection, therefore $\Delta\alpha_j = 0.0977$ is the deflection increment.

Increment of the displacing force is:

$$\Delta H_j = N_{AB} \cdot \Delta\alpha_j = 0.13$$

its value related to the original displacing force is:

$$\frac{\Delta H_j}{Q_j} = \frac{0.13}{0.5} = 0.26 \approx 0.25$$

According to the foregoings it is sufficient to calculate the internal forces by first–order analysis, but magnified, by levels, with the displacing force

$$\overline{Q}_j = \frac{1}{1 - \frac{\Delta H_j}{Q_j}} Q_j = 0.688\frac{k}{\ell}$$

Solution of the equilibrium equations

$$e_B = 0.1237\ell$$
$$\Theta_B = 0.1329$$

$\overline{M}_{AB} = -0.4765k$ in the most loaded section of column AB, which is 2% lower than that of calculated by analysis b).

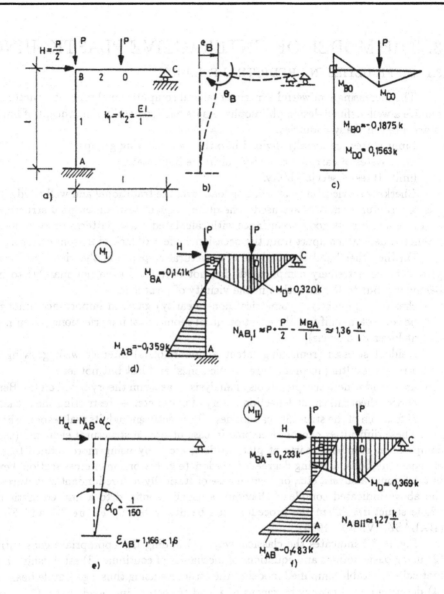

Figure 2.13

3. THE MODEL OF "INTERACTIVE PLASTIC HINGE"

3.1. DIFFICULTIES IN PREDICTING FAILURE

The increasingly powerful experimental and computational tools of structural design require a well–defined design philosophy. As its basis generally the concept of limit states is accepted in many countries.

Limit states are usually divided into the two following groups:
- limit states of carrying capacity /ultimate limit states/
- limit states of serviceability.

Checking serviceability at working load level the traditional and well tried out methods of structural analysis are used. The quite frequent tests on original structures show as a rule a relatively good accordance with calculated stress patterns or even better with predicted deflections apart from the occasional effect of lack of fit connections.

On the other hand the analysis of structural response in the vicinity of peak load proved to be extremely complicated, due not only (and even not mainly) to inelastic behaviour, but to the fact, that in the vicinity of peak load
- change in geometry (geometrical non–linearity) gains in importance among others because of magnifying the effect of initial geometrical imperfections (often negligible at lower load levels),
- residual stresses (remaining latent at lower loads) interact with growing active stresses resulting in premature plastic zones, and last but not least
- usual and widely accepted tools of analysis — as beam theory based on the Bernoulli–Navier theorem; small deflection theory of plates etc. — restricting the actual degree of freedom of the structure cannot describe exactly enough its real response at failure.

These difficulties can be overcome in case of simple structural elements (separated compression members, parts of plate girders, etc.) by using more refined (e.g. finite element) methods, allowing degrees of freedom (e.g. distortion of cross sections) excluded in traditional analysis, etc., or even in case of statically non-redundant structures, where the above indicated complex behaviour is usually confined to a limited section of the whole structure. Then the procedure can be illustrated by Figure 3.1 and Figure 3.2 (HALÁSZ, IVÁNYI 1985).

Figure 3.1 indicates the classical way: (1) finding the appropriate constititive law; (2) using basic notions and equations of mechanics of continua; (3) establishing a mathematically treatable simplified model of the structure using thus e.g. simple beam theory; (4) describing load history by means of a load-trajectory in a load-space; (5) computing the corresponding change of primary parameters (trajectory of load-actions, stresses, deflections) describing structural response; (6) selecting out of them the so-called quality parameters (BOLOTIN 1952) (e.g. maximum moments or stresses) playing decisive role in judging the onset of limit states; which again can be defined by the intersection of the trajectory of quality parameters with a given limit surface.

If the simplified model is not elaborate enough to reflect real structural behaviour, a secondary, more detailed local model is inserted (see Figure 3.2) to depict the mostly critical part of the structure, by which more realistic quality parameters (and limit surface) can be deduced from the already known primary parameters.

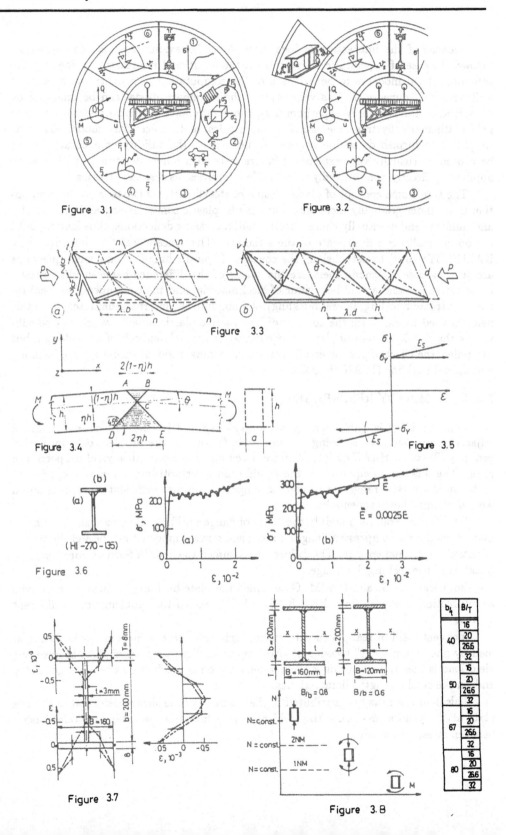

Figure 3.1

Figure 3.2

Figure 3.3

Figure 3.4

Figure 3.5

Figure 3.6

Figure 3.7

Figure 3.8

Because of the interaction of local and global behaviour this pattern cannot be followed in case of hyperstatic structures, as the additional information gained by the secondary, local model is to be fed back to the computation of primary parameters as well. For this purpose, if – as very often – the secondary model can be analysed by numerical methods or only experimentally, the results have either be re–interpreted to gain mathematically treatable, simple enough rules, or the secondary model has to be simplified to furnish digestible results. In both cases the validity or accuracy has to be proved by (usually very expensive) failure tests with full–scale strucures. The same applies for directly non or hardly measurable quantities, as residual stresses.

The traditional concept of plastic design of steel structures is based on the assumption that, under gradually increasing static loads, plastic zones develop and grow in size and number, and eventually cause unrestricted, increasing deflections; thus loading until the onset of ultimate limit state of the structure. The concept was first introduced by KAZINCZY (1914) by establishing the concept of "plastic hinge". Some basic questions are still discussed. Among them are the effects of the difference between ideal–plastic constitutive law and actual behaviour of steel material and the consequence of local instability (plate buckling; lateral buckling). Joining into the international research in this field we tried to introduce the concept of "interactive plastic hinge", which can be substitute the classic concept of plastic hinge in the traditional methods of limit design, but can reflect the effect of phenomena like strain-hardening, residual stresses, plate buckling and lateral buckling (IVÁNYI 1983).

3.2. PRELIMINARY RESEARCHES

Plate buckling is influenced by plate geometry and by the supporting action of adjacent plate parts, these being the two basic factors for the problem of deformation capacity. Tests by HAAIJER (1957) were based on the discontinuos yield properties of steel. The differential equation of the equilibrium of orthotropic plates was applied to plates in the plastic range with different supports; various coefficients in the equation were determined experimentally.

LAY (1965) examined mainly the effect of flange buckling. He established a mathematical model for compressed flanges under yield stress supported by the web by means of fictitous torsional springs. The failure was assumed to occur in form of pure torsional buckling of the rectangular flange.

Both Haaijer, G. and Lay, M. G. assumed the plate buckling problem to be solved as bifurcation of equilibrium. BEN KATO (1965) tackled the problem from a different aspect.

Incipient yield of real plates with rather high width-to-thickness ratios is accompanied by fine "crumplings" (waviness). No drop in load capacity takes place at once but yield load is maintained during a certain deformation depending on the plate geometry and on the steel strength characteristics.

This behaviour may be attributed to the possibility of arbitrary deformations in the plate at the yield stress, since the yield plateau exhibits no unambiguous relationship between stress and strain.

"Crumpling" itself may be considered as a yield mechanism where in the plastic hinges not constant but increasing "ultimate moments" develop due to strain hardening. Load capacity — or rather deformation capacity — gets exhausted where the quoted "ultimate moments" ceases to increase. This was assumed by Ben Kato to occur at the ultimate tensile stress rather than at the yield stress in the plastic hinges. This approach permits to establish a relation between maximum compression and plate geometry. His tests involved separate analysis of flanges and web plate of the I-section, assuming separate yield mechanism for each constituting plate, omitting the effect of elastic deformations as well. For the yield mechanism, the strain energy consisting of the bending moment and shear force energies at the plastic hinges, and the potential energy, product of compressive force by the compression, where established.

His relationship permit to put restriction on plate geometry ensuring adequate deformation capacity for the member.

Shapes assumed for the yield mechanism of the flanges and web are seen in Figure 3.3a and 3.3b, respectively.

3.3. BASIC METHOD AND A SIMPLE EXAMPLE

In the course of plate experiments, if the width-to-thickness ratio is high, the plate does not lose its load bearing capacity with the development of plastic deformations but is able to take the load causing yield until a deformation characteristic of the plate occurs and is even able to take a small increase in load. In the course of the process "crumplings" (bucklings) can be observed on the plate surface. These "crumplings" form a yield mechanism with the plastic moments acting in the linear plastic hinges (peaks of waves) not constant but ever increasing due to strain-hardening. The yield mechanism formed by the "crumplings" extends to the component plates of the bar. The description of its behaviour is obtained, from among the extreme values theorems of plasticity, with the aid of the *theorem of kinematics.*

Thus, in the course of our investigations an upper limit of the load bearing has been determined, however, to be able to assess the results, the following have to be taken into consideration: on one hand the yield mechanisms are taken into account through the "crumpling" forms determined by experimental results and on the other hand, the results of theoretical investigations are compared with those of the experimental investigations.

Thus, a yield mechanism will be adopted such as to correspond to the geometrical conditions and the assumed yield criterion. In addition to Ben Kato, this method has been applied by ALEXANDER (1960), PUGESLEY and MACAULAY (1960) for examining cylindrical shells, while by CLIMENHAGA and JOHNSON (1972) for the analysis of buckling due to the negative moment of the steel beam component of composite beams.

The method will be illustrated by a simple example (Figure 3.4). Let us take a bar of an isotropic, incompressible, rigid-plastic material (Figure 3.5), subject to the Tresca yield condition. Rigid and plastic zones include an angle of 45° with the bar axis. For a small rotation Θ, in the plastic hinge (yield mechanism) deformations of identical magnitude but opposite sign develop in directions x and y, while incompressibility requires zero strain in direction z $(\varepsilon_z = 0)$. According to the yield criterion, stresses in directions

y and z are zero ($\sigma_y = \sigma_z = 0$) while in direction x they are equal to the yield point ($\sigma_x = \pm \sigma_Y$).

Centre of rotation C is located as a function of depth h.

Writing deformation energy of the plastic hinge (yield mechanism):

$$W_a = V \int_{(\varepsilon)} \sigma \, d\varepsilon$$

where

$$V = \left(\frac{2\eta^2 h^2}{2} + \frac{2(1-\eta)^2 h^2}{2} \right) a = \left(1 - 2\eta + 2\eta^2 \right) h^2 a.$$

$V = $ minimum for $\eta = 0.5$, hence

$$V = \frac{ah^2}{2}.$$

For a small rotation Θ, because of the 45° limit, the strain becomes

$$\varepsilon_x = \pm \frac{\Theta}{2}$$

$$W_a = V \int_0^{\varepsilon_x} \sigma_x \, d\varepsilon = \frac{ah^2}{2} \int_0^{\Theta/2} \sigma_Y \, d\varepsilon = \frac{ah^2}{2} \cdot \Theta \cdot \frac{\sigma_Y}{2}$$

The potential energy:

$$W_h = \int_{\Theta} M \, d\Theta$$

Equating deformation and potential energies:

$$\int_{\Theta} M \, d\Theta = \frac{ah^2}{4} \sigma_Y \Theta$$

Deriving this equation with respect to Θ:

$$M = \frac{ah^2}{4} \sigma_Y$$

In case of a strain-hardening material (Figure 3.5):

$$\sigma_x = \sigma_Y + \varepsilon_x E_s$$

Substituting these equations:

$$M = \frac{ah^2}{4} \left(\sigma_Y + \frac{1}{2} \Theta E_s \right)$$

The presented method suits also more complicated cases.

3.4. YIELD MECHANISM OF STRUCTURAL MEMBERS

3.41. Yield Mechanism Forms Based on Experimental Results

The different form of yield mechanisms can be determined on the basis of experimental results.

Experimental investigations are essentially expected partly to supply physical background (often inspiration) needed to establish models for theoretical examinations, and partly to delimit the range of validity.

Our experiments involved measurements to determine:

a) plastic material characteristics of steel;
b) residual deformations;
c) load carrying capacities of members in compression, bending and eccentric compression.

ad a) Plastic material characteristics of steel have been determined in tensile tests on a total of 76 standard specimens of different plate thicknesses. Material testing results for specimen No 21 are seen in Figure 3.6.

ad b) Residual strains or stresses of welded I-sections have been determined by the sectioning method. Residual strains affect both strength and stability phenomena. Tests comprised determination of residual deformations of 32 different cross sections in all. Distribution of residual stresses in specimen No 21 is seen in Figure 3.7.

ad c) Load carrying capacities of members in compression, bending and eccentric compression have been tested by experimental methods. The testing program covered I-sections of two different outlines ($B/b = 0.6; 0.8$). Either of the two cases had four different web slenderness, each of them with four flange widths (Figure 3.8). Thereby a series comprised 32 specimens in all, each type was exposed to four different distributions of stresses. The experimental program involved testing of 132 specimens.

The yield mechanism forms of an I-section bar can be classified according to the following criteria:

a.) according to the manner of loading,
b.) according to the positions of the intersecting lines of the web and the flanges, the so-called "throat-lines", thus
    bi.) the evolving formation is called a planar yield mechanism if the two "throat-lines" are in the same plane also after the development of the yield mechanism.
    bii.) the evolving formation is called a spatial yield mechanism if the two "throat-lines" are not in the same plane after the development of the yield mechanism.

3.411. Yield mechanism of compression members

(i) Planar yield mechanism
The buckled form of the compression member and the formation of the chosen yield mechanism is shown in Figure 3.9.

Plastic deformation occurs in the shaded regions.

As an effect of compressive force N a compression part develops.

The symbol of the yield mechanism is $(N)_P$ where P stands for the planar yield mechanism formation.

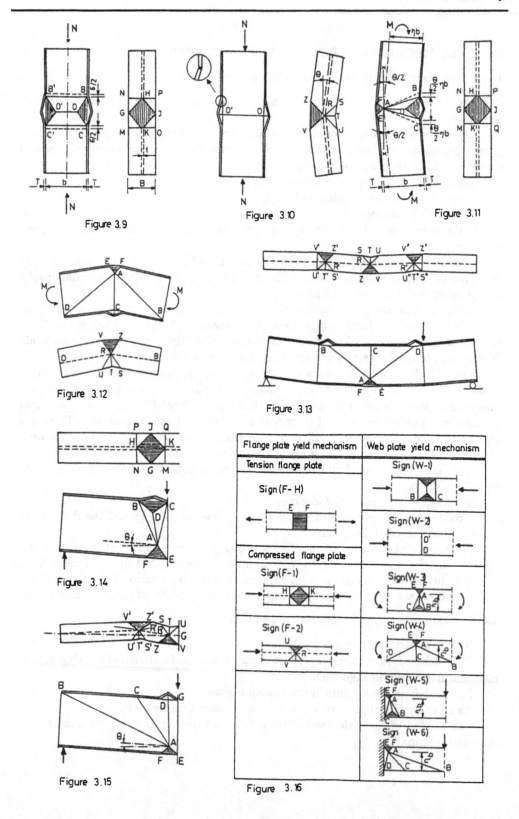

Figure 3.9

Figure 3.10

Figure 3.11

Figure 3.12

Figure 3.13

Figure 3.14

Figure 3.15

Figure 3.16

Flange plate yield mechanism	Web plate yield mechanism
Tension flange plate	Sign (W-1)
Sign (F-H)	
Compressed flange plate	Sign (W-2)
Sign (F-1)	Sign (W-3)
Sign (F-2)	Sign (W-4)
	Sign (W-5)
	Sign (W-6)

(ii) Spatial yield mechanism

The formation of the spatial yield mechanism for a compression member is shown in Figure 3.10. The ends of the member are assumed to be hinge–supported in both main inertia directions. The yield mechanism that has occurred is called spatial, the phenomenon models the planar buckling of the compression member or the buckling of the component plates in the course of buckling.

The symbol of the yield mechanism is $(N)_S$ where S stands for the spatial yield mechanism formation.

3.412. Yield mechanisms of bent members

a.) In case of a bending moment constant along the members axis:

ai) Planar yield mechanism

The buckled form of the bent specimen and the chosen yield mechanism formation is shown in Figure 3.11. As an effect of moment M a rotation Θ develops.

As an effect M, tension and compression parts develop.

The symbol of the yield mechanism is $(MC)_P$, where C stands for the constant bending moment.

aii) Spatial yield mechanism

The form of the spatial yield mechanism in the case of a bent rod is indicated in Figure 3.12. The rod ends are assumed to be hinge–supported in both main inertia directions. The yield mechanism models the buckling of the component plates of the bent member, the lateral buckling of the beams as well as their interaction.

The buckled formation of a simple–supported beam specimen and the chosen yield mechanism is shown in Figure 3.13.

In case of the yield mechanism formations in Figure 3.13 the effect of neighbouring supports (the effect of ribs) has also been taken into account. As an effect of the moment, a rotation Θ develops.

The symbol of the yield mechanism is $(MC)_S$.

b.) In case of a bending moment varying along the rod axis.

In case of a varying bending moment along the member axis it is assumed that the "crack" of the web plate of the I-section in the cross section of the concentrated force is hindered by the suitable thickness of the web plate or by the ribs.

CLIMENHAGA and JOHNSON (1972) assumed yield mechanism forms similar to those introduced in point (b) for the investigation of bucklings occuring at the steel beam parts of composite steel–concrete construction.

bi) Planar yield mechanism

The buckled form of a bent specimen and the selected yield mechanism is shown in Figure 3.14. As an effect of the moment a rotation Θ develops.

Because of the clamping of the cross-section EC, the yield mechanism loses its symmetric character.

The symbol of the yield mechanism is $(MV)_P$ where V stands for the varying moment.

bii) Spatial yield mechanism

The form of the spatial yield mechanism in case of a bending moment varying along the rod axis is shown in Figure 3.15.

As an effect of the moment a rotation Θ develops.

The symbol of the yield mechanism is $(MV)_S$.

3.42. Yield Mechanism of the Component Plates of an I-section Member

Yield mechanism formations have been determined for different stresses. On the basis of the experimental results it is expedient to decompose these yield mechanism formations to the yield mechanism formations of the component plates of an I-section rod as certain component plate formations appear in other yield mechanism, too.

To classify the yield mechanisms of component plates, the following division has been used:

a.) flange plate, if the plate is supported along the central line,

b.) web plate, if the plate is supported at the unloaded ends.

Figure 3.16 shows the yield mechanisms of the component plates where F is the flange plate, W is the web plate, the odd numbers refer to the planar yield mechanisms while the even ones to the spatial yield mechanisms.

3.43. "Joining" the Yield Mechanisms of Component Plates

The "joining" of the yield mechanisms of component plates depends on the positions of the so-called "throat-lines" of the yield mechanism chosen on the basis of the experimental results.

In cases pertaining to planar yield mechanisms this "joining" is to be realized in a linear way, with a linear plastic hinge: the length of the linear plastic hinge is governed — due to the properties of the chosen yield mechanism — by the length of the yield mechanism of the compression flange plate (F-1). In case of spatial yield mechanisms the "joining" should be realized in one point or more points.

The relationships of the component plate yield mechanisms and the "joining" of the component plates have been given by IVÁNYI (1983); IVÁNYI (1979a and 1979b) show the basic relationships of partial cases.

In addititon to the above model, a suitable yield mechanism was considered for cyclic bending, such as to correspond to the geometrical conditions and the assumed yield criterion (Figure 3.17) (GIONCU et al., 1989).

The cyclic load–deflection curve of beam S-3 is shown in Figure 3.18.

3.44. Yield Mechanism Curves of Compression Members

The yield mechanism adopted according to experimental data is seen in Figure 3.19.

Shaded areas undergo plastic deformation; if their — undeformed — boundaries include an angle of 45° with bar axis, then the model may be applied, valid also for the heavy lines in the figure, considered as linear plastic hinges. Yield points of web and flanges are assumed not be equal but in any case the strain–hardening modulus is E_S.

Plastic strain in the web plate develops in zones BCD and $B'C'D'$, with the development of plastic hinges BD, $B'D'$, CD, $C'D'$, BB', CC' and DD'.

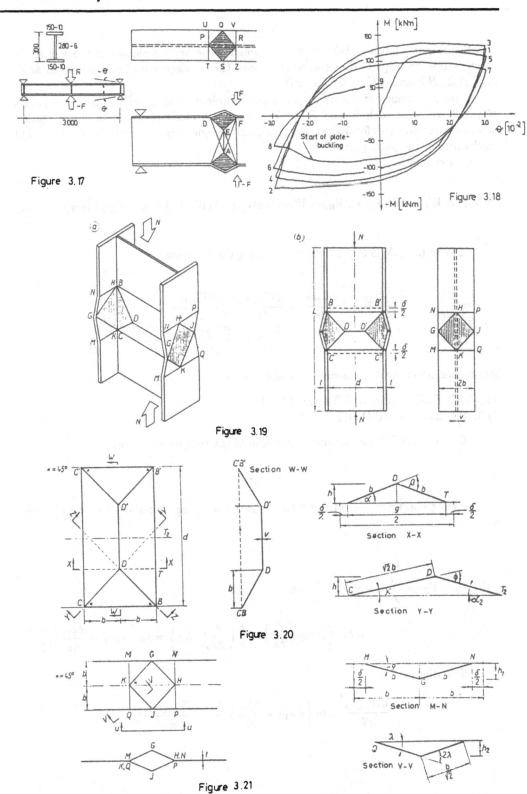

Figure 3.17

Start of plate-
buckling

Figure 3.18

Figure 3.19

Figure 3.20

Figure 3.21

In the web plate (its behaviour is assumed to be symmetric about the bar axis), plastic strain develops in the zone $KGHJ$, with the development of plastic hinges MK, KQ, NH, HP and GK, GH, JK, JH.

Yield mechanism of the web plate is compatible with those of the flanges if points B and C coincide with H and K, making the yield mechanism of the entire section determinate. Now, deformation and potential energies may be written for the yield mechanism of the plate parts.

As a conclusion:

$$\int_\delta N\, d\delta = W_a = 2W_{BCD} + 2W_{BB'} + W_{DD'} + 4W_{BD} + 2(W_{KJ} + W_{NH} + W_{GG} + W_{BC}) \qquad (1)$$

Value of force N is found by differentiating the equation:

$$N = \frac{dW_a}{d\delta} = 2\frac{dW_{BCD}}{d\delta} + 2\frac{dW_{BB'}}{d\delta} + \frac{dW_{DD'}}{d\delta} + 4\frac{dW_{BD}}{d\delta} +$$

$$+ 2\frac{dW_{KJ}}{d\delta} + 2\frac{dW_{NH}}{d\delta} + 2\frac{dW_{GG}}{d\delta} + 2\frac{dW_{BC}}{d\delta} \qquad (II)$$

Strain energies of yield mechanism of plate parts are:

N1. WEB PLATE (Figure 3.20a and 3.20b)
N1.1. Plastic zones BCD and $B'C'D'$

BCD and $B'C'D'$ are strains in direction of the compressive force:

$$\varepsilon = \frac{\delta}{2a}.$$

Utilizing characteristics of the rigid – strain-hardening material, the strain energy becomes:

$$W_{BCD} = V_{BCD} \int_0^{\delta/2b} (\sigma_{YW} + \varepsilon \cdot E_S)\, d\varepsilon =$$

$$= ta^2 \left\{ \sigma_{YW}\left(\frac{\delta}{2a}\right) + \frac{1}{2}\left(\frac{\delta}{2a}\right)^2 E_S \right\} = ta\left\{ \sigma_{YW} + \frac{\delta E_S}{4a} \right\}\frac{\delta}{2}.$$

Differentiating:

$$\frac{dW_{BCD}}{d\delta} = ta\left\{ \sigma_{YW} + \frac{\delta E_S}{2a} \right\}\frac{1}{2} = \frac{dW_{B'C'D'}}{d\delta}$$

N1.2. Plastic hinges BB' and CC'

First, the hinge rotation angle due to the assumed yield mechanism has to be determined.

Section $X - X$ indicated in Figure 3.20a is seen in Figure 3.20c showing hinges CC' and BB' to rotate by α:

$$g = 2a - \delta$$

$$\cos \alpha = \frac{g}{2a} = 1 - \frac{\delta}{2a}$$

$$\sin \alpha = \sqrt{1 - \left(1 - \frac{\delta}{2a}\right)^2}$$

$$\alpha = \arccos \left(1 - \frac{\delta}{2a}\right)$$

To obtain the derivatives of α with respect to δ:

$$\frac{d\alpha}{d\delta} = \frac{1}{2a} \cdot \frac{1}{\sin \alpha}$$

For a small rotation ε, the axial strain was as demonstrated in Section 3.3:

$$\varepsilon = \frac{\alpha}{2}$$

The strain energy:

$$W_{BB'} = V_{BB'} \int_0^{\alpha/2} (\sigma_Y w + \varepsilon\, E_S)\, d\varepsilon = \frac{1}{2}t^2 b \left\{\sigma_Y w + \frac{\alpha\, E_S}{4}\right\} \frac{\alpha}{2}$$

The derivate with respect to δ:

$$\frac{dW_{BB'}}{d\delta} = \frac{1}{4}t^2 b \left\{\sigma_Y w + \frac{\alpha\, E_S}{2}\right\} \left\{\frac{d\alpha}{d\delta}\right\} = \frac{dW_{CC'}}{d\delta}$$

N1.3. Plastic hinge DD'

The plastic hinge rotates by β, according to Figure 3.20c: $\beta = 2\alpha$ the strain being:

$$\varepsilon = \frac{\beta}{2} = \alpha$$

and the strain energy:

$$W_{DD'} = V_{DD'} \int_0^{\alpha} (\sigma_Y w + \varepsilon\, E_S)\, d\varepsilon = \frac{1}{2}t^2 (b - 2a) \left\{\sigma_Y w + \frac{\alpha\, E_S}{2}\right\} \alpha$$

Its derivative with respect to δ:

$$\frac{dW_{DD'}}{d\delta} = \frac{1}{2}t^2(b-2a)\{\sigma_{YW} + \alpha\,E_S\}\left\{\frac{d\alpha}{d\delta}\right\}$$

N1.4. Plastic hinges BD; CD and $B'D'$; $C'D'$

Section $Y - Y$ and $Z - Z$ in Figure 3.20a are seen in Figure 3.20d where the plastic hinge rotates by:

$$\Phi = \kappa + \alpha_2$$

Due to geometry: $\kappa = \alpha_2$

$$\sin\kappa = \frac{h}{\sqrt{2}a} = \frac{\sin\alpha}{\sqrt{2}}$$

since $h = a\sin\alpha$ (from section $X - X$):

$$\Phi = 2\arcsin\left\{\frac{\sin\alpha}{\sqrt{2}}\right\}$$

In need of the derivative of $\sin\alpha$ with respect to δ:

$$\frac{d(\sin\alpha)}{d\delta} = \frac{\cos\alpha}{\sin\alpha} \cdot \frac{1}{2a}$$

and the derivate of Φ with respect to δ:

$$\frac{d\Phi}{d\delta} = \frac{1}{\sqrt{2}} \cdot \frac{1}{\cos\alpha} \cdot \frac{d(\sin\alpha)}{d\delta}$$

Plastic hinge strain:

$$\varepsilon = \frac{\Phi}{2}$$

The strain energy:

$$W_{BD} = V_{BD}\int_0^{\Phi/2}(\sigma_{YW} + \varepsilon\,E_S)\,d\varepsilon = \frac{\sqrt{2}\,at^2}{2}\left\{\sigma_{YW} + \frac{\Phi E_S}{4}\right\}\frac{\Phi}{2}$$

The derivative with respect to δ:

$$\frac{dW_{BD}}{d\delta} = \frac{\sqrt{2}}{4}at^2\left\{\sigma_{YW} + \frac{\Phi E_S}{2}\right\}\left\{\frac{d\Phi}{d\delta}\right\} = \frac{dW_{CD}}{d\delta} = \frac{dW_{B'D'}}{d\delta} = \frac{dW_{C'D'}}{d\delta}$$

N2. FLANGE PLATE (Figure 3.21)

N2.1. Plastic zone $KGHJ$:

The strain energy:

$$\varepsilon = \frac{\delta}{2a}$$

The strain energy:

$$W_{KJ} = V_{KJ} \int\limits_0^{\delta/2a} (\sigma_{YW} + \varepsilon\, E_S)\, d\varepsilon = 2a^2 T \left\{ \sigma_{YW} \left(\frac{\delta}{2a} \right) + \frac{1}{2} \left(\frac{\delta}{2a} \right)^2 E_S \right\} =$$

$$= aT \left\{ \sigma_{Yf} + \frac{\delta E_S}{2a} \right\}$$

Differentiating:

$$\frac{dW_{KJ}}{d\delta} = aT \left\{ \sigma_{Yf} + \frac{\delta\, E_S}{2a} \right\}$$

N2.2. Plastic hinges NH, HP, QK and KM

Section $U - U$ in Figure 3.21a is seen in Figure 3.21b where plastic hinges NH, HP, QK and KM rotate by:

$$h_1 = \sqrt{a\delta - \frac{\delta^2}{4}}$$

$$\sin \rho = \frac{h_1}{a} = \frac{\sqrt{a\delta - \frac{\delta^2}{4}}}{a}$$

$$\rho = \arcsin \left\{ \frac{\sqrt{a\delta - \frac{\delta^2}{4}}}{a} \right\}$$

The derivate of ρ with respect to δ:

$$\frac{d\rho}{d\delta} = \frac{1}{\cos \rho} \cdot \frac{1}{2a} \cdot \frac{a - \frac{\delta}{2}}{\sqrt{a\delta - \frac{\delta^2}{4}}}$$

Strain in the plastic hinge:

$$\varepsilon = \frac{\rho}{2}$$

The strain energy:

$$W_{NH} = V_{NH} \int\limits_0^{\rho/2} (\sigma_{Yf} + \varepsilon\, E_S)\, d\varepsilon = 2T^2 a \left\{ \sigma_{Yf} + \frac{\rho E_S}{4} \right\} \frac{\rho}{2}$$

The derivative with respect to δ:

$$\frac{dW_{NH}}{d\delta} = aT^2 \left\{ \sigma_{Yf} + \frac{\rho E_S}{2a} \right\} \left\{ \frac{d\rho}{d\delta} \right\} = \frac{dW_{HP}}{d\delta} = \frac{dW_{QK}}{d\delta} = \frac{dW_{KM}}{d\delta}$$

N2.3. Plastic hinges GH, HJ, JK and KG

Section $V - V$ in Figure 3.21a is seen in Figure 3.21c with the plastic hinge rotating by:

$$2\lambda$$

Due to geometry:

$$h_2 = \frac{h_1}{2}$$

$$\sin \lambda = \frac{\sqrt{2} h_2}{a} = \frac{\sqrt{2}}{2} \sin \rho$$

$$\lambda = \arcsin \left\{ \frac{\sqrt{2}}{2} \sin \rho \right\}$$

The derivate of λ with respect to ρ:

$$\frac{d\lambda}{d\delta} = \frac{1}{\cos \lambda} \cdot \frac{\sqrt{2}}{2} \cos \rho \left\{ \frac{d\rho}{d\delta} \right\}$$

Plastic hinge strain:

$$\varepsilon = \frac{2\lambda}{2} = \lambda$$

Strain energy for the four hinges:

$$W_{GG} = V_{GG} \int_0^\lambda (\sigma_{Yf} + \varepsilon E_S) \, d\varepsilon = 2\sqrt{2} aT^2 \left\{ \sigma_{Yf} + \frac{\lambda E_S}{2} \right\} \cdot \lambda$$

Derivative with respect to δ:

$$\frac{dW_{GG}}{d\delta} = 2\sqrt{2} aT^2 \left\{ \sigma_{Yf} + \frac{\lambda E_S}{2} \right\} \left\{ \frac{d\lambda}{d\delta} \right\}$$

N3. Plastic hinges BC and $B'C'$

Plastic hinge developing between the web and the flange — or assumed in the web plate — is seen in Figure 3.22.

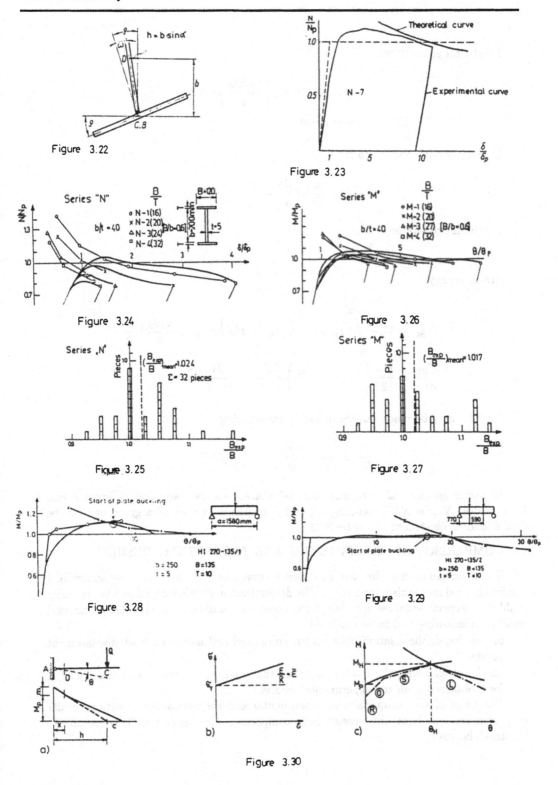

Figure 3.22

Figure 3.23

Figure 3.24

Figure 3.26

Figure 3.25

Figure 3.27

Figure 3.28

Figure 3.29

Figure 3.30

Plastic hinge rotation:

$$\omega = \rho - \arcsin \left\{ \frac{a \sin \alpha}{a} \right\}$$

$$\omega = \rho - \alpha$$

Derivative with respect to δ:

$$\frac{d\omega}{d\delta} = \frac{d\rho}{d\delta} - \frac{d\alpha}{d\delta}$$

Plastic hinge strain:

$$\varepsilon = \frac{\omega}{2}$$

Strain energy:

$$W_{BC} = V_{BC} \int_0^{\omega/2} (\sigma_Y w + \varepsilon E_S)\, d\varepsilon = t^2 a \left\{ \sigma_Y w + \frac{\omega E_S}{4} \right\} \frac{\omega}{2}$$

$$\frac{dW_{BC}}{d\delta} = \frac{t^2 a}{2} \left\{ \sigma_Y w + \frac{\omega E_S}{2} \right\} \left\{ \frac{d\omega}{d\delta} \right\} = \frac{dW_{B'C'}}{d\delta}$$

Eq. (II) may be made dimensionless by introducing:

$$\frac{N}{N_P} = \frac{1}{N_P} \cdot \frac{dW_a}{d\delta} \qquad (III)$$

Knowing geometrical data and material characteristics, assuming different δ and δ/δ_Y values, N and N/N_p defining the yield mechanism curves of a given bar can be determined as illustrated in Figure 3.23.

3.5. COMPARISON OF EXPERIMENTAL AND THEORETICAL RESULTS

The determination of the yield mechanism curve has been carried out by assuming a particular yield mechanism formation. The determination of this formation was primarily enabled by experimental results, thus, for the comparison of theoretical and experimental results the following had to be analyzed:

- on one hand, the relation between the computed and measured load–displacement curves,
- on the other hand, the relation between the assumed formation and the buckling form obtained from the experimental results.

The basis of the comparison of experimental and theoretical investigation results were compression, bent and eccentrically compressed members as well as so–called "control"–beams.

3.51. Investigation of the Results of Compression Members

Because of the arrangement of the specimens, the end–support is clamped from the point of view of bending and torsion to the so–called "weak" axis and the evolving yield mechanism is a planar one, its symbol is $(N)_P$.

The four load–displacement curves determined on the basis of the experimental results are shown in Figure 3.24.

Here:

$$N_p = 2 \cdot B \cdot T \cdot \sigma_{YF} + b \cdot t \cdot \sigma_{YW} \qquad \delta = \frac{N_p \cdot L}{E \cdot A}$$

The relation and coincidence of the experimental and theoretical results is acceptable.

Figure 3.25 indicates the results of measurement of forms evolving as an effect of plate buckling that served to disclose the form of buckling of the flange plate. The bucklings of the component plates are in interaction with each other and thus it is sufficient to investigate the buckling formation of the flange plate. The assumed buckling form of the flange plates has a length B the diagram compares the measured buckling wave lengths $(B)_{exp}$ to the width B of the flange plate. It can be observed that the assumed yield mechanism formation coincides with that obtained from the experimental results.

3.52. Investigation of the Results of Bent Specimens

The buckling investigation of component plates of bent beams can also be carried out with the aid of the yield mechanism curve.

Because of the arrangement of the specimens the evolving yield mechanism is a planar one, its symbol is $(M)_P$.

The load–displacement curves determined on the basis of the experimental results fit to the curve determined with the aid of the yield mechanism (see Figure 3.26).

Here:

$$\Theta_p = \frac{M_p \cdot L}{2 \cdot E \cdot I_z}$$

Figure 3.27 shows the results of measurement of the form developing as an effect of plate buckling. The diagram contains the results of measurements realized on the buckling form of the compression flange plate and it can be seen that the assumed yield mechanism formation coincides with the obtained from the measurement results.

3.53. Investigation of the Results of Welded I–section Beams ("Control" Beams)

3.531. Investigation of the results of beams bent with varying moment

In case of a varying moment diagram plate buckling appear first of all. The beam section has been selected to be short enough to obtain a planar yield mechanism, i.e. the so–called "throat–lines" are in the same plane after loading, too. The symbol of the yield mechanism is $(MV)_P$.

Figure 3.11 indicates the yield mechanism formations while Figure 3.28 indicates the theoretical and experimental results (load–displacement curves).

3.532. Investigation of the results of beams bent with constant moment

In case of beams bent with a constant moment, the so–called beam rotation occurs, first of all, because of the support conditions, namely the two "throat–lines" were not in the same plane after loading and the cause of the final deterioration was plate buckling in this case, too.

Figure 3.13 shows the selected yield mechanism formation, its symbol is $(MC)_S$.

Figure 3.29 indicates the experimental results as well as the results of theoretical investigations obtained by assuming a yield mechanism.

It can be observed that the yield mechanism curves determined on the basis of the experimental results describe well the "descending" part of the load–displacement curve and also that the yield mechanism formation coicides with the one obtained from the experimental results. Hence, the yield mechanism curves are suitable to determine the stress–displacement relationships of the member elements by taking into account both effect of plate buckling and lateral buckling.

3.6. MODEL OF THE INTERACTIVE HINGE

The plastic load–bearing investigation assumes the development of rigid–ideally plastic hinges, however, the model describes the inelastic behaviour of steel structures but with major constraints and approximations. There are some effects with the consideration of which the behaviour of the steel material and the I-section member can be taken into account in a more realistic way.

i) When determining the load–displacement relationship of an I-section member, the symbol of the elastic state is E and if the so–called "rigid" state is assumed instead of the elastic one, the symbol of the rigid state is R.

ii) The effect of residual stress and deformation is characterized by a straight line for the sake of case of handling. The symbol used when taking the residual stress and deformation into consideration is O.

iii) Strain–hardening is one of the important features of the steel material, S indicates that it had been accounted for.

iv) In Section 3.5. the effect of buckling of the I-section member component plates on the rod element load–displacement relationship has been investigated, this is indicated by L.

The models that take the above effects into consideration at the investigation of load–displacement (relative displacement) relationship of I-section are called "interactive" ones.

The model of the interactive plastic hinge taking into consideration the effect of rigid — residual stress — strain hardening — plate bucklings can be described with the aid of the "equivalent beam length" suggested by HORNE (1960) (Figure 3.30a). The material model employed at the investigations is shown in Figure 3.30b. The effect of the residual stresses and deformations is substituted by a straight line. The effect of strain–hardening can be determined with the help of the rigid–hardening $(R - S)$ model. The buckling of the I-section member component plates is described by the yield mechanism curve which is substituted by a straight line.

Figure 3.30c indicates the load–displacement relationship belonging to the $(R$–O–S–$L)$ interactive plastic hinge. The substitution by straight lines is justified to simplify the investigations. In the $(R$–O–$S)$ sections the intersections are connected while in section L the moment-rotation relationship is substituted by a tangent that can be drawn at the apex.

4. PREDICTION OF ULTIMATE LOAD OF STEEL FRAMES

4.1. INTRODUCTION

Questions relating to the effects of the plastic properties of steel material on the load bearing capacity of steel structures have long been of interest to researchers, a number of problems have been solved. The application of the theory of plasticity to designing has been enabled by the introduction of plastic load capacity investigations. These take into account that in steel structures an increasing number of extensive plastic zones are brought about by gradually increasing so-called "static" load until in the end, at the limit of load bearing, the structure, without further increase in load, is able to undertake continuous displacements. Calculations of plastic load bearing capacity are based on the research work of Gábor KAZINCZY.

Several questions remained unanswered in connection with plastic load bearing capacity investigations. Thus, for instance, the difference between the behaviour of the ideally plastic model and of real steel material, as well as problems connected with loss of stability (plate buckling, flexural–torsional buckling). Extensive research, especially in the experimental field, tried to answer these questions determining the conditions the adherence to which allows the maintance of the validity of plastic load bearing capacity investigations.

In present designing practice the plastic load bearing capacity investigation is carried out and thereupon complementary investigations are made to check the effects of loss of stability (plate buckling, flexural buckling and flexural–torsional buckling, etc.). The need for a separative investigation is due to the lack of a plastic hinge model that could have reflected these effects.

Our aim is a theoretical and experimental investigation that studies steel structures in steps known from traditional methods with the aid of a hinge model suitable to describe "more refined" properties, to embrace more phenomena (strain–hardening, residual deformation, plate buckling, flexural–torsional buckling, etc.).

4.2. EFFECT OF LOCAL INSTABILITY

The final collapse of steel structures is mostly caused by instability phenomena (THÜRLIMANN 1960). These instability phenomena may be:
- disadvantageous change in the steel structure geometry,
- disadvantageous change in the cross section geometry.

The effect of disadvantageous change in structure geometry can be — traditionally — grouped in the field of plastic instability. It was Ottó HALÁSZ (1976) who treated the problem in a doctoral thesis and, over and above theoretical studies, he also introduced a method suitable for practical design work.

The disadvantageous changes in cross section geometry are mainly plate bucklings. Plate buckling causes a change in the behaviour of the plastic hinge, too, and thus for the plastic investigations of statically undeterminate steel structures, methods have to be elaborated that take also the effect of plate buckling into consideration.

The case of ideally plastic material is an assumed case and it should be taken while investigating the effects of the evolving plastic hinges, lest the previously formed plastic hinges should "close".

Studying the effect of strain–hardening and plate buckling one should keep in mind that the load–displacement diagram of the structure may be of an ascending type even if the characteristic curves of the given member section or sections are of a descending type in individual cross sections because of the effect of plate buckling.

In the theory of plasticity, when deriving the condition of plasticity or some other physical relationships, Drucker's postulate for stability is applied, by assuming stable materials (DRUCKER 1951).

It should be noted that Drucker's postulate is not a natural law but a criterion of classification (DRUCKER 1964), the materials very often do not correspond to the assumptions of stable materials, or structural elements may behave in an unstable way, while, at the same time, their material is of a stable state.

MAIER (1961) was the first to treat the problem of the effect of the unstable state of certain members on the behaviour of a triangulated structure. Again it was Maier who in 1966 re-introduced the subject and investigated a structure consisting of compressed members and rigid beams where the load–displacement diagram of individual members contained stable and unstable parts.

MAIER and DRUCKER (1966) re-examined the original Drucker postulate applied when determining the condition of plasticity since the original postulate is suitable for the determination of the convexity and normality of the condition of plasticity in case of stable materials only.

When studying the load bearing capacity of steel structures, the problem of unstable material or softening material, according to Drucker's postulate does not appear since the strain–hardening of the steel material may increase in a major way the plastic load bearing capacity of steel structure. However, as it has been known for a long time, the final collapse of steel structures is caused — in a high percentage of cases — by instability (plate buckling, flexural-torsional buckling) phenomena that may occur in the cross section or in a structural unit. (Figure 4.1)

Concerning steel structures the properties of plastic hinges over and above the usual elastic–ideally plastic–hardening behaviour may be complemented with the effect of instability (flexural–torsional buckling) developing in the given structural unit (environment of the plastic hinge).

This type of inelastic or interactive hinge describes the behaviour of the structural unit and at the same time, also satisfies the criteria of unstable or softening structural unit, according to Maier–Drucker's postulate.

When determinig the plastic load bearing capacity of steel structures the interactive hinge of softening has so far not been considered or applied. The effects of the stability phenomena causing the softening character (flexural–torsional buckling, plate buckling) can be taken into account indirectly with the aid of construction rules. In principle, mathematical programming allows the investigation of more complex steel structures,

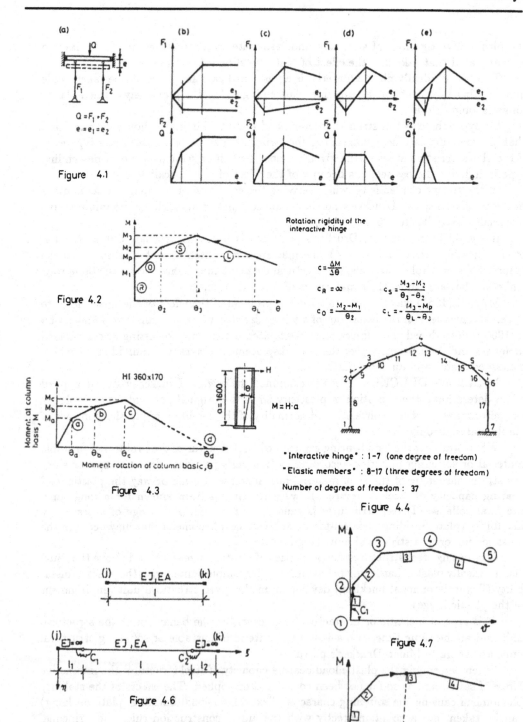

(a)

$Q = F_1 + F_2$
$e = e_1 = e_2$

Figure 4.1

Figure 4.2

Rotation rigidity of the interactive hinge

$$c = \frac{\Delta M}{\Delta \theta}$$

$$c_R = \infty \qquad c_S = \frac{M_3 - M_2}{\theta_3 - \theta_2}$$

$$c_O = \frac{M_2 - M_1}{\theta_2} \qquad c_L = -\frac{M_3 - M_P}{\theta_L - \theta_3}$$

HI 360×170

$M = H \cdot a$

Moment at column basis, M

Moment rotation of column basic, θ

Figure 4.3

"Interactive hinge" : 1-7 (one degree of freedom)
"Elastic members" : 8-17 (three degrees of freedom)
Number of degrees of freedom : 37

Figure 4.4

(j)   EJ,EA   (k)

Figure 4.6

Figure 4.7

Figure 4.8

too, however, it is less suitable for designing practice. The author (IVÁNYI 1980) has suggested a procedure that is taking into account the softening character of the inelastic hinge in the form of an interactive zone. The softening character of the interactive zone is caused by the buckling of the component plates, a phenomenon that can be studied with the help of the yield mechanism.

4.3. COMPUTER PROGRAMS FOR STEEL FRAMES

4.31. Analysis of Steel Frames with Conditional Joints

In engineering practice the plastic load bearing investigation of plane frame structures is carried out with the aid of the so-called plastic hinge. The investigation for computing statically undeterminate structures of elastic-plastic material, using the principles of matrix-calculation has been elaborated — among the first — by TASSI and RÓZSA (1958).

In the theory of structures the wide application of matrix calculation was introduced by SZABÓ and ROLLER (1971). It gave the possibility to treat the theoretical and computational investigations of rod structures in a uniform way, allowed the application of computer technique. The results most important for our investigations are those concerned with the application of kinematic load. In their work Kaliszky, S. and Kurutz, M. elaborated in detail the computerized computation of structures containig conditional joints with the kinematic load (KALISZKY 1978), (KURUTZ 1975). The procedure was based on the solution of structures by the displacement method. The investigation imitates the effect of the hinges with a rigidity changing stepwise as a kinematic load without alterning the original rigidity matrix of the structure continually softening due to the gradually developing interactive hinges.

The investigation of the change of state of the frame structure is carried out by matching two model-parts:
a.) an ideal linearly elastic member
b.) an interactive hinge: $R-O-S-L$ model containing the effects of rigid — residual stress — strain hardening — plate buckling.

The ideally elastic member was studied by usual methods (SZABÓ, ROLLER 1971).

4.311. The interactive hinge

(i) In Chapter 3. a description of the interactive hinge has been introduced that takes into consideration the effects of rigid — residual stress — strain-hardening — plate buckling and is called a $R-O-S-L$ hinge. The model of the $R-O-S-L$ hinge can be described with the aid of the "equivalent beam length". In the course of the investigations, the length "h" of the "equivalent beam" was determined from the moment diagram established during plastic load bearing investigations and this length h does not change — according to assumptions — with the increase of the load.

Figure 4.2 shows the model of the $R-O-S-L$ hinge. Rigidity of the hinge for different phases is expressed by the rigidity factor, $c = \Delta M/\Delta\Theta$.

Figure 4.2 also shows the kinematic loads valid for different rotation phases.

Compatibility relations of rotation sections:

R section:

$$c = \infty; \quad 0 \leq M \leq M_1; \quad \Theta = 0$$

$$\text{kinematic load}: t = 0$$

O section:

$$c = c_0; \quad M_1 \leq M \leq M_2; \quad 0 \leq \Theta \leq \Theta_2$$

$$\Theta = \frac{1}{c_0}(M - M_1); \quad 0 = -\Theta + \frac{1}{c_0}M + \left(-\frac{1}{c_0}M_1\right);$$

$$\text{kinematic load}: t = -\frac{1}{c_0}M_1$$

S section:

$$c = c_s; \quad M_2 \leq M \leq M_3; \quad \Theta_2 \leq \Theta \leq \Theta_3$$

$$\Theta - \Theta_2 = \frac{1}{c_s}(M - M_2); \quad 0 = -\Theta + \frac{1}{c_s}M + \left(\Theta_2 - \frac{1}{c_s}M_2\right)$$

$$\text{kinematic load}: t = \Theta_2 - \frac{1}{c_s}M_2$$

L section:

$$c = c_L; \quad M > M_3; \quad \Theta \geq \Theta_3$$

$$\Theta - \Theta_3 = \frac{1}{c_L}(M - M_3); \quad 0 = -\Theta + \frac{1}{c_L}M + \left(\Theta_3 - \frac{1}{c_L}M_3\right)$$

$$\text{kinematic load}: t = \Theta_3 - \frac{1}{c_L}M_3$$

It was assumed that interactive hinges do not "close" with an increase in load, no elastic–type unloading occurs (MAJID 1972).

(ii) Anchoring of the columns has not been developed in the form of a mechanical hinge or of complete clamping and therefore, primarily on the basis of experimental results, a "hinge" has been assumed in the cross sections of the anchoring of the columns where an increase in moment involves a rotation increasing in a changing measure (Figure 4.3).

4.312. Model of planar frame structures

The analysis of the Conder–system frame structure figuring in the experimental program has been selected as basis for our investigations. The selected model is shown in Figure 4.4.

A potentially occurring interactive hinge has been assumed in the cross section of possible maximal moment. Interactive hinges 1 through 7 have one degree of freedom while elastic members 8 through 17 have three degrees of freedom and thus the number of degrees of freedom of the model is 37. The state equation is:

$$\mathbf{G}^* \cdot \mathbf{s} + \mathbf{q} = 0$$
$$\mathbf{G} \cdot \mathbf{u} + \mathbf{F} \cdot \mathbf{s} + \mathbf{t} = 0$$

where:

G — geometric matrix

s — vector of load actions

q — load vector

u — displacement vector

t — vector of "kinematic load"

F — flexibility matrix

The solution of the equations is possible in the usual form (SZABÓ, ROLLER 1971).

4.313. Changes of state with a uni-parameter load system taken into account

The load vector q is increased by a value Δq from a suitable starting value till the moment M in the cross section of a potential interactive hinge attains the value M_1. At this point the relevant kinematic load t is inserted into the hinge and the hinge rotation rigidity is changed to the new value. Further increasing the amount of q, the modifications are undertaken until the determinant of the rigidity matrix becomes negative. This means that the size of the further steps is: $-\Delta q$.

In the course of the changes of state an interactive hinge may be encountered, where because of a plate buckling development, a descending curve characterizes the hinge behaviour, however, the frame structure can take up further loads.

The computations have been carried out with aid of a **PDP 11/34** computer.

4.32. Analysis of Steel Frames with Global Bar Elements

Recent development of engineering structures imposed to increase the accuracy of existing computing methods, to develop new methods for computing the novel structures — in fact, increasing dimensions and even complex structural design require lengthy, highly exact computations. The fast generalization of computers and rapid development of their technical parameters permitted to replace "manual computation" by computer analysis methods, procedures such as finite element, finite strip, bar system programs, shells, lattice girders, trusses and frameworks. Application of these methods has the double advantage of making the "inaccessible" structures computable: and of increasing the accuracy of determining the stress pattern of structures computed earlier in a different way — generally approximated — improving the economy of design.

Analyses concerned more exact determination of the stress pattern of steel plane frameworks in the plastic range (by a more realistic approximation of the given technical parameters) are possible in either of the following ways:

a) Non–linear model of the material and the bar

For structural analyses concerning static loads, essential strength and deformation characteristics of steel are supplied static characteristic curves obtained in laboratory tests. σ–ε diagrams of steel are commonly known but too intricate for practical computations, hence subject to different idealizations such as linear elastic, rigid-plastic, linear-elastic — linear-strain-hardening, etc. σ–ε curves. Selection of the material model is not determined by the required computation accuracy or the admissible computation complexity alone, since in several kinds of structures, certain circumstances prohibit the development of important deformations or yielding. In such cases, the load capacity increase between the tensile strength and the yield point of steel is a *priori* to be ignored.

Based on theoretical and experimental research (IVÁNYI 1983), introduction of the interactive plastic hinge model closer approximating the σ–ε diagram permitted to reckon with the elastic range, the effect of yield and strain hardening, of residual stresses of the structure, and member buckling phenomena.

b) Increased accuracy of geometry description, i.e. applying second- order theory (BAKSAY, IVÁNYI, PAPP 1985)

Also reckoning with geometrical non–linearity increases the accuracy, though at a significant increase of the volume of computations.

The program writes equilibrium equations for deformed rather than undeformed members; thus it ignores the principle of stiffening.

Computations according to the second–order theory permit to study how exactly the first–order theory describes the behaviour of the tested frameworks. Second–order computations are also justified by the important deformations of steel at and beyond the yield point, not to be omitted by applying the principle of stiffening.

Meeting these two kinds of requirements leads to a much more realistic image of the behaviour of structures.

4.321. Fundamentals of bar system computation

Matrix methods are available for computer determination of stresses in plane bar systems (SZABÓ, ROLLER 1971). These methods are relying either on the force or on the displacement method, this latter has been applied in the program.

A method well-known and proved in the literature has been followed. As a first step, the structure has been decomposed into nodes and rectilinear bars. Bar ends have three degrees of freedom in displacement: two displacements in the structure plane (u_x and u_y) and rotation in this plane (Φ_z). Accordingly, in any bar, three different stresses can be determined, i.e., due to normal force N, shear T, and bending moment M. In knowledge of bar stiffness matrices $\hat{K}$, overall stiffness matrix K of the structure can be compiled. Also reckoning with supporting conditions followed the known procedure: cancelling from K block rows and columns corresponding to support numerals, thus, reckoning with a stiffness matrix of reduced size. For a given load, also load vector q is known, hence, after solving the equilibrium equation system

$$K \cdot u = q \qquad \text{to} \qquad u = K^{-1} \cdot q,$$

reaction forces are obtained from the system of nodal displacements.

To increase the accuracy of geometry description, the second–order theory has been applied. Second–order computation of bars loaded by normal forces may apply the known stability functions (HORNE, MERCHANT 1965). As a consequence, bar stiffness matrix K will not be constant (like in the first–order theory) but function of the bar force parameter $\frac{N}{N_E}$ (where N_E is the Euler critical load).

Thereby all the computation becomes iterative, solutions $u_1, u_2, \ldots, u_n$ of the equilibrium equations system $K \cdot u = q$ converge after a number of computation cycles to the second–order solution.

4.322. The applied bar element

Simpler cases involve the bar element in Figure 4.5 permitting fast, easy computations mainly on an elastic material model. Bar stiffness matrix $\hat{K}$ is common knowledge; stiffness values are obtained by solving basic problems of hyperstatic beams.

Our goal seemed to be better achieved by applying complex bar element (Figure 4.6):

Two end parts of the bar, of lengths ℓ_1 and ℓ_2, are infinitely rigid (maybe $\ell_1 = \ell_2 = 0$); the middle part is elastic. Rigid and elastic bar parts are connected by a rotation spring each, able to rotation ϕ in the structure plane alone. Stiffnesses, i.e., spring constants are c_1 and c_2.

This assumption of the bar element had the following motivations:
- A plastic second–order analysis was the goal. Applying the usual bar elements, development of plastic hinges would require intercalation of a spring of stiffness c instead of a plastic hinge each hence to increase the stiffness matrix K of the structure. This assumption of the bar element avoids this computational problem, by permitting simulation of the development of plastic hinges by changing the stiffness of the inner spring.
- Assumption of an ideal clamping or ideal hinge exact enough for closer computations. This bar element permits also to reckon with elastic clamping.
- Structural design may produce infinitely stiff bar parts (gussets). Although it can be reckoned with by assuming EI sufficiently high for conventional bar elements, stiffness matrix $\hat{K}$ becomes thereby poorly conditioned, causing sometimes numerical errors. The presented bar element helps to avoid this problem.

Stiffness matrix K of the bar element can be written in knowledge of bar reaction forces resulting from unit nodal displacements.

4.323. Calculation of reaction forces of complex bars

Assumptions were the following:
- normal force arises exclusively from displacement along the bar axis (u_ξ^j and u_ξ^k);
- displacements normal to it (u_η^j and u_η^k), and bar end rotations (φ^j and φ^k) produce shear force and bending moment.

Thereby four computations are needed to see bar reaction forces arising from $\varphi^j = 1, \varphi^k = 1, u_\eta^j = 1, u_\eta^k = 1$.

At last, four vectors result:

$$\mathbf{s}_{\varphi j} = \begin{bmatrix} \overline{M}_k[\varphi_j] \\ \overline{M}_j[\varphi_j] \\ F[\varphi_j] \end{bmatrix} \qquad \mathbf{s}_{\varphi k} = \begin{bmatrix} \overline{M}_k[\varphi_k] \\ \overline{M}_j[\varphi_k] \\ F[\varphi_k] \end{bmatrix}$$

$$\mathbf{s}_{uj} = \begin{bmatrix} \overline{M}_k[u_j] \\ \overline{M}_j[u_j] \\ F[u_j] \end{bmatrix} \qquad \mathbf{s}_{uk} = \begin{bmatrix} \overline{M}_k[u_k] \\ \overline{M}_j[u_k] \\ F[u_k] \end{bmatrix}$$

where M_j and M_k are bending moments on bar ends j and k, and F is the shear force.

Physical purport of the stiffness matrix of the bar supplies matrix stiffness value true to sign:

$$\hat{\mathbf{K}} = \begin{bmatrix} \hat{K}_{AA} & \vdots & \hat{K}_{AB} \\ \cdots\cdots\cdots \\ \hat{K}_{BA} & \vdots & \hat{K}_{BB} \end{bmatrix} =$$

$$\begin{bmatrix} \frac{EA}{\ell} & 0 & 0 & \vdots & -\frac{EA}{\ell} & 0 & 0 \\ 0 & F[u_j] & F[\varphi_j] & \vdots & 0 & -F[u_k] & -F[\varphi_k] \\ 0 & \overline{M}_j[u_j] & \overline{M}_j[\varphi_j] & \vdots & 0 & \overline{M}_j[u_k] & \overline{M}_j[\varphi_k] \\ \cdots\cdots\cdots\cdots\cdots\cdots\cdots\cdots\cdots\cdots\cdots\cdots \\ -\frac{EA}{\ell} & 0 & 0 & \vdots & \frac{EA}{\ell} & 0 & 0 \\ 0 & -F[u_j] & -F[\varphi_j] & \vdots & 0 & F[u_k] & F[\varphi_k] \\ 0 & \overline{M}_k[u_j] & \overline{M}_k[\varphi_j] & \vdots & 0 & \overline{M}_k[u_k] & \overline{M}_k[\varphi_k] \end{bmatrix}$$

The vectors $\mathbf{s}_{\varphi j}, \mathbf{s}_{\varphi k}, \mathbf{s}_{uj}$ and $\mathbf{s}_{uk}$ are obtained by solving four equation systems:

(1)
$$\begin{bmatrix} 1 + \frac{s \cdot k}{c_2} & \frac{s \cdot c \cdot k}{c_1} & -\frac{s \cdot k \cdot \ell_2}{c_2} - \frac{s \cdot c \cdot k \cdot \ell_1}{c_1} - \ell_2 \\ \frac{s \cdot c \cdot k}{c_2} & 1 + \frac{s \cdot k}{c_1} & -\frac{s \cdot k \cdot \ell_1}{c_1} - \frac{s \cdot c \cdot k \cdot \ell_2}{c_2} - \ell_1 \\ s(1+c)\frac{k}{\ell} \cdot \frac{1}{c_2} & s(1+c)\frac{k}{\ell} \cdot \frac{1}{c_1} & 1 - s(1+c)\frac{k}{\ell}\left(\frac{\ell_2}{c_2} + \frac{\ell_1}{c_1}\right) \end{bmatrix} \begin{bmatrix} \overline{M}_k \\ \overline{M}_j \\ F \end{bmatrix} =$$

$$= \varphi_j \begin{bmatrix} s \cdot c \cdot k\left(1 - \frac{p \cdot \ell_1}{c_1}\right) + s(1+c)\frac{k}{\ell}\ell_1 \\ s \cdot k\left(1 - \frac{p \cdot \ell_1}{c_1}\right) + s(1+c)\frac{k}{\ell}\ell_1 - p\ell_1 \\ s(1+c)\frac{k}{\ell}\left(1 - \frac{p \cdot \ell_1}{c_1}\right) + \frac{2s(1+c)}{m} \cdot \frac{k}{\ell^2}\ell_1 \end{bmatrix}$$

$$\mathbf{F}^*_{\varphi j} \cdot \mathbf{s}_{\varphi j} = \varphi_j \cdot \mathbf{b}_{\varphi j}, \qquad \mathbf{s}_{\varphi j} = \varphi_j \cdot (\mathbf{F}^{*-1}_{\varphi j} \cdot \mathbf{b}_{\varphi j})$$

that is, $s_{\varphi j}$ is expressible from the equation in dependence of φ_j.

(2)
$$
\begin{bmatrix}
1+\frac{s\cdot k}{c_2} & \frac{s\cdot c\cdot k}{c_1} & \frac{s\cdot k\cdot \ell_2}{c_2}+\frac{s\cdot c\cdot k\cdot \ell_1}{c_1}+\ell_2 \\[2mm]
\frac{s\cdot c\cdot k}{c_2} & 1+\frac{s\cdot k}{c_1} & \frac{s\cdot k\cdot \ell_1}{c_1}+\frac{s\cdot c\cdot k\cdot \ell_2}{c_2}+\ell_1 \\[2mm]
-s(1+c)\frac{k}{\ell}\cdot \frac{1}{c_2} & -s(1+c)\frac{k}{\ell}\cdot \frac{1}{c_1} & 1-s(1+c)\frac{k}{\ell}\left(\frac{\ell_2}{c_2}+\frac{\ell_1}{c_1}\right)
\end{bmatrix}
\begin{bmatrix}
\overline{M}_k \\[2mm]
\overline{M}_j \\[2mm]
F
\end{bmatrix}
=
$$

$$
=\varphi_k
\begin{bmatrix}
s\cdot k\left(1-\frac{p\cdot \ell_2}{c_2}\right)+s(1+c)\frac{k}{\ell}\ell_2-p\ell_2 \\[2mm]
s\cdot c\cdot k\left(1-\frac{p\cdot \ell_2}{c_2}\right)+s(1+c)\frac{k}{\ell}\ell_2 \\[2mm]
-s(1+c)\frac{k}{\ell}\left(1-\frac{p\cdot \ell_2}{c_2}\right)-\frac{2s(1+c)}{m}\cdot \frac{k}{\ell^2}\ell_2
\end{bmatrix}
$$

$$
\mathbf{F}^*_{\varphi k}\cdot \mathbf{s}_{\varphi k}=\varphi_k\cdot \mathbf{b}_{\varphi k},\qquad \mathbf{s}_{\varphi k}=\varphi_k\cdot \left(\mathbf{F}^{*-1}_{\varphi k}\cdot \mathbf{b}_{\varphi k}\right)
$$

that is, $s_{\varphi k}$ is expressible from the equation in dependence of φ_k.

(3)
$$
\begin{bmatrix}
1+\frac{s\cdot k}{c_2} & \frac{s\cdot c\cdot k}{c_1} & -\frac{s\cdot k\cdot \ell_2}{c_2}-\frac{s\cdot c\cdot k\cdot \ell_1}{c_1}-\ell_2 \\[2mm]
\frac{s\cdot c\cdot k}{c_2} & 1+\frac{s\cdot k}{c_1} & -\frac{s\cdot k\cdot \ell_1}{c_1}-\frac{s\cdot c\cdot k\cdot \ell_2}{c_2}-\ell_1 \\[2mm]
s(1+c)\frac{k}{\ell}\cdot \frac{1}{c_2} & s(1+c)\frac{k}{\ell}\cdot \frac{1}{c_1} & 1-s(1+c)\frac{k}{\ell}\left(\frac{\ell_2}{c_2}+\frac{\ell_1}{c_1}\right)
\end{bmatrix}
\begin{bmatrix}
\overline{M}_k \\[2mm]
\overline{M}_j \\[2mm]
F
\end{bmatrix}
=
$$

$$
=u_j
\begin{bmatrix}
s(1+c)\frac{k}{\ell}\ell_1 \\[2mm]
s(1+c)\frac{k}{\ell}\ell_1 \\[2mm]
\frac{2s(1+c)}{m}\cdot \frac{k}{\ell^2}
\end{bmatrix}
$$

$$
\mathbf{F}^*_{uj}\cdot \mathbf{s}_{uj}=u_j\cdot \mathbf{b}_{uj},\qquad \mathbf{s}_{uj}=u_j\cdot \left(\mathbf{F}^{*-1}_{uj}\cdot \mathbf{b}_{uj}\right)
$$

that is, s_{uj} is expressible from the equation in dependence of u_j.

(4)
$$
\begin{bmatrix}
1+\frac{s\cdot k}{c_2} & \frac{s\cdot c\cdot k}{c_1} & \frac{s\cdot k\cdot \ell_2}{c_2}+\frac{s\cdot c\cdot k\cdot \ell_1}{c_1}+\ell_2 \\[2mm]
\frac{s\cdot c\cdot k}{c_2} & 1+\frac{s\cdot k}{c_1} & \frac{s\cdot k\cdot \ell_1}{c_1}+\frac{s\cdot c\cdot k\cdot \ell_2}{c_2}+\ell_1 \\[2mm]
-s(1+c)\frac{k}{\ell}\cdot \frac{1}{c_2} & -s(1+c)\frac{k}{\ell}\cdot \frac{1}{c_1} & 1-s(1+c)\frac{k}{\ell}\left(\frac{\ell_2}{c_2}+\frac{\ell_1}{c_1}\right)
\end{bmatrix}
\begin{bmatrix}
\overline{M}_k \\[2mm]
\overline{M}_j \\[2mm]
F
\end{bmatrix}
=
$$

$$
=u_k
\begin{bmatrix}
-s(1+c)\frac{k}{\ell}\ell_1 \\[2mm]
-s(1+c)\frac{k}{\ell}\ell_1 \\[2mm]
\frac{2s(1+c)}{m}\cdot \frac{k}{\ell^2}
\end{bmatrix}
$$

$$F^*_{uk} \cdot s_{uk} = \varphi_k \cdot b_{uk}, \qquad s_{uk} = \varphi_k \cdot \left(F^{*-1}_{uk} \cdot b_{uk}\right)$$

that is, s_{uk} is expressible from the equation in dependence of u_k.
Legend for the equation systems:

ℓ, ℓ_1 and ℓ_2 bar element length data

c_1 and c_2 torsional spring stiffness factor

$k = EI/\ell$ bar stiffness

p normal force on the bar

$s = s(\rho)$ stiffness (stability) function

A area of cross section

$$s(\rho) = \frac{(1 - 2\alpha \cdot \cot 2\alpha)\alpha}{\tan \alpha - \alpha} \qquad \text{where} \qquad \alpha = \frac{\pi}{2}\sqrt{\rho} \;\; \text{if} \;\; \rho = \frac{N}{N_E} \geq 0$$

$$s(\rho) = \frac{(1 - 2\gamma \cdot \coth 2\gamma)\gamma}{\tanh \gamma - \gamma} \qquad\qquad\qquad \gamma = \frac{\pi}{2}\sqrt{-\rho} \;\; \text{if} \;\; \rho < 0$$

$$\left(N_E = -\frac{\pi^2 EI}{\ell^2}\right)$$

$c(\rho) =$ carry-over function

$$c(\rho) = \frac{2\alpha - \sin 2\alpha}{\sin 2\alpha - 2\alpha \cdot \cos 2\alpha} \qquad \text{or} \qquad c(\rho) = \frac{2\gamma - \sinh 2\gamma}{\sinh 2\gamma - 2\gamma \cdot \cosh 2\gamma}$$

$$m = m(\rho) = \frac{2s(1+c)}{2s(1+c) - \pi^2 \rho}$$

4.324. Spring characteristics

Spring characteristics are of the general form in Figure 4.7. Sections have different spring constants $c = \frac{\Delta M}{\Delta \Theta}$ indicating the given section of the elasto–plastic behaviour or of the stability condition of the bar part. The characteristic is strictly monotonous for Θ but not for M. Namely there is a peak followed by a descending path of the curve.

Application of spring characteristics lessens the validity of the theorem that a hyperstatic beam with n redundances fails at the development of the $(n + 1)$-th plastic hinge.

It is true only for computations relying on rigid–plastic not strain–hardening models. It is both logically obvious and demonstrated by computations that plastic reserves after yield permit more than $n + 1$ interactive hinges to develop. Final failure depends on a complex interaction between structural design, loss of stability phenomena and yield mechanism.

4.325. The computation method

Also plastic analysis is set of computations to be considered as elastic where the structural problem is solved successive load increments. In each load increment, the program performs the following computations:

- elastic–type computation for a load increment of assumed value (calculating with the actual c value for each spring), determining the moment at each spring;
- examination of each spring for eventual spring characteristic changes, hence for the need of turning to a section of direction tangent;
- computation by linear interpolation or extrapolation of the least value of load increment (in second–order computation by several iterative cycles) leading exactly to the end of a straight section on a spring characteristic. Adding characteristics belonging to this load increment (e.g. moment level of springs, overall displacement vector of the structure, vector of normal load parameters of the bars, load increment value, etc.) to values obtained in the former step yields state characteristics for the complete load level,
- in the spring affected by the change, the program reduces the spring characteristic value according to the Θ–M function, underlying computation of the next load increment.

Details of the computation procedure have been described in (BAKSAI 1983).

This program lends itself to solve bar system with one–parameter loads. It has generally the consequence that spring rotations monotonously increase with the load level. It can, however, be realized that in certain structures the developing interactive hinges change the stress pattern in the structure so as to unload certain springs, reducing the arisen rotation. If at that place no interactive hinge has developed to then, hence section 1 of the spring characteristic prevails, also unloading follows according to spring constant c_1 corresponding to this section (Figure 4.8)

In section 2, 3, maybe 4, however, unloading follows (Figure 4.8).

Clearly, since an interactive hinge behaves elastically in opposite rotation. Thereby direction opposite to the former one, the needed data have to be modified accordingly, and load increment be recomputed.

In course of loading, the bar can carry further load, although with ever less increments, by developing futher interactive hinges. Meanwhile, however, stability loss phenomena arise at these spots. After a given level, loading cannot be further increased, thus, increasing displacements arise at decreasing load levels. This phenomena has to be reckoned with in computation, possible by examining the sign of the determined of stiffness matrix $\mathbf{K}$. Namely until $\det |\mathbf{K}| > 0$, the beam can be loaded further, while for $\det |\mathbf{K}| < 0$, there is unloading.

4.4. A SIMPLE APPROXIMATE METHOD

Numerous approximate engineering methods are introduced in the literature (HORNE, MORRIS 1981) from which as one of the possibilites we are going to deal with the extension of the Mechanism Curve Method. The Mechanism Curve Method — above the determination of the plastic load bearing capacity — can be applied to take the effect of finite deformations and strain hardening of steel into consideration.

4.41. Mechanism Curve Method

(i) Plastic collapse loads are idealizations of the failure loads of elastic-plastic structures. In these idealizations the collapse load refers to infinitely small differences from the undeformed states with infinitely small plastic hinge rotations.

Based on previous consideration an adequate yield mechanism (pattern of plastic hinges) is to be chosen (Figure 4.9). Denoting the displacements of the external forces in the yield mechanism by u_i and the hinge rotations by θ_j (Figure 4.9), the virtual work equation furnishes:

$$\lambda \sum_i Q_i u_i = \sum_j M_p \theta_j$$

This gives the collapse mechanism load factor λ_p (Figure 4.9).

(ii) In order to determine the rigid–plastic mechanism line it is necessary to follow the variation in the load factor λ at increasing finite values of the rotation ϕ. At finite deformations, the mechanism itself becomes non–linear, and consideration of the exact geometry changes becomes laborious. HORNE (1960) has shown that a simple treatment gives a value for λ which is correct up to the first power of ϕ. The work equation for the incremental deformation may be written as:

$$\lambda \sum_i Q_i \, du_i + \sum_k N_k L_k \phi_k \, d\phi_k = \sum_j M_{pj} \, d\theta_j$$

The second term of the equation is due to the additional external work arising from finite deformations.

The axial thrusts N_k may be obtained with sufficient accuracy by proportion from the values they have in the simple collapse state. The axial thrusts are $\lambda(N_{kp}/\lambda_p)$. A further approximation is that the total and incremental rotations and displacements are all in the same proportion as those for the same mechanism during an infinitely small deformation from the undeformed state. Hence

$$\frac{du_i}{u_i} = \frac{d\Phi_k}{\Phi_k} = \frac{d\Theta_j}{\Theta_j}$$

and the previous equation becomes

$$\lambda \left(\sum_i Q_i u_i + \sum \frac{N_{kp}}{\lambda_p} L_k \phi_k^2 \right) = \sum_j M_{pj} \Theta_j$$

This gives the relationship AC shown in Figure 4.9.

(iii) HORNE (1960) proposed the use of the simple rigid–plastic–rigid relationship in order to take into account the effect of strain–hardening on the collapse load of a structure.

Change of geometry due to elastic–plastic deformations tends to decrease the ultimate load bearing capacity of steel frames in comparison with the plastic collapse load. This tendency is counteracted by the strain–hardening properties of steel. The rigid–plastic–rigid theory of structural behaviour is found to be an adequate mean to assess

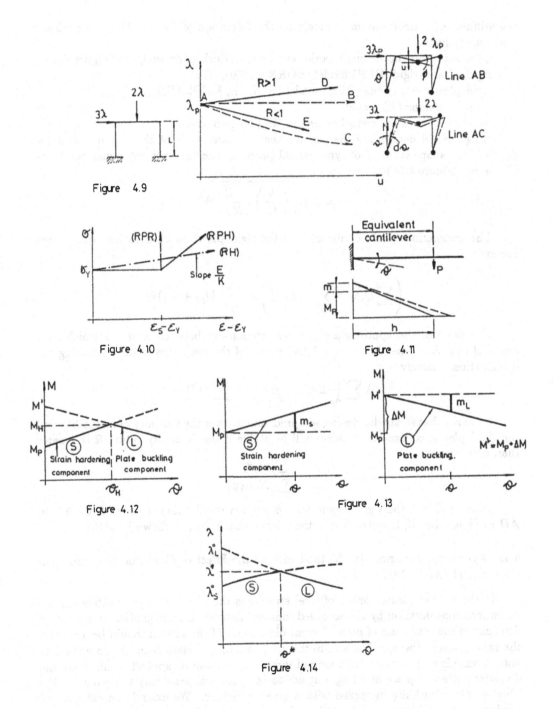

Figure 4.9

Figure 4.10

Figure 4.11

Figure 4.12

Figure 4.13

Figure 4.14

the stiffness of a structure immediately on the formation of the last hinge in a plastic hinge mechanism.

Different strain–hardening theories can be used during the analysis: (Figure 4.10)
- rigid–plastic–rigid (RPR) model (HORNE 1960),
- rigid–plastic–hardening (RPH) model (HORNE, MEDLAND 1966),
- rigid–hardening (RH) model.

This treatement uses the rigid–hardening (RH) hinge model.

Hinge model determines the end moments of cantilevers of the length h and the depth b, the shape factor f of symmetrical I-section, the end moment comes out to be $(M_p + m)$ (Figure 4.11)

$$m = M_p \sqrt{\left(\frac{b}{fh}\right) \cdot \frac{E}{K\sigma_Y} \Theta}$$

The summations could be included in the rigid–plastic work equation, which then becomes

$$\lambda \left(\sum_i Q_i u_i + \sum_k N_k L_k \phi_k^2 \right) = \sum_j (M_{pj} + m_j) \Theta_j$$

The slope of the approximately linear relationship between λ and u which as a result of equation depends on the relative values of the two extra terms containing the deformations, namely

$$\lambda \sum \frac{N}{\lambda_p} L \phi^2 \qquad \text{and} \qquad \sum m\Theta.$$

It follows that strain–hardening will predominate over the tendency of instability and the rigid–plastic collapse mechanism will be stable if the "stability ratio" R is greater than unit where

$$R = \frac{\sum_j m_j \theta_j}{\lambda_p \sum_k N_k L_k \phi_k^2}$$

Thus, if $R > 1$ the rigid–plastic load–deflection relationship is raised from AC to AD in Figure 4.9. If, however $R < 1$, the relationship drops, as shown by AE.

4.42. Approximate Engineering Method to Take the Effect of Plate Buckling into Consideration (IVÁNYI 1983)

In the field of plastic design of steel structures the effect of plate buckling can be taken into consideration by the so called indirect method. During analysis it should be determined that the ratio of plate element dimensions of the section should be less than the ratio given in the specification; in this case buckling of plate elements do not occur until mechanism formation. Such kind of direct method can be applied to eliminate the disturbing effect of plate buckling, but not to analyze — at least only to predict — the effect of plate buckling in regard with a given structure. We extend the category of hardening plastic hinges by taking the effect of plate buckling into consideration. Such hinge model can be the basis of an Approximate Engineering Method, that — without analyzing the full load history —, with simple methods can directly take the effect of plate buckling into consideration.

Figure 4.12 shows the linear interaction of moment–rotation of interactive hinge that contains the effects of strain hardening and plate buckling. The essence of Approximate Engineering Method is that the two effects are separated and the interactive hinge of the structure is put together from two separate components: (Figure 4.13)

(1) Strain-hardening component: (S)

(2) Plate buckling component: (L)

With the assumed two hinge components the values of the load parameter for the chosen mechanism of the framework can be determined as a function of finite deformations.

The expression for (1) strain-hardening component is:

$$\lambda_{(S)} \left(\sum Qu + \sum NL\phi^2 \right) = \sum M_p\theta + \sum m_S\theta$$

$$\lambda_{(S)} = \frac{\sum M_p\theta + \sum m_S\theta}{\sum Qu + \sum NL\phi^2}$$

To write down the expression for the (2) plate buckling component it should be assumed that the interactive hinge characteristic curve contains rigid, plate buckling effects, so rigid behaviour goes up to the value of $M' = M_p + \Delta M$ first, then a linearly decreasing change is taken into consideration due to the effect of plate buckling. Because of the shape of the characteristic curve belonging to the (2) plate buckling component, external and internal capacities and works are written similarly to the (1) strain–hardening component — except the sign of the increment $m\, d\theta$

$$\lambda_{(L)} \left(\sum Qu + \sum NL\phi^2 \right) = \sum M'\theta - \sum m_L\theta$$

$$\lambda_{(L)} = \frac{\sum M'\theta - \sum m_L\theta}{\sum Qu + \sum NL\phi^2}$$

Load parameter $\lambda_{(S)}$ takes the effect of strain–hardening into consideration, while load parameter $\lambda_{(L)}$ that of plate buckling.

From the displacement given by the intersection of the two curves; the reduction–like change of state is due to the effect of plate buckling (Figure 4.14). In connection with the results it should be emphasized, that — similarly to plastic load bearing capacity analysis — the expression — taking the two separate components into consideration — assumes the structure motionless till the moments M_P and M' in the hinges form.

The axial forces in bars are assumed to be proportional to the external loading. Equivalent cantilever length h for the interactive hinge can be determined by the moment diagram from plastic load bearing capacity analysis.

5. EVALUATION OF LOAD BEARING CAPACITY OF STEEL FRAMES

5.1. TEST PROGRAM

The experimental research project was carried out in the Laboratory of the Department of Steel Structures, Technical University, Budapest.

Three main series of experiments were covered by the test program.

The first part of the program contained the additional tests on stub columns, frame corners, plates elements, simple beams (HALÁSZ, IVÁNYI 1979). A brief summary of this serie is given in Figure 5.1.

5.11. Full-scale Tests of Frames

In the second part of the program the full–scale tests of frames have been included. Figure 5.2 gives a brief summary of the full–scale tests and dimensions of the specimens, indicating the loads and the characteristics of the loading process.

Test frame C–3 had rafters with a slope of 30% (16.7°), welded column sections (Figure 5.3a).

Rafter–to–column and mid–span connections have been end–plated ones with high–strength prestressed bolts (Figure 5.3b).

Different types of lateral supports were applied (Figure 5.4) to the frames to prevent or decrease out–of–plane displacements and/or rotations of selected sections.

Vertical loads at the purlin supports were applied to the upper flange of rafter, so web and bottom flange were not restrained laterally. To make horizontal displacement (sideway) unrestricted, jacks were fastened not directly to the floor–slab, but through a so–called gravity load simulator (Figure 5.5). This latter one consisted of three elements: two bars, and a rigid triangle. The two bars had pin–joints at both ends, resulting in a one–degree–of–freedom mechanism. Hydraulic jacks were joined to the rigid triangle. This mechanism produced a vertical load acting upon the intersection of the two bar axes. Characteristic curve of the simulator illustrating the ratio of horizontal and vertical load component as function of horizontal displacement is given in Figure 5.5.

A remark in connection with the character of loading seems rather important.

At mathematical (theoretical) investigations, virtual disturbances have been assumed for the equilibrium state analysis of the structure so that these disturbances do not influence loading (HOFF 1956). However, in case of experimental investigations, these disturbances are, quite naturally, real ones thus their effect does not only manifest itself on the structure but also in the loading system. Therefore the interaction of the structure and the loading system has also to be determined at experimental investigations.

The significant majority of steel structures in engineering practice is loaded with dead or gravity load, thus the highest point of the load–displacement diagram of the structure also indicates the loss of stable equilibrium state. However, this general observation has become a hindrance to the cognitive process, since, on the basis of the above observation, not only the complete structure but also individual structural elements have mainly been analyzed experimentally by gravity load. Therefore, according to the character

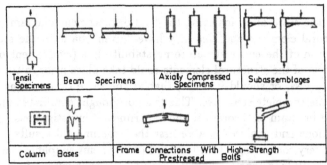

Figure 5.1

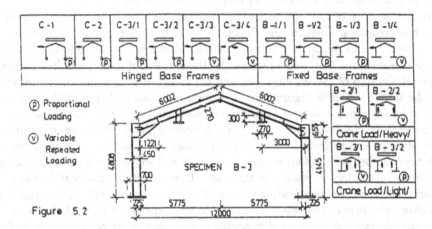

Figure 5.2

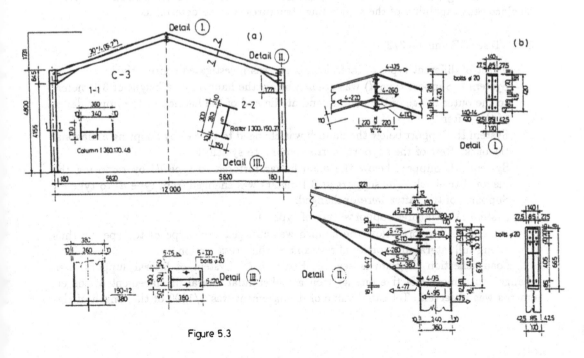

Figure 5.3

of gravity load, the state after achieving the load-displacement diagram peak was not known in structural elements and this may have "given" basis to the statement that at the investigation of the entire structure no stability loss (evolvement of descending load–displacement) of the structural elements could take place prior to the development of global stability loss, or should it occur, that at the same time, means the collapse, stability loss of the complete structure. This train of thought eliminated major problem spheres from the program of theoretical and experimental investigations. In the course of our investigations and analyses it were first the experimental results that indicated and then proved very convincingly that this type of viewpoint simplifies the behaviour of the structure, gravity load type loading "covers up" the exhaustion of the load bearing capacity of individual structures and its effect on the structure.

A full recognition of the behaviour of the structure also involves the knowledge of the behaviour of structural elements and thus it is not only demand from the viewpoint of "comfort" that when investigating the supporting structures also the descending section should be revealed, but this is also required by the demand of a complete knowledge.

The relationship of loading character and the model of supporting structure can be seen in the simple model in Figure 5.6a. The state indicated in Figure 5.6b develops as an effect of gravity load character loading. The state shown in Fig. 5.6c — taking also into consideration the descending section of the load–displacement curves of individual units — develops as an effect of deformation type loading. The maximum load capacity characterizes the behaviour of structures sufficiently well so the use of deformation-type loading is not absolutely necessary in this case. However, it is expedient to carry out the experiments usual for investigating structural elements, beams, columns, connections, column bases, etc. first and foremost with deformation–type loading if both the ascending and descending parts of the load–displacement curve are to be considered and if the displacement capability of the supporting structures is to be determined.

5.12. Test of Frame C–3/2

Effect of different types of lateral supports was investigated consecutively.

- System I. Support type d) was applied below the haunch, at a height of 3.1 meters to the outer flange of the column and at the end of the haunch to the upper flange of the rafter.
- System II. Support below the haunch was changed to type e) by adopting a diagonal tie–back. Rest of the supports corresponding to system I.
- System III. Support below the haunch was restored to type d) by removing the diagonal tie–back. The support at 3.1 meters was changed from type d) to type e). Supports of the rafter were unchanged.
- System IV. All three supports were of type e).
- System V. Support below the haunch was changed from type e) to type f), thus giving full lateral and torsional restraint at this cross section.

Load-deflection diagram is seen in Figure 5.7. System of lateral supports was changed to a more effective one as soon as substantial lateral displacement of the elements was observed. Tolerated value of displacement was 1/1000 of the length of ele-

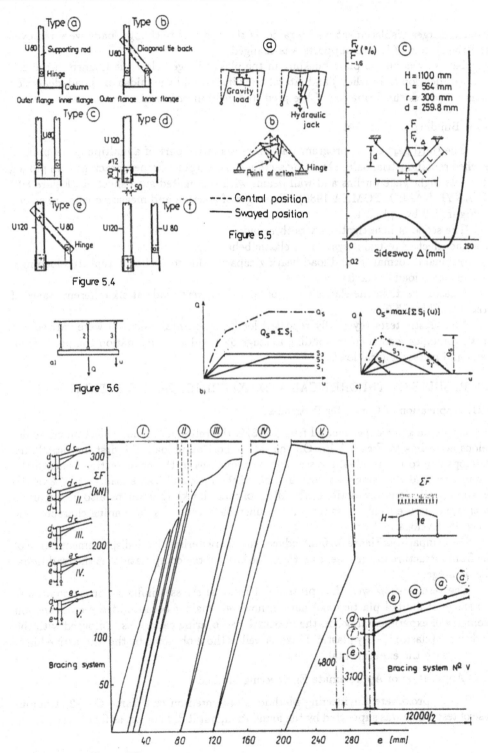

Type (a)

U80 | Supporting rod

Hinge

| Column

Outer flange Inner flange

Type (c)

U80

Type (e)

U120 — U 80

Hinge

Figure 5.4

Type (b)

U80 | Diagonal tie back

Outer flange Inner flange

Type (d)

U 120

Ø12

50

Type (f)

U120 — U 80

(a)

Gravity
load

(b)

Hydraulic
jack

Point of action   Hinge

---- Central position
—— Swayed position

Figure 5.5

$\frac{F_V}{F}$ (°/o)   (c)
-1.6

H = 1100 mm
L = 564 mm
r = 300 mm
d = 259.8 mm

F
F_V Δ

d   L

r

0   250
                      Sidesway Δ [mm]
-0.2

1 2 3

a)   Q

Figure 5.6

Q
                                          Q_s
          $Q_s = \Sigma S_i$
                                          S_3
                                          S_2
                                          S_1
                                      u
b)

Q $O_s = max.[\Sigma S_i (u)]$
                              a_s
      S_1 S_i
                              u
c)

Figure 5.7

ments. Larger displacement was regared as the onset of buckling: loads were removed and the system of lateral supports was changed.

Failure was due to plate buckling in the plastic hinge below the haunch in the column (to develop the earliest) and lateral buckling around the midspan. Load–deflection diagram proves the formation of the predicted yield mechanism.

5.13. Building Section Test

The third part of the program was a representative part of a multipurpose, pinned, pitched roof industrial hall: a building section consisting of 3 frames, bracings with pinned elements, light gage purlins and wall beams with corrugated steel sheeting (Figure 5.8) (IVÁNYI, KÁLLÓ, TOMKA 1986). Six stages of erection and measurements are signed in Figure 5.9 by a, b,... f.

The scope of investigations was threefold:
- the effect of restraint system on elastic behaviour,
- residual deformation and load bearing capacity due to cyclically repeated loading,
- ultimate load of the frames.

Measurements in the elastic range of behaviour were made at six different stages of erection (Figure 5.9).

Non–elastic tests (cyclically repeated load, incremental collapse) were carried out on the building section corresponding to stage 6 using load combination composed from dead and meteorological loads.

5.2. RESULTS OF THEORETICAL AND EXPERIMENTAL INVESTIGATIONS

5.21. Application of Computer Programs

Concerning the experimental frame C-3/2, the relation of load–deflection curve develops according to Figure 5.10. On the side of horizontal load, the first inelastic hinge develops due to the residual stresses and deformations in the cross section beneath the frame knee and this hinge develops at 52% of the maximal frame load. At 97% of the maximal load, Zone L describing the effect of plate buckling develops also in this cross section, i.e. in the frame cross section an "unstable" state — a descending characteristic curve — develops.

For comparison Figure 5.10 introduces the characteristic load–displacement curve of the frame structure in the case, too, when the basis of the computations is the traditional plastic hinge.

The results show well that presence of residual stresses influence in a major way the range of limited plastic deformation, however, mainly because of the cross section geometry of experimental beam, the maximal load bearing capacities computed with the traditional (elastic-ideally plastic) hinge as well as those obtained by the interactive hinge coincide with the experimental results.

5.22. Application of Approximate Engineering Method

The Approximate Engineering Method is presented on test frame C-3/2. Column base of test frame was supported by the foundation, but it did not act as fix end, so plastic

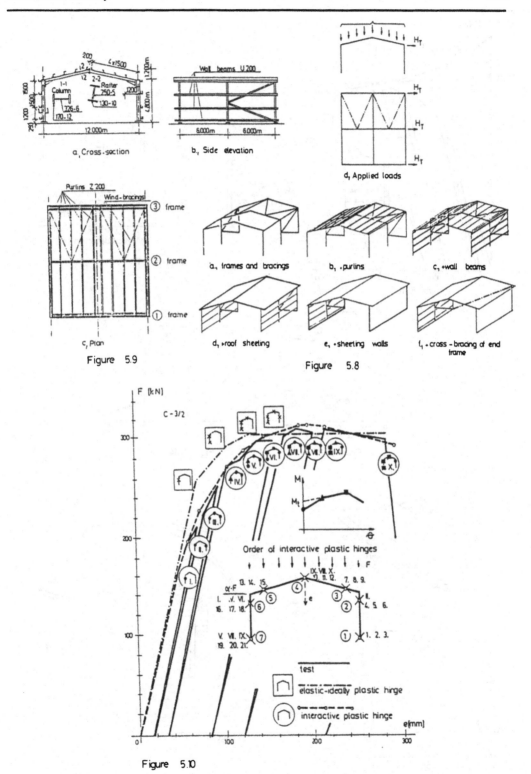

a, Cross-section

b, Side elevation

d, Applied loads

Figure 5.9

a., frames and bracings   b, .purlins   c, +wall beams

d, »roof sheeting   e, ·sheeting walls   f, · cross - bracing of end frame

Figure 5.8

Order of interactive plastic hinges

test

elastic-ideally plastic hinge

interactive plastic hinge

Figure 5.10

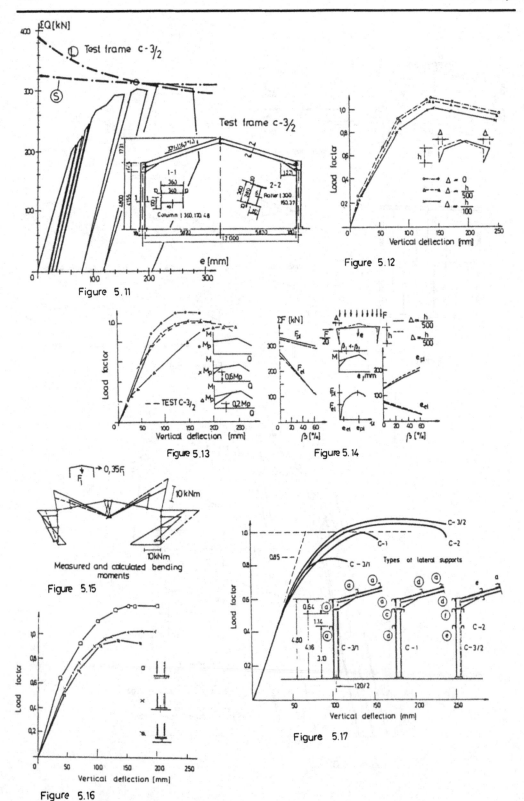

Figure 5.11

Figure 5.12

Figure 5.13

Figure 5.14

Measured and calculated bending moments

Figure 5.15

Figure 5.16

Figure 5.17

load bearing capacity of column base was determined by experiments; these values were used in the calculations.

Approximate Engineering Method described in Section 4.4 is suitable for analyses of the strain hardening and the effect of plate buckling. The most important steps of the analysis of frame C–3/2 given in Figure 5.11 are as followes:

(i) Plastic load bearing capacity analysis: by mechanism chosen by test results,

(ii) Analysis of effects of finite deformations with the help of the chosen mechanism,

(iii) Effect of strain–hardening: moment–rotation relationship of inelastic zones can be replaced by straight lines, therefore rigid–hardening model can be applied to determine m_S,

(iv) Effect of plate buckling: characteristic curves of interactive hinges can be replaced by straight lines too, therefore the rigid–hardening model can be applied to determine M' and m_L.

Figure 5.11 compares test result and the results of Approximate Engineering Method taking strain–hardening of steel and the effect of the plate buckling into consideration. The comparison shows that the Approximate Engineering Method gives a satisfactory result for the maximum loads and the unstable equilibrium state path of whole structure as well; and at the same time the analysis can be done at the "desk of the designer".

5.3. ANALYSIS OF THE INFLUENCE PARAMETERS

5.31. Effect of Fabrication and Erection

Effect of incorrect geometry was investigated by introducing different initial lateral displacements. The consequences are illustrated in Figure 5.12.

The effect of residual stresses of different intensities is shown in Figure 5.13. The medium curve was in coincidence with test results.

The presented method for the complex analysis of frameworks takes several effects into consideration (Figure 5.14).

5.32. Effect of Structural Details

It seems worthwhile to draw attention to the occasional decisive effect of minor differences in structural details on failure as well. Some of the results are reproduced below.

(i) Column bases

Column bases were fixed or hinged. The hinges were not ideal: columns could have been supported by larger base–plates. Figure 5.15 compares the measured bending moment due to vertical and horizontal load with the calculated ones assuming pinned (dashed line) and fixed (solid line) frame. The corresponding moment–rotation diagram of column base was checked experimentally, its adaption to an interactive plastic hinge indicating the load-deflection diagrams obtained by different end conditions (Figure 5.16).

(ii) Lateral supports

Spacing and efficiency of lateral supports proved to be of basic importance. Their effect is illustrated in Figure 5.17.

The importance of adequate spacing of lateral supports and their efficiency in preventing the rotation of cross section around the bar axis has to be emphasized as purlins and rails connected to tension flanges often cannot regarded fully effective in case of thin webs. Not only the load carrying capacity can thus be substantially reduced (as by elastic lateral buckling in case of frame C-3/1 in Figure 5.17), but the yield plateau in the load–deflection diagram can be too short (as in the case of frame C-1 in Figure 5.17), rendering the structure sensitive againts initial imperfections.

Figure 5.18 illustrates the role of the two fundamental structural requirements of plastic design. The diagrams onthe left show the typical responses of members with different width to thickness ratios. Here full lateral support is assumed. On the right the load histories of frames with various type of lateral support can be seen. The similarity is striking.

5.33. Frame Structure under Variable Repeated Load

By repeated cycles of variable loads — for instance by those involving the subsequent application of a light crane–load D and uniformly distributed vertical load $P1$ as indicated in Figure 5.19, incremental collapse can be produced by a load–factor surpassing slightly the shake–down load predicted by a first–order ideally elastic–plastic analysis. The difference between test and analytical values was similar to that observed in proportional loading (due probably to strain–hardening), so the gap between limit loads in proportional and cyclic loading (about 10% in the case indicated in Figure 5.19) is the same in test and computation for both loading cases.

Surprising was the quick progression of residual deflections after just a few load cycles (see Figure 5.19),
- to be attributed possibly to the effect of axial loads connected with remarkable changes in geometry;
- to gradual increase of imperfections (both lateral deflections of beams and curvature of plates); thus to work–softening effects overcoming the work–hardening ones.

5.34. Building under Proportional Loading

(i) Cross Bracing of the End–Frame

Measured deflections from uniform horizontal loads are shown in Figure 5.20, representing the effect of both semi–rigid cross bracings of the end frame.

(ii) Horizontal and Verical Bracing System

Load-displacement diagrams of incremental collapse tests are shown in Figure 5.21. Ultimate loads are influenced by local loss of stability, previous loadings and the layout of frame–horizontal and vertical bracing connections.

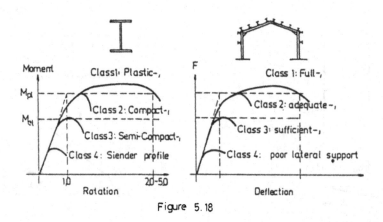

Figure 5.18

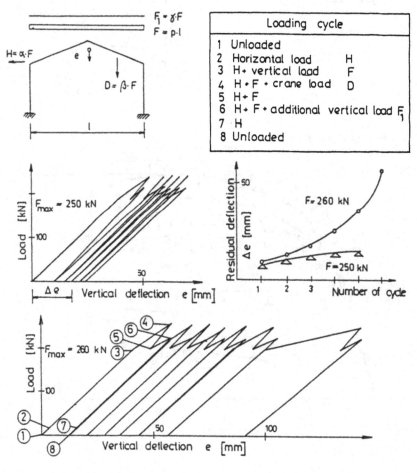

Figure 5.19

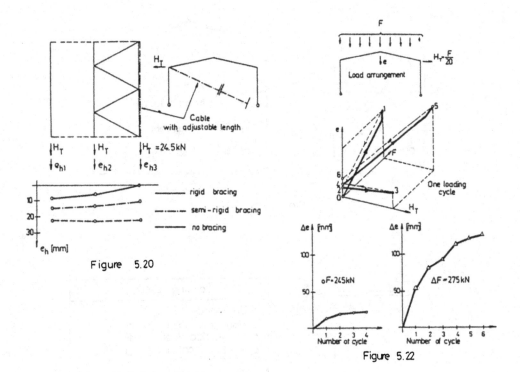

Figure 5.20

Figure 5.22

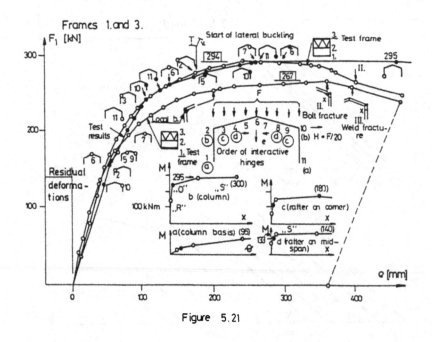

Figure 5.21

5.35. Building under Cyclically Repeated Meteorological Loads

Schematic diagram of a cycle of repeated meteorological loads can be found in Figure 5.22.

Analysing the results of the experiments in the different phases of construction and in the fully completed state, the following conclusions can be drawn:

The actual behaviour of hall stuctures is more favourable than that generally taken into consideration in design practice. In strength analysis, and in stability analysis in plane of the frame the effects of semi–rigid column bases, roof sheeting and end walls are generally neglected.

The measurement results showed the limit to which these effects advantageously affect both internal forces and the rigidity (mostly that of the horizontal displacement) of the stucture.

The actual rigidity of the hall structures, the semi–rigid connection of the structural elements advatageously affect the value of load intensity belonging to the loss of stability perpendicular to the plane of the frame.

It also should be mentioned that, beside advantageous effects, also effects disadvantageously affecting the load–bearing capacity of the structure are occurring. These are the residual stresses due to welding, imperfections due to manufacturing and assembling, eccentricity.

No doubt these effects must be taken into consideration according to their significance. The method and the values of factors (such as fictious eccentricity) applied in calculations are mostly contained in our present specifications. In cases not regulated or not adequately known, the decision of the designer governs. It should be mentioned that economic design can not be imagined such as taking the disadvantageous circumstances, practically in full, into consideration, and totally neglecting the advantageous ones, or taking effects into considerably underestimated.

5.4. CONCLUSION

Experimental and theoretical investigations have been carried out in connection with the elastic and the plastic load–bearing study of frame and hall structures, with strain–hardening of steel material, the residual stresses and plate buckling taken into account.

A method has been presented for the investigation of frame structures applying the steps of known, traditional methods so that the structural behaviour can be analyzed during the entire loading process. Certain effects determining the structural behaviour (e.g. residual deformation, steel material strain–hardening and plate buckling) have been taken into consideration with the aid of the interactive plastic hinge. The interactive hinge was incorporated into an investigation method operating with the structure matrix–calculation method. The results of the elaborated method has been compared with the experimental investigation of full–scale structures.

REFERENCES

ALEXANDER, J. M. 1960.

An approximate analysis of the collapse of thin cylindrical shells under axial loading, Quarterly Jrl. of Mechanical and Applied Mathematics, Vol. 13.

BAKSAI, R. 1983.

Plastic analysis by theoretical methods of the state change of steel frameworks. Diploma work, TUB, Budapest, (In Hungarian)

BAKSAI, R. – IVÁNYI, M. – PAPP, F. 1985.

Computer program for steel frames taking initial imperfections and local buckling into consideration, Periodica Politechnica, Civil Enginnering, Vol. 29. No. 3-4. pp. 171-185.

BEN KATO, 1965.

Buckling strength of plates in the plastic range, Publications of IABSE, Vol. 25.

BJORHOVDE, R. 1972.

Deterministic and probabilistic approach to the strength of steel columns. PhD. dissertation. Lehigh University

BOLOTIN, V. V. 1952.

Reliability theory and structural stability. University of Waterloo, Study No.6. 385

CLIMENHAGA, J. J. – JOHNSON, P. 1972.

Moment-rotation curves for locally buckling beams. Jrnl. of Struct. Div. ASCE, Vol. 98 ST6.

DRUCKER, D. C. 1951.

A more fundamental approach to plastic stress-strain relations. Proceedings, 1st U. S. Natl. Congress of Applied Mechanics, ASME, 487

DRUCKER, D, C. 1964.

On the postulate of stability of material in the mechanics of continua. Journal de Mechanique, Paris, 3, 235

GIONCU, V. – MATEESCU, G. – ORASTEANU, S. 1989.

Theoretical and experimental research regarding the ductility of welded I-sections subjected to bending, Stability of Metal Structures, Proceedings of Fourth International Colloquium on Structural Stability, Asian Session, Beijing, China, pp. 289-298.

HAAIJER, G. 1957.

Plastic buckling in the strain-hardening range, Jrl. ASCE, Vol. 83. EM2

HALÁSZ, O. 1976.

Limit design of steel structures. Second-order problems (In Hungarian). DSc. Thesis, Budapest

HALÁSZ, O. – IVÁNYI, M. 1979.

Tests with simple elasic-plastic frames. Periodica Polytechnica, Civil Engineering, Vol. 23, 157

HALÁSZ, O. – IVÁNYI, M. 1985.

Some lessons drawn from tests with steel structures, Peridica Polytechnica, Civil Engineering, Vol. 29. No. 3-4. pp. 113-122.

HOFF, N. J. 1956.

The analysis of structures, John Wiley and Sons, New York

HORNE, M. R. 1960.

Instability and the plastic theory of structures. Transactions of the EIC. 4, 31

HORNE, M. R. – MEDLAND, J. C. 1966.

Collapse loads of steel frameworks allowing for the effect of strain-hardening, Proc. of Inst. of Civil Engineers, Vol. 33. pp. 381-402.

HORNE, M. R. – MERCHANT, W. 1965.

The stability of frames, Pergamon Press

HORNE, M. R. – MORRIS, L. J. 1981.

Plastic design of low-rise frames, Granada, Constrado Monographs

IVÁNYI, M. 1979a.

Yield mechanism curves for local buckling of axially compressed members, Periodica Polytechnica, Civil Engineering, Vol. 23. No 3-4. pp. 203-216.

IVÁNYI, M. 1979b.

Moment-rotation characteristics of locally buckling beams, Periodica Polytechnica, Civil Engineering, Vol. 23. No 3-4. pp. 217-230.

IVÁNYI, M. 1980.

Effect of plate buckling on the plastic load carrying capacity of frames. IABSE 11th Congress, Vienna

IVÁNYI, M. 1983.

Interaction of stability and strength phenomena in the load carrying capacity of steel structures. Role of plate buckling. (In Hungarian). DSc. Thesis, Hung. Ac. Sci., Budapest

IVÁNYI, M. 1985.

The model of "interactive plactic hinge" Periodica Politechnica, Vol. 29. No. 3-4. pp. 121-146.

IVÁNYI, M. 1987.

Complementary report for the session of frames, Second Regional Colloquium on "Stability of Steel Structures", Final Report, pp. 149-156.

IVÁNYI, M. – KÁLLÓ, M. – TOMKA, P. 1986.

Experimental investigation of full-scale industrial building section, Second Regional Colloquium on Stability of Steel Structures, Hungary, Final Report, pp. 163-170.

KALISZKY, S. 1978.

The analysis of structures with conditional joints, Jrl. of Structural Mechanics, 6., 195.

KAZINCZY, G. 1914.

Experiments with fixed-end beams. (In Hungarian). Betonszemle, 2, 68

KURUTZ, M. 1975.

Mechanical computation of structures containing conditional joints under kinematic loads (In Hungarian), Magyar Épitőipar 24, 455.

LAY, M. G. 1965.

Flange local buckling in wide-flange shapes, Jrl. ASCE, Vol. 91. ST6.

MAIER, G. 1961.
Sull'equilibrio elastoplastico delle strutture reticolari in presenza di diagrammi forze-elongazioni a trotti desrescenti. Rendiconti, Istituto Lombardo di Scienze e Letture, Casse di Scienze A, Milano 95, 177

MAIER, G. – DRUCKER, D. C. 1966.
Elastic-plastic continua containing unstable elements obeying normality and convexity relations. Schweizerische Bauzeitung 84, No. 23., Juni. 1

MAJID, K. 1972.
Non-linear structures. Matrix methods of amnalysis and design by computers. Butterworths, London

MCGUIRE, W. 1984
Structural Engineering for the 80's and Beyond. Engineering Journal, AISC. Second Quarter, pp.77–88.

MSZ 15024-85 (Hungarian Standard)
Design of steel constructions for buildings, Budapest

NETHERCOT, D.1985.
Utilisation of experimentally obtained connection data in assessing the performance of steel frames, "Connection Flexibility and Steel Frames", ed. Wai-Fah Chen pp. 13-37.

PUGESLEY, A. – MACAULAY, M. 1960.
Large-scale crumpling of thin cylindrical columns. Quarterly Jrl. of Mechanical and Applied Mathematics, Vol. 13.

SZABÓ, J. – ROLLER, B. 1971.
Theory and analysis of bar systems. (In Hungarian) Műszaki Könyvkiadó, Budapest

TASSI, G. – RÓZSA, P. 1958.
Calculation of elasto-plastic redundant systems by applying matrix theory (In Hungarian), ÉKME Tud. Közlemények, 4.21.

THÜRLIMANN, B. 1960.
New aspects concerning the inelastic instability of steel structures. Journal of the Structural Division, ASCE, 86, STI, Jan.

STABILITY OF STEEL FRAMES WITH SEMI-RIGID JOINTS

F. M. Mazzolani
University of Naples, Naples, Italy

Abstract

The load carrying capacity of steel frames is deeply influenced by the inelastic behaviour of their connections.

The introduction of the concept of "semi-rigid joints" leads nowadays to interesting applications, because of their convenience from the point of view of both constructional aspects (simplicity means economy) and behavioural aspects (damping due to slippage effects).

The subject of stability of semi-rigid frames is developed in this part, which is subdivided into five chapters.

Chapter 1 gives general definitions of steel frames and their connections. Chapter 2 and 3 are dealing with the behaviour of connections and its analytical interpretation. Chapter 4 is devoted to the evaluation of the influence of the semi-rigidity on the buckling behaviour of elastic and inelastic frames. Finally, Chapter 5 gives a new interpretation of the "industrial frames", in which the behaviour of semi-rigid joints is considered as a special kind of imperfection.

1. FRAMES AND THEIR CONNECTIONS IN STEEL STRUCTURES

1.1 Frames

A frame is a composition of linear members joined together by connections (fig. 1.1); both members and connections are not perfect [1]. Referring to the flexural behaviour, the rilevant properties of members and connections can be derived from the moment-rotation relationship from which it is possible to evaluate (fig. 1.2):
- *Strength*, corresponding to the ultimate moment;
- *Stiffness*, which is the slope of the M-Φ curve in the elastic region;
- *Deformation capability*, which is the capacity to undergo plastic deformations up to rupture and, in one word, is called "ductility";

According to EC3 [2], frame structures can be classified both on the basis of the presence of bracing systems (braced and unbraced) and of the possibility to have lateral displacements (sway and non sway). As a consequence we can define (fig. 1.3):

Braced frame, when it is acceptably accurate to assume.that all horizontal loads are resisted by bracing systems;

Non sway frames, when its in-plane lateral stiffness is sufficient to make acceptably accurate to neglect the geometrical 2nd order effects (i.e. the so called P-Δ effect).

Sway frame, when condition 2 is not fulfilled (P-Δ effect is very important) whether for braced or for unbraced frames. It means that also if the frame is braced, it can be considered a sway frame when the bracings don't give enough rigidity to make the P-Δ effect negligible.

1.2 Sections

An important classification given by EC3 [2] is dealing with the plastic behaviour of sections, depending on their capability to go into plastic range and according to the shape of the cross-section which composes the bars. Four categories of cross-sections are considered (fig. 1.4):

Plastic section (class 1), when there is no local buckling effect and the section can express its whole load carrying capacity. The ultimate moment reaches the value of the fully plastic moment Mp.

Compact section (class 2) can reach the same ultimate moment as class 1, but the rotation capacity is less than the one of perfectly plastic sections.

Semi-compact sections (class 3) means that the fully plastic moment can not be reached and the load bearing capacity is limited to the conventional elastic limit moment only.

Slender sections (class 4) in which the local buckling effects are determinant and decrease the load carrying capacity below the elastic limit, according to the definition of the so-called "effective section" (it is the case of cold-formed thin-walled sections).

A cross-section is considered belonging to one of the above mentioned classes according to the b/t ratio. EC3 gives a lot of limitations which are referred to the different parts of the cross-sections. They are mainly flange and web elements which can be loaded in bending or in compression.

In case of class 1 and 2, the plastic behaviour is allowed and, therefore, the stress distribution is constant, while in the semi-compact sections (class 3) only the elastic behaviour is allowed, so the stress distribution is linear. For slender sections (class 4) the stress distribution is limited to the effective part of the reduced cross-section.

1.3 Connections

Going to the connections, we can observe that the use of *"semi-rigid connections"* is spreading more and more in the practice of steelwork. One of the reasons is that, because of the increase of the cost of labour, the workmanship cost increases in the years. The consequence is that when we design a constructional detail we prefer to eliminate the stiffners and to increase the thickness of the connection parts, because the cost of material is less important than the cost of labour. This tendency leads to more simplified solutions, because they are more economic. The consequence is that the behaviour of connections becomes very far from the ideal rigid one, so they behave as semi-rigid connections.

On the contrary, in the last century, rigid connections were not used in the former steel structures. They preferred to use perfect hinges coming from the mechanical engineering. These mechanical hinges can work as a perfect hinge provided that good manteinance and lubrification are assured. If not, the perfect hinge might be transformed into a built-up restraint and the stress distribution in the structure completely change.

The traditional classification of connections leads to three main categories (fig. 1.5) [3]:

Nominally pinned connection (it is the case of a very short end plate or of angles as cleats);

Perfectly rigid connection (where a lot of stiffeners are introduced in the web of the column, that correspond to a very expensive solution);

Semi-rigid connection (which is the most economic way to connect members and, therefore, the most used in practice).

This classification is used in case of elastic design. When we go into plastic design, different definitions must be done. In this case it is necessary to make distinction between connections and joints. According to the up-to-dated definitions, _connection_ means the parts which mechanically fasten the connected members (i.e. beam to column), while _joint_ means the connection plus the corresponding zone of interaction between the connected members.

From the plastic point of view, joints are classified as follows, according to their M-Φ curve (fig. 1.6):

Full strength joints, when the plastic hinge takes place in the member adjacent to the connection, but *not in the connection*. No rotation capacity is required for the connection (cases A and B). The joint reaches an ultimate moment bigger than the fully plastic one of the connected beam. Of course, they can have different degrees of ductility even if the stiffness is the same.

Partial strength joints, when the plastic hinge takes place in the connection, because the moment capacity of the connection is less than the moment capacity of the connected member. In this case the rotation capacity can vary according to the different kinds of section (cases C,D, and E), but sufficient rotation capacity is required for the connection.

In one word, we have *full strength joints* when the connection is stronger than the member, *partial strength joints* in the opposite case.

EC3 [2] introduces a link between the type of framing, the metodology of global analysis and the type of connections. The type of framing can be simple, continuous or semi-continuous. Semirigid connections are used only in the case of semi-continuous frames.

All together, the different types of connections can be classified as simply pinned, rigid or semi-rigid, as well as they can belong to full strength or partial strength joints. There is an intersection between these two ways of classification. It is not necessary that the full strength connection should be rigid; alternatively a partial strength connection can be rigid, but not necessarily semirigid [4].

This is a very significant aspect which has been developed for the first time in EC3, where a combination of typologies, calculation methods and behaviour of connections is assessed in a rational way.

Summing up, connections can be classified by using a nondimensional diagram (fig. 1.7) which gives different zones. The lower part corresponds to very low stiffness and very low moment capacity; curves falling in this zone cover the nominally pinned joints, which behave as simple hinges.

On the contrary if the curve belongs to the left part, the behaviour is rigid. The intermediate zone, which belongs to the semi-rigid joints, is the widest.

By means of this diagram, the classification of joints can be made according to rigidity:

Nominally pinned, when the joint can not develope significant moments.

Rigid, when its deformation has no significant influence.

Semi-rigid, when it provides a predictable degree of interaction between members, based on the design moment-rotation characteristics of the joints.

According to strength, they can be classified as:

Nominally pinned, when transmit the design force only without moments.

Full-strength, when its design resistance is at least equal to the one of the member.

Partial-strength, when its design resistance is less than the one of the member.

1.4 Ductility of joints

The ductility of joints is the capability to develope their strength in plastic range without premature collapse due to excessive strain.

If we compare the experimental results (fig. 1.8) of a simple tensile test of a member connected to a gusset plate by means of welds or bolts [5], it can be observed that in case of welds the joint reaches the full strength conditions together with a very good ductility (curve I).

In case of bolted connections, the result of curve II corresponds to a very brittle behaviour without possibility to have full-strength capacity and with very low ductility. This happens when the design is not based on the ultimate resistance of the bar, but simply on the axial force given by the elastic calculation. This is important to avoid this kind of behaviour in the design of braced structures (generally not-redundant) in seismic zones, when the collapse of a joint produces the collapse of the whole structure.

This behaviour can be improved by using the same number of bolts located in different way (curve III), that allows to eliminate the eccentricity due to the spacing between bolt holes and the centroid axis of the bar. In this way the flexural phenomena, which are responsible of the brittle behaviour of the previous case, are eliminated and the ductility is increased.

Let us now consider the case of a beam with span L connected to two columns at the ends A and B and uniformely loaded by q (fig. 1.9). Every equilibrated solution is accaptable, provided [5]:

$$M_A = \frac{1}{8} qL^2 - M_B \le M_b$$

$$M_b \le M_c$$

$$\Theta = \frac{qL^3}{24EI} - \frac{M_B L}{2EI} \le \Theta_c$$

being:

M_h the design resistance of the beam;
M_c the design resistence of the beam-to-column connection;
Θ_c the maximum rotation capability.

At the collapse condition, when a plastic hinge takes place at mid span of the beam, the ductility condition becomes:

$$\frac{L}{6EI} (2M_b - M_c) \le \Theta_c$$

By considering both strength and ductility limitations, we can built up a domain which is limited by the design moment M_b of the cross-section of the beam (horizontal line) and by the sloped straigthline corresponding to the moment-rotation relationship of the beam (fig. 1.10).

By comparing the M-Θ curves of these three types of connections, we can found that:

-case I: the fully welded joint with stiffeners is a full-strength joint with sufficient ductility;

-case II: the joint is still welded but the stiffeners in the column web are eliminated; there are two possibilities: partial strength joint with enough ductility (IIb) or partial strength joint with insufficient ductility (IIa); in both cases the design moment capacity is not reached;

-case III: the joint with angles only in the web is considered as a nominally pinned joint with sufficient ductility due to the large allowable rotations, but it is unable to resist bending moment.

1.5 Beam-to-column connections

Different types of beam-to-column connections are given in fig. 1.11:

1 and 2 - Single or double web cleat (they represent the simplest way to connect);

3 - Web side plate (in this case the angle is not bolted but welded to the column flange);

4 - Flange cleats (made by means of two angles bolted to both column and beam flanges);

5 - Bottom flange cleat and web cleat (always by means of angles connecting web and bottom flange of the beam);

6 - Flexible end plate (with a very small plate limited to the center of beam web);

7 - Flush end plate (when the plate has the same depth of the beam);

8 - Extended end plate (it is the best way because it increases the arm of the opposite forces which equilibrates the external moment);

9 - Web + flange cleats (it is a combination of cases 2 and 4);

10 - Tee-stubs (an alternative to case 4, by using tees instead of angles);

11 - Top plate with a seat angle (the top angle of case 4 is substituted by a plate);

12 - Tee-stubs and web cleats (it is a combination of cases 2 and 10).

Some typical beam-to-column joints for continuous or semi-continuous framing are shown in fig. 1.12 [5].

Case a) is the typical rigid joint which is completely welded and integrated by horizontal
         stiffeners in the column web.

Case b) is the same as a) without stiffeners and it behaves as rigid or semirigid according
         mainly to the stiffness of the column web.

Case c) is the typical end plate joint with extended end plate, which is usually considered as
         semirigid.

Case d) gives a different solution in which top and bottom plates are introduced. In this case
         bolts work in shear and not in tensile, leading to a more compact behaviour especially
         in the case of small clearance between holes and bolts.

Case e) is an alternative solution in which a seat is located in the compressed part instead of
         a plate.

In order to avoid the possible damage (cases d and e) during transportation and erection, the welded plates can be substituted by bolted T or L stubs together with web angles (cases f and g).

In case of pinned structures integrated by bracings, some possibilities are shown in fig. 1.13. They are [5]:

a) continuos beam-interrupted column with bolted end plates;
b) web angles to column web connection;
c) web angles to column flange connection;
d) end plate with shear bolts;
e) f) g) bottom seat and upper connecting angles with different degrees of eccentricities;
h) typical solution with diagonal bracing.

Usually the floor structure is made of corrugated sheets and concrete slab which is built on the steel beams and connected by means of studs [6]. In this case the beam-to-column joint is a composite joint (fig. 1.14).

This composite joint can be both rigid and pinned (fig. 1.14a and b). In the last case it is necessary to cut the slab in order to allow a simple behaviour as a hinge.

A semi-rigid joint can be obtained by the combination of a very flexible connection in the steel part and a continuos concrete slab on the top, which contributes to increase the rigidity of the joint. The case of fig. 14c shows one of the most common types of semi-rigid composite joint, nowday under investigation.

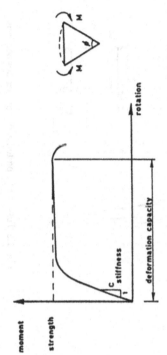

Fig. 1.2 Relevant properties
of members and connections.

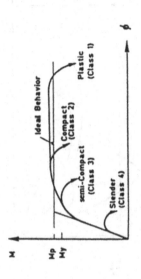

Fig. 1.4 Categories of sections.

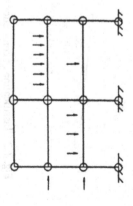

Fig. 1.1 Frame: linear members
joined together by connections.

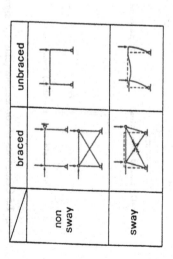

Fig. 1.3 Classification criteria for frames
(according to EC3)

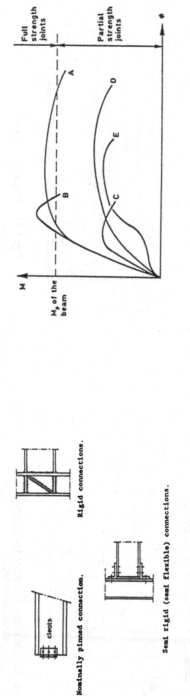

Nominally pinned connection.   Rigid connections.

Semi rigid (semi flexible) connections.

Fig. 1.5 Categories of connections (elastic design)

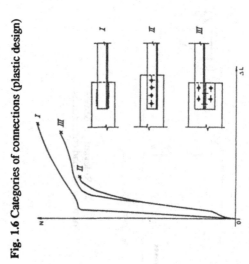

Fig. 1.6 Categories of connections (plastic design)

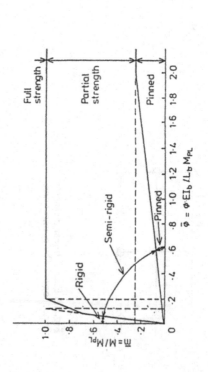

Fig. 1.8 Test results on tension member connections.

Fig. 1.7 Classification of connections (according to EC3).

$$\bar{\phi} = \phi \cdot EI_b / L_b M_{PL}$$

$$\bar{m} = M/M_{PL}$$

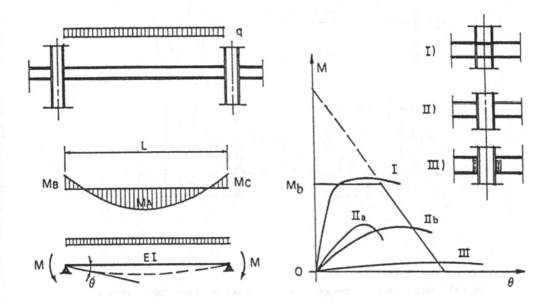

Fig. 1.9 Evaluation of rotation capability.   Fig. 1.10 Minimum rotation capacity connections.

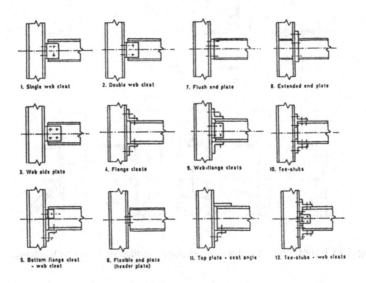

Fig. 1.11 Classification of beam-to-column connections.

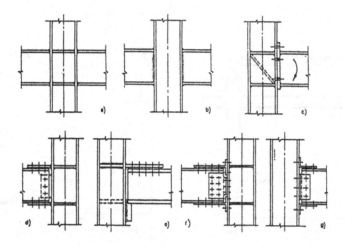

Fig. 1.12 Typical beam-to-column joints (for continuous or semi-continuous framing).

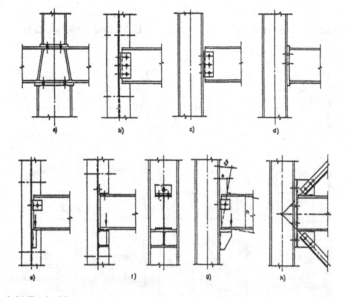

Fig. 1.13 Typical beam-to-column joints (for pin-ended structures with bracings).

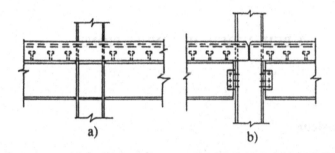

Beam-to-column composite joint: (a)   rigid joint; (b)   simple joint.

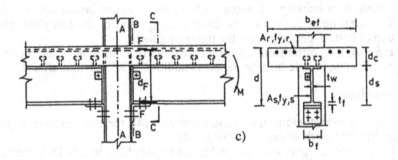

c)   semi-rigid joint.

Fig. 1.14 Composite joints.

2. BEHAVIOUR OF CONNECTIONS

2.1 Monotonic behaviour

The behaviour of connections can be analyzed in simplified way by referring to three typical families of joints [4], which cover all the range of semi-rigidity (fig. 2.1):

1. *welded joints*; their behaviour is different from the ideal rigid joint because the nodal panel deformation due to the highly concentrated forces produces local buckling, which is responsible of the decrease of sfiffness;

2. *end plate joints*; their behaviour is more far from the ideal one because of the additional effects of flexural deformations of the column flange and of the end plate, plus axial deformation of bolts depending upon the tightening force;

3. *shear bolts joints*; they differ from the ideal case more than the other ones, because of the slipping of bolts due to hole-bolt clearance, in addition to all previous effects.

2.2 Cyclic behaviour

If we pass from the monotonic to the cyclic behaviour [4], the three examined joints can produce three different kinds of behaviour (fig. 2.2):

a) The joints of the first category exhibit a stable behaviour characterized by hysteresis loops having the same area inside the curve which remains constant as far as the number of cycles increases.

b) The joints of the second category, instead, exhibit an unstable behaviour due to permanent deformations in holes and bolts, thus reducing the stiffening effect of the tightening force. Therefore, for constant amplitude cyclic loading, we observe hysteresis loops with increasing deformations up to the complete collapse. The slope of the hysteresis curves characterizing the stiffness of the i-th cycle is continuosly decreasing.

c) The joints of the third category exhibit also an unstable behaviour characterized by slipping of bolts. This phenomenon significantly modifies the shape of the curve by reducing the dissipated energy for the same values of deformations.

Figure 2.3 shows the result of some experimental tests on welded joint (a) and on end plate joint (b): the first has a stable behaviour, while the second one has an unstable behaviour.

2.3 Behavioural parameters

The problem to define the main parameters, which are determinant in order to interprete the cyclic behaviour of joints, has been considered in the field of activity of the Seismic Design Commitee TC13 of ECCS. At the first glance, the following parameters were considered as fundamental for characterizing the behaviour of joints at a given cycle i:

a) ductility parameter $\Delta_i = \delta_i/\delta_y$

defined as the ratio between the maximun deflection and the deflection at conventional elastic limit;

b) the energetic efficiency parameter $E_i = \dfrac{A_i}{F_y(\delta_i - \delta_y)}$

is the area included in the loop over the area of an ideally elastic-perfectly plastic loop;

c) duct-energy parameter (combination of ductility and energy) $D_i = \dfrac{A_i \delta_i}{\delta_y F_y(\delta_i - \delta_y)}$

defined as the product of the ductility and the energetic efficiency parameters;

d) rigidity parameter $R_i = \dfrac{F_i \delta_y}{\delta_i F_y}$

defined as the ratio between the secant slope at the end of the i-th cycle and the slope corresponding to elastic conditions;

e) strength parameter $S_i = \dfrac{F_i}{F_y}$

is the maximum force reached in the cycle over the elastic limit.

These parameters can be represented in a sheet (fig. 2.4), which contains all data necessary to define the cyclic behaviour of a joint, by means of the complete hysteresis loops, the full ductility and the variation of the main parameters (peak strength, secant modulus, energy efficiency, ductility) as far as the number of cycles increases during the test.

In case of stable behaviour (fig. 2.4a), the strength increases, the secant modulus (i.e. the stiffness) decreases, the energy efficiency parameter first increases and then decreases, the duct-energy parameter always increases.

In case of unstable behaviour (fig. 2.4b), the variation of the main parameters is very different, being characterized by a sudden decreasing of strength and energy.

From the analysis and the comparison of these parameters the main problem is how to define the best joint: the best joint can not be defined in general and it must be chosen according to the requested performances (for instance we can need enough ductility, or a given amount of energy absorption, or a guaranteed strength, or a combination of more requirements). In addition it is very difficult to define what ultimate conditions means for a given joint according to its behaviour and how to decide from the test results the condition in which the joint reaches its maximum load carrying capacity.

We also observe that when analyzing the results of cyclic tests coming from different laboratories in the world, every researcher uses a different way to perform the test on his joint, so it is quite impossible to compare each other the collected results. It means that this effort can be unuseful in order to obtain generalized results.

For this reason in the field of activity of ECCS-TC13 we decided to propose an unified testing procedure by means of a Recommendation published in 1986: **"Recommended Testing Procedure for Assessing the Behaviour of Structural Steel Elements under Cyclic Loads"**, which gives a methodology for measuring the fundamental parameters and for interpreting the obtained results [7].

In these Recommendations the definition of the elastic limit is given in different conventional ways (fig. 2.5):

a) the classic definition is to identify the end of the elastic behaviour;

b) in case of a quite brittle behaviour the straight elastic line can be considered up to the top force without significant error;

c) in case of round-house curves the elastic limit can be defined as the one for which the elastic deformation is equal to the plastic one: the horizontal line cuts the straight line (elastic) and the actual curve in two equal portions; this assumption is particularly convenient when the Ramberg-Osgood law is used to interpret the load-deflection curve.

d) finally, the elastic limit can be given by the intersection between the prolungation of the elastic line and a tangent line with a given slope.

The parameters which interprete the single cycle of the test are (fig. 2.6):

- the extremes of displacement v_i^+ and v_i^-;
- the values of the force F_i^+ and F_i^- corresponding to the extremes of displacement v_i^+ and v_i^-;
- the extremes of displacement in the positive and negative range of the applied force Δv_i^+ and Δv_i^-;
- the tangent modulus corresponding to the change of the sign of the applied load, $tg\alpha_i^+$ and $tg\alpha_i^+$;
- the areas A_i^+ and A_i^- of the positive and negative half cycle.

The cyclic behavioural parameters of the ECCS Recommendations are quite similar to the ones defined before with small changes only of symbols, but the meaning is the same. They are:

- *partial ductility*: $\mu_{oi}^+ = v_i^+/v_y^i$, $\mu_{oi}^- = v_i^-/v_y^-$
- *full ductility*: $\mu_i^+ = \Delta v_i^+/v_y^+$, $\mu_i^- = \Delta v_i^-/v_y^-$
- *full ductility ratios*: $\psi_i^+ = \Delta v_i^+/(v_i^+ + (v_i^- - v_y^-))$, $\psi_i^- = \Delta v_i^-/(v_i^- + (v_i^+ - v_y^+))$
- *resistance ratios*: $\varepsilon_i^+ = F_i^+/F_y^+$, $\varepsilon_i^- = F_i^-/F_y^-$
- *rigidity ratios*; $\xi_i^+ = tg\alpha_i^+/tg\alpha_y^+$, $\xi_i^- = tg\alpha_i^-/tg\alpha_y^-$
- *absobed energy ratios*: $\eta_i^+ = A_i^+/(v_i^+ + v_i^- - v_y^+ - v_y^-) \cdot F_y^+$, $\eta_i^- = A_i^+/(v_i^+ + v_i^- - v_y^+ - v_y^-) \cdot F_y^-$

2.4 Testing procedures

The first application of the ECCS Recommendations has been done in 1987 [8] in the field of a common research program on cyclic behaviour of beam-to-column joints between the University of Naples and the Politecnico of Milan. Tests were performed by means of the apparatus of the laboratory in Milan, which is composed by (fig.2.7):

1. horizontal base beam for the anchorage of the specimen;
2. braced contrast frame;
3. elastic screw jack in order to impress horizontal displacements (100 KN load and 300 mm stroke);
4. jack to apply axial loads to the horizontal member;
5. specimen.

The specimen corresponds to a beam-to-column joint in which the column is horizontal and an axial force is applied there; the beam is vertical and at the top cyclic deflections are applied. The choice of specimens was oriented to some typical beam-to-column joints belonging to the market of steel structures. Four series A, B, C, D of joints (in total 14 joints) were selected (fig. 2.8). For each series the number 1 indicates the basic case. A1 is the basic case of the series A characterized by the top plate and the bottom plate with shear bolts.

Series B has angles instead of plates at top and bottom flanges.

Series C is the typical case of an extended end plate, symmetrical in order to undergo cyclic alternative loads. In this case the column has been reinforced with stubs in order to find exactly the behaviour of the flanges not influenced by the behaviour of the column.

The last series D is the classical rigid joint, completely built up by welding with horizontal stiffeners in the column web.

Cases 2,3 and 4 introduce some variations to the main items.

Starting from the basic case A1, it is possible to introduce a diagonal stiffener in the column web (A2), or a cover plate in the inside part of the flanges (A3), as well as both of these variations (A4).

For the type B, the variations are: in B2 stiffeners are located in the column webs behind both angles; in B3 angles are stiffened by transverse triangular plates in order to prevent their deformation; the last case B3 is the combination of the previous ones: stiffeners in the angles and stiffeners in the web.

In case of end plates (C), two stiffeners in the web of the beam (C2) can be introduced; in C3 the thickness of the end plates has been increased but web stiffeners are not complete; finally case C4 is a combination of all previous cases . Nothing has been varied in the column because it is always more rigid than the beam in order to prevent additional effects in the column.

For the last case D, the variation consists only in stiffening of the web of the column with two plates in addition to the horizontal stiffeners.

2.5 Analysis of experimental results

Fig. 2.9 shows how so different histeresis loops have been derived from tests for the analized beam-to-column joints. From the observation of these resuls, the problem is how to find the influences of the different kinds of detail variation on the cyclic behaviour.

When considering the first case A, we can find that the presence of the diagonal stiffeners produces the effect to increase the load carrying capacity, but to reduce both ductility and energy absorption. The elimination of the stiffeners produces less strength but more ductility and the area of the loop is increased.

This behaviour can lead to a general rule: if we want to introduce something that increases strength we must be careful because ductility and energy absorption can be reduced; it means that if the design is in seismic areas particular attention must be payed to this problem.

The results of second type B supplie a confirmation of what stated before. When we consider the last case compared to the first one, the difference is due to the introduction of two types of stiffeners; the consequence is still an increment of strength and a reduction of ductility with no important difference in energy absorption.

The cases of end plates extended in both directions (C) give better results. Also here the last case C4 is less ductile than the first one C1, where on the contrary there is a lowering of the maximum force. In the intermediate cases the maximum force at each cycle is always increased. Anyway in all cases C the loop area is very satisfactory and larger than in the previous cases.

As expected, the best combination of strength and ductility characterizes the fully welded joints (cases D). The stiffening of the web of the column just increases the strength at the beginning but not at the end of the test (cases D2), where only in the intermediate cycles a maximum value bigger than in the first case is reached.

An attempt to correlate the different types of joint by considering the main parameters referred to the last cycle is given in fig. 2.10. Here, the loop area (Ω) (corresponding to the absorbed energy), the maximun load F_{max}, the collapse load F_u (which corresponds to the presence of very evident collapse mechanism in the specimen) and the maximum diplacement v_u reached in the test are compared togheter.

It is evident the strong difference between A1, A3 (without diagonal stiffeners) and A2, A4 (with diagonal stiffeners) as regard to the absorbed energy.

For joints of type B we observe pratically the same difference between stiffened and unstiffened cases. In case D2 the introduction of additional stiffeners reduces the amount of absorbed energy.

The diagrams of the maximun load, which is a strength parameter, are quite complementary to the previous ones, leading to the confirmation that when strength increases energy absorption decreases.

The maximum displacement v_u is a ductility parameter which goes in the same direction of absorbed energy. The stiffening of some part of the joint produces a reduction of the ductility of the overall joint.

In conclusion, the stiffeners must be used very carefully, because we risk to deteriorate the behaviour of the joint. In addition, do not forget that stiffening operations increase the cost of the joint.

For the above reasons, the choice of semi-rigid joints seems to be the most rational both from structural and economical points of view.

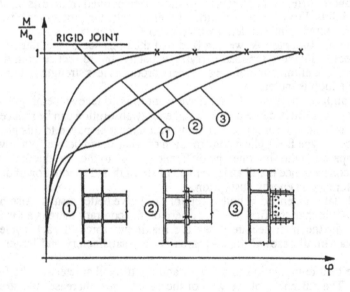

Fig. 2.1 Monotonic behaviour.

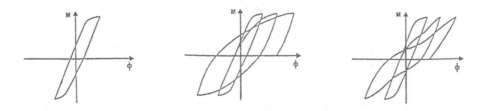

Fig. 2.2 Cyclic behaviour.

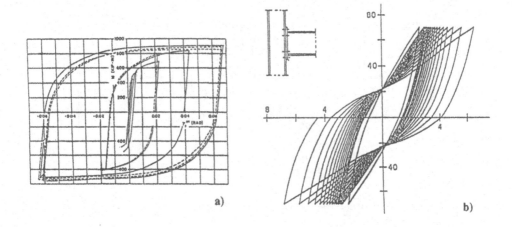

Fig. 2.3 Typical hysteresis loops;
a) Stable behaviour; **b)** Unstable behaviour.

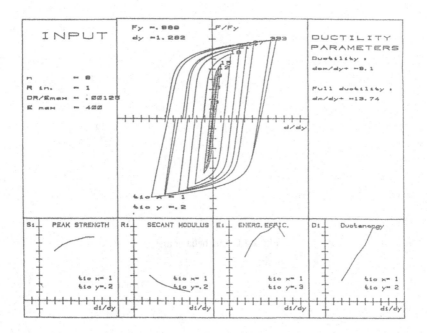

Fig. 2.4 a) Evaluation of cyclic parameters.

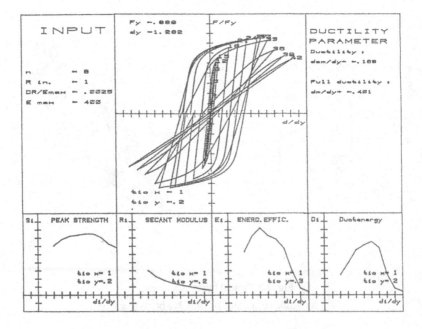

Fig. 2.4 b) Evaluation of cyclic parameters.

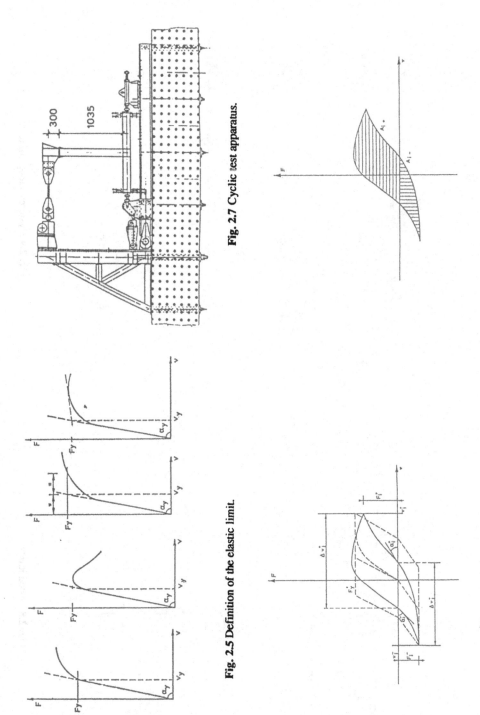

Fig. 2.7 Cyclic test apparatus.

Fig. 2.6 b) Parameters of interpretation of one cycle.

Fig. 2.5 Definition of the elastic limit.

Fig. 2.6 a) Parameters of interpretation of one cycle.

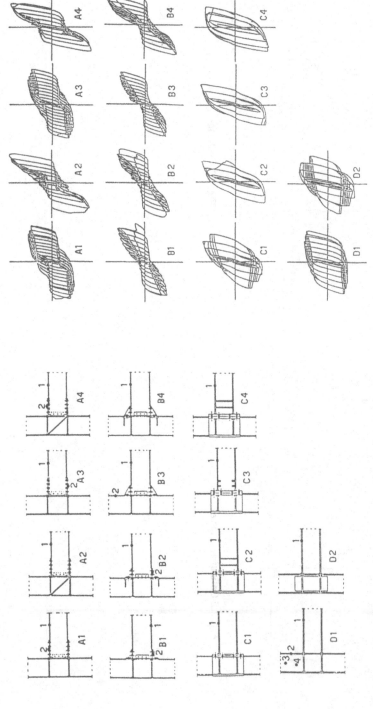

Fig. 2.9 Hysteresis loops of tested joints.

Fig. 2.8 Tested joints.

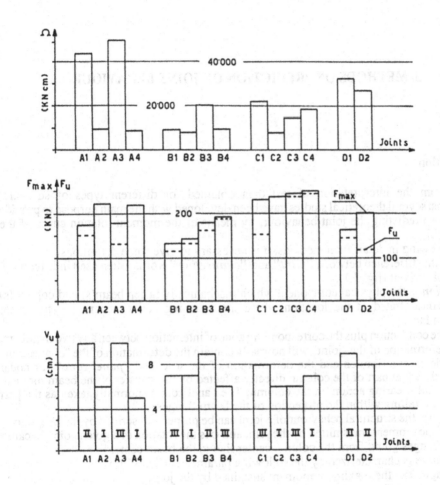

Fig. 2.10 Comparison among behavioural parameters.

3. METHODS OF PREDICTION OF JOINT BEHAVIOUR

3.1 Definitions

Starting from the interpretation of test data obtained for different types of structural connections, several theoretical studies have been developed with the main purpose to provide methods for predicting the joint behaviour, by means of the moment-rotation curve of the joint [9],[10].

Before reviewing in detail such methods, it is necessary to give some definitions.

First of all the difference between *"joint"* and *"connection"* words, often used indifferently, must be pointed out (fig. 3.1).

A *connection* is the physical component which mechanically fasten beam(s) and column(s) and it is concentrated at the location where the fastening action occurs, i.e. the beam end/column face.

A *joint* is the connection plus the corresponding zone of interaction between any two members. Thus the deformation of the "joint" will normally contain the deformation of the "connection" as a significant contribution and the deformation of the so-called *"panel zone"* (or nodal zone), which is that part of the column directly affected by the transfer of the beam moment as a horizontal shearing action on the column. The panel zone is normally taken as the part of the column falling within the projection of the beam flanges.

As said before, the structural behaviour of a joint can be properly described by the M-φ curve, where M is the moment transmitted by the joint and φ is the rotation of a given point, because the angle changes going from the column center line to the beam center line.

The M-φ curve is characterized by the following parameters (fig. 3.2):

Moment capacity - the maximum moment sustained by the joint.

Rotation capacity - the rotation achieved at a given value of moment; frequently the rotation achieved at the design moment capacity, i.e. before any significant reduction in moment occurs.

Stiffness - a measure of the slope of the M-φ curve.

Secant stiffness - the slope of a straight line joining two points of the M-φ curve; its value depends upon the points (defined in terms of M or φ) chosen.

Tangent stiffness - the istantaneous slope of the M-φ curve at a particular value of M (or φ); tangent stiffness decreases with increasing moment (and rotation).

According to the definition given for the moment and for the rotation, the choice of the point on the center line of the beam which to refer for the evaluation of the M-φ curve is multiple (fig. 3.3): the moment-rotation curve can be evaluated at the ideal intersection beam-column point, or at the end beam point, or at a given point; of course, the choice of each of these points corresponds to the choice of a different idealized model, which must adequately represent the real arrangement.

As said before, the joint deformation includes the contributions of the panel zone and of the connection, therefore the M-φ diagram of the joint can be obtained as the sum of the curves of the panel zone and of the connection (fig. 3.4).

3.2 Classification of semi-rigid joints

Obviously the structural behaviour of the joint is strongly influenced by its own typology, so different M-φ curves correspond to different joint details (fig. 3.5). Generally such curves are non-linear and present remarkable differences in terms of stiffness (slope of curve), of strength (moment capacity), and of ductility (rotation capacity); for istance an extended end plate joint (fig. 3.5, case a) has the maximum strength, but has a very poor rotation capacity, while a web cleats joint presents a low moment capacity, but a rather high ductility (fig. 3.5, case e); between these two bounding cases there is a series of structural joints with intermediate behaviour between extended end plate and web cleat joints; they are: flush end plate (case b), web and seat cleats (case c), flange cleats (case d), which give intermediate values of strength and stiffness.

At the light of this, a way to classify the joint may be based on the analysis of its M-φ curve, when a request of rotational capacity and a plastic moment of the beam are assigned.

The joint, with reference to these data, may be (fig. 3.6):

case A) full strength, clearly suitable for all applications, because the joint is stronger than the beam and possesses adequate rotational capacity.

case B) full strength, stronger than the beam but without adequate rotation capacity, so that the load carrying capacity of the structure will be governed by the development of plastic hinges in the beam.

case C) partial strength with inadequate rotational capacity; it can only be used with elastic design methods.

case D) partial strength, but possessing sufficient rotation capacity to permit redistribution of moment, leading to the formation of a collapse mechanism.

The Eurocode n.3 provides two limit M-φ diagrams, respectively for braced and unbraced frames, representing the boundary between rigid and semi-rigid joint behaviour, in order to classify any joint according to the position of its M-φ curve with respect to the limit one (fig. 3.7).

So, from the Recommendations point of view, the joint classification is strictly codified; therefore a joint may be defined as *rigid - full strength*, with different values of rotation and moment capacity, or *rigid - partial strength*, or *semi-rigid - full strength*, or *semi-rigid - partial strength* (see fig. 3.8), according to its position on the basic diagram.

3.3 Mathematical expressions

In order to include the semi-rigid joints in the frame structural analysis, the joint M-φ relationship must be analytically expressed [9]. Starting from 1933 up to nowadays, many models have been proposed by different authors (see table of fig. 3.9).

In a more synthetic form, the proposals coming from literature can be grouped together as regard to the "order" of the model (linear, bilinear, multilinear, non-linear) (see table of fig. 3.10).

The next step in the study of joints deals with the cyclic M-φ behaviour. Also in this framework a lot of proposals more or less sophisticated were provided by different authors (fig. 3.11). In this field the most up-to-dated models are the ones of Altman $et\ al.$ (1982) (fig. 3.11a) and of De Martino, Faella & Mazzolani (1988) (fig. 3.12). The former is a kind of polynomial model, extension of Moncarz & Gerstle trilinear model, while the latter provides a continuous M-φ law, which allows to interprete very closely the experimental results. In this model [10] the rotation φ is given as sum of three contributions: $\phi = \phi_0 + \phi_1 + \phi_2$, where the basic term ϕ_1 is given by a Ramberg-Osgood type law, which interprets the stable behaviour of the joint; the incremental rotation $\Delta\phi$ due to slip, which corresponds to the unstable behaviour of the joint, is taken into account by the ϕ_2 term; the residual rotation of the previous cycle is expressed by ϕ_0. The simulation of experimental tests performed on different typologies of joints has been made by means of this model and the results obtained look quite satisfactory [8]: the cycles are very close to the experimental results and the scatter between simulated values and experimental data for strength and energy absoption are very small (less than 10%) (see fig. 3.13 a, b, c, d).

3.4 Simplified analytical models

The so-called simplified analytical models are based on the following general statements [9]:
1. Close observation of test behaviour to identify the major sources of deformation in the connection.
2. Elastic analysis of the initial loading phase, concentrating on the key component(s) to predict initial connection flexibility.
3. A plastic mechanism analysis for the key component(s) to predict ultimate moment capacity.
4. Verification of the resulting equations against test data.
5. Description of connection M-φ behaviour by curve fitting, by using the evaluated initial stiffness and ultimate moment capacity values in suitable expressions.

The results of the application of such principles lead to the construction of a semi-empirical M-φ curve, as given in fig. 3.14.

3.5 Mechanical models

In order to obtain the actual M-φ curve without the need to constrain it to follow predetermined patterns, mechanical models have been developed by several researchers for the single connection as well as for the whole joint [9]. The complexity of the connection/joint behaviour, however, makes it quite difficult to set up simple but comprehensive models. Generally the joint is conceived as a set of rigid and deformable components, representing the behaviour of "elemental" parts. The nonlinearity of its response is then accounted for by inelastic constitutive laws adopted for the deformable elements, obtained from test data or from analytical models. In literature there are rather sophisticated mechanical models (Tschemmernegg, 1988) in which horizontal springs account for the bending behaviour of the joint and diagonal springs interprete the shearing behaviour. Such a model allows also to account for possible stiffner elements by means of the introduction of infinitely rigid elements (fig. 3.15).

3.6 Finite element analysis

The finite element analysis represents, in principle, the most suitable tool presently available for conducting a really exhaustive investigation on joint behaviour of complex shape (fig. 3.16).
The main aspects of joints behaviour which can be modelled by means of the finite element numerical analysis are the following:
1. geometrical and material nonlinearities of the various plate components of the members and of the connection;
2. bolt pre-load and behaviour under a general stress distribution resulting in a combination of moment, shear and tension;
3. bolt interaction with the plate components of the members and of the connection;
4. compressive interface stresses due to bolt pre-tensioning and consequent friction resistance;
5. possibility of slip due to bolt-to-hole clearance;
6. variability of the zones of contact between component elements under loading;
7. behaviour of weldments under complex loading;
8. presence of initial imperfections (residual stresses, lack of fit, etc.).

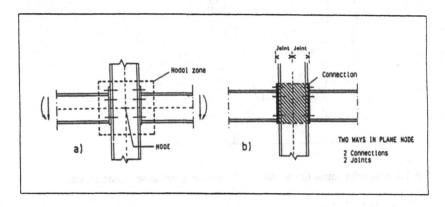

Fig. 3.1 Definition of joint and connection.

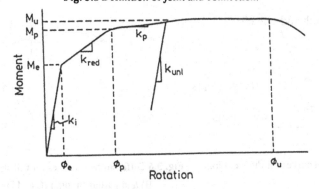

Fig. 3.2 Definition of M-φ curve.

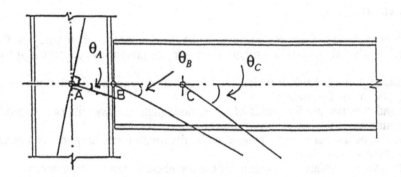

Fig. 3.3 Choice of the point where to evaluate M-φ curve.

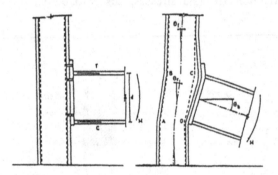

Fig. 3.4 Joint deformation (including contributions of panel zone + connection).

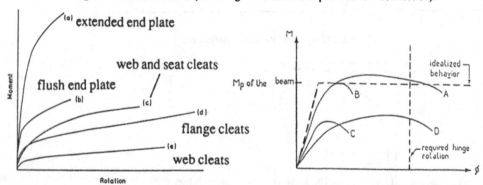

Fig. 3.5 Moment-rotation curves for different joint details.

Fig. 3.6 Different M-φ curves for full strength (A and B) and partial strength (C and D) joints.

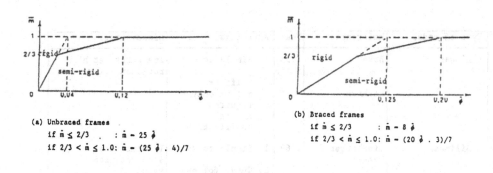

(a) Unbraced frames
if $\dot{m} \leq 2/3$,   : $\dot{m} - 25 \, \dot{\phi}$
if $2/3 < \dot{m} \leq 1.0$: $\dot{m} - (25 \, \dot{\phi} . 4)/7$

(b) Braced frames
if $\dot{m} \leq 2/3$: $\dot{m} - 8 \, \dot{\phi}$
if $2/3 < \dot{m} \leq 1.0$: $\dot{m} - (20 \, \dot{\phi} . 3)/7$

$$\dot{m} - \frac{M}{M_{pl;beam}} \quad : \quad \dot{\phi} - \frac{\phi \, E \, I_b}{I_b \, M_{pl;beam}}$$

Fig. 3.7 Boundaries for the classification of beam-to-column connections according to Eurocode 3.

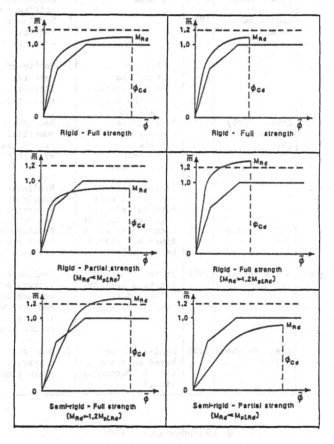

Fig. 3.8 Different cases compared with EC3 definition.

Type of model	References	Year	Advantages	Disadvantages
- Linear	Baker	1933	1. Simple to use	Inaccurate at high rotation values
	Rathbun		2. Stiffness matrix only requires initial modification	
- Bilinear	Lionberger & Weaver	1969	1. Simply to use	Inaccurate at some rotation values
	Romstad & Sumbramanian	1970	2. Curve follows M-ϕ curve more closely than linear model	
- Polynomial	Sommer	1970	Produce a close approximation to the shape of the M-ϕ data	1. Can produce inaccurate (even negative connection tangent stiffness values
	Frye & Morris [5]	1975		
	Radziminski et.al.	1982		2. Nonlinear requires iterative evaluation
- Cubic B-spline			1. Produces a very close approximation to the M-ϕ data shape	1. Nonlinear requires iterative evaluation
	Jones, Kirby & Nethercot [6]	1980	2. Produces accurate values of connection stiffness	2. Requires special numerical procedures for evaluation
- Richard formula			Produces a good fit to the test data for single angle connections untried for other types but should be suitable	1. Nonlinear requires iterative evaluation
	Richard et.al. [8]	1980		2. Requires weighted least squares evaluation
- Ramberg-Osgood	Ang & Morris [7]	1984	Produces a good fit to a variety of test data similar to type 3	1. Nonlinear requires iterative evaluation
				2. Requires weighted least squares evaluation
	Yee & Melchers [9]	1986	Semi-empirical based on theoretically determined limits	1. Nonlinear requires iterative evaluation

Fig. 3.9 Mathematical models.

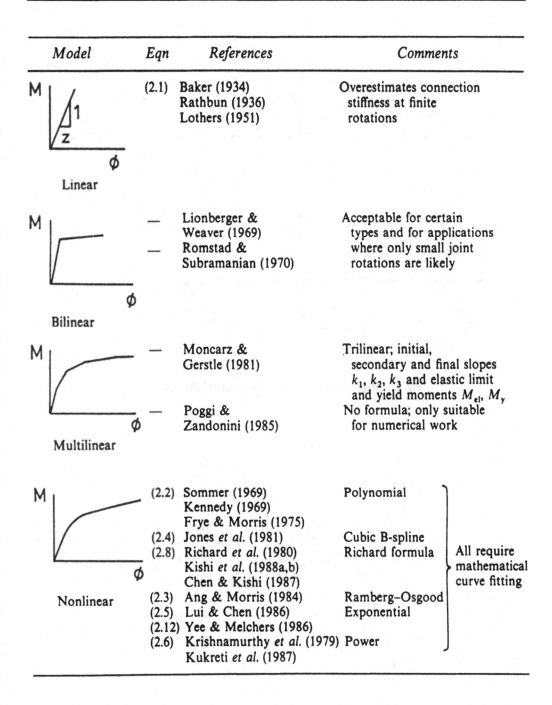

Model	Eqn	References	Comments
M ... Linear	(2.1)	Baker (1934) Rathbun (1936) Lothers (1951)	Overestimates connection stiffness at finite rotations
M ... Bilinear	— —	Lionberger & Weaver (1969) Romstad & Subramanian (1970)	Acceptable for certain types and for applications where only small joint rotations are likely
M ... Multilinear	— —	Moncarz & Gerstle (1981) Poggi & Zandonini (1985)	Trilinear; initial, secondary and final slopes k_1, k_2, k_3 and elastic limit and yield moments M_{el}, M_y No formula; only suitable for numerical work
M ... Nonlinear	(2.2) (2.4) (2.8) (2.3) (2.5) (2.12) (2.6)	Sommer (1969) Kennedy (1969) Frye & Morris (1975) Jones et al. (1981) Richard et al. (1980) Kishi et al. (1988a,b) Chen & Kishi (1987) Ang & Morris (1984) Lui & Chen (1986) Yee & Melchers (1986) Krishnamurthy et al. (1979) Kukreti et al. (1987)	Polynomial Cubic B-spline Richard formula Ramberg–Osgood Exponential Power All require mathematical curve fitting

Fig. 3.10 Representations of M-φ curve: static behaviour.

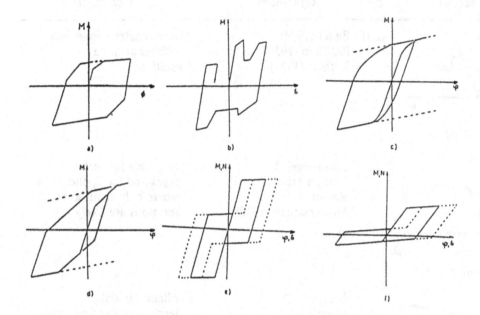

Fig. 3.11 Representations of M-φ curve: cyclic behaviour.

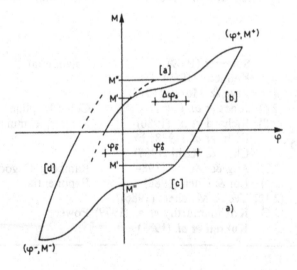

Fig. 3.12 The four branches of the analytical model.

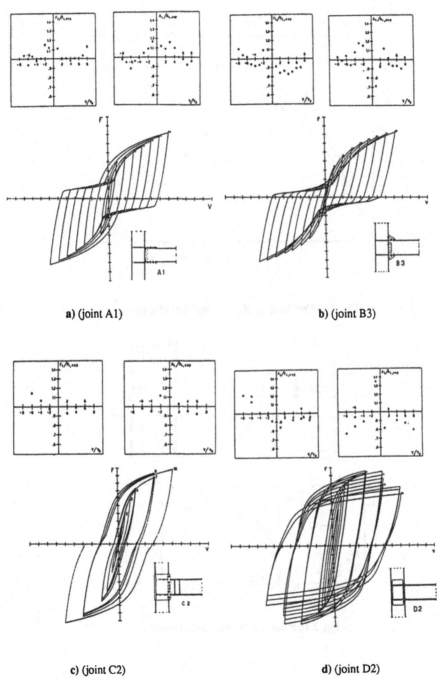

a) (joint A1)   **b)** (joint B3)

c) (joint C2)   **d)** (joint D2)

Fig. 3.13 Numerical model application.

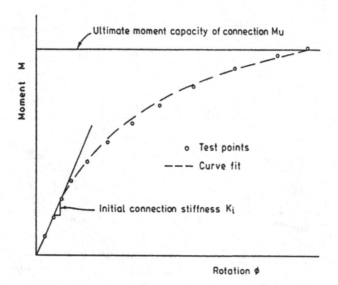

Fig. 3.14 Construction of semi-empirical M-φ curve.

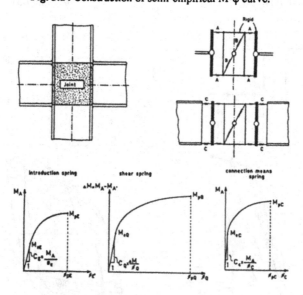

Fig. 3.15 Prediction by mechanical models.

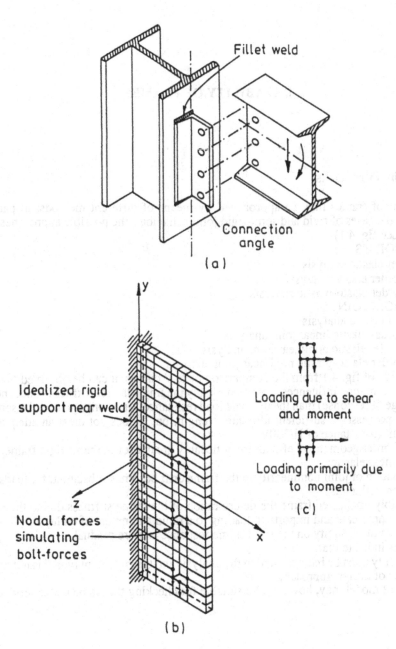

Fig. 3.16 Prediction by finite element analysis.

4. STABILITY OF FRAMES

4.1 Preliminary remarks

The analysis of frames may be approached by means of different methods; in particular, referring to the cases of rigid and semi-rigid joint behaviour, the possible approaches are the following (see fig. 4.1):
1. RIGID JOINTS
a - first order elastic analysis
b - second order elastic analysis
c - second order elasto-plastic analysis
2. SEMI-RIGID JOINTS
a - first order elastic analysis
b - second order elastic linear joint analysis
b'- second order elastic nonlinear joint analysis
c - second order elasto-plastic nonlinear joint analysis
In the example of fig. 4.1 from the comparison of α-V curves, it can be observed that:
- the joint deformation influences more remarkably the frame lateral stiffness (since the initial elastic range of response) than the ultimate load resistance. As a consequence, the semi-rigid frame still possesses a sufficient ultimate strength, but it does not meet an adequate drift service limit (for istance of H/300);
- the second order geometrical effects are noticeably greater for the semi-rigid frame, even if joints are linear elastic;
- the contribution of joint nonlinearity to the frame drift becomes substantial for loads higher than the nominal loads;
- in the flexibly-connected frame the deterioration of the frame stiffness due to the member yielding is immediate and important, leading to a rapid attainment of collapse;
- the influence of plasticity on the rigid frame response is more gradual.
These results indicate that:
- joint flexibility must be incorporated in the analysis and the rigid joint model is not adequate for all types of design approach;
- a linear joint model may, however, be suitable for checking the frame under serviceability loads.

4.2 Elastic buckling

As far as the elastic analysis is concerned, the M-φ linear relation can be assessed in the form M=Kφ, being K the stiffness of the connection [11].

In fig.4.2, the ultimate moment versus connection stiffness relationship is presented for different types of joints.

With reference to a simple portal frame (fig. 4.3a) a parametrical analysis has been carried out by Cosenza, De Luca and Faella [12], considering semi-rigid joints both for beams and columns; of course, by using symmetry considerations, it has been possible to define a semi-portal scheme (fig. 4.3b). The following step has been the computation of the fictitious stiffness of the top spring leading to an equivalent cantilever scheme (fig. 4.3c). The non-dimensional parameters used in the analysis are defined in fig. 4.3.

The evaluation of the load carrying capacity has been performed for different values of the ratio T1 between girder to column stiffness and the non-dimensional connection stiffness K. The results for fixed base portal frames are provided in fig.4.4 where the curves relate the non-dimensional connection stiffness of the semi-rigid joint to the non-dimensional critical load (fig. 4.4a) and to the non-dimensional critical multiplier (fig. 4.4b).

The analysis has been extended to the case of semi-rigid base portal frame (fig. 4.5). In this case the curve corresponding to $T1^* = 0$ starts from 0 because for pinned connections (K'=0) the structure is not able to withstand loads (fig. 4.5a).

In fig. 4.5b the critical multiplier of semi-rigid base frame non-dimensionalized with respect to fixed base frame is provided as a function of the non-dimensional connection stiffness for different values of the girder to column stiffness ratio. It can be observed that in the practical range of interest ($T1^* = 0.5 - \infty$) the scatters are not significant and only one curve can be approximatively adopted.

This kind of approach has been also extended to case of multi-story frames (fig. 4.6). The possibility to provide an equivalent one column frame has also been considered by using symmetry considerations as in the portal frame case. The influence of the variation of the columns stiffness along the height has been examined for a single bay frame. The parametrical analysis has required the introduction of a new parameter T2 which defines the variation of the column stiffness (fig. 4.7).

The results of the analysis have been represented in diagrams characterized by different values of T2 (from 0 to 1), in which the critical multiplier of the frame versus the girder to column stiffness ratio $T1^*$ is provided for different values of the number of stories n (from 1 to 10) (see fig. 4.8).

4.3 Simplified methods

The results of the above parametrical analysis [12] are useful to point out simplified methods for the evaluation of the critical multiplier of both portal and multi-story frame, due to the linear relationship between a representation of the critical multiplier α_k - through the $(\alpha_k-\alpha_0)/(\alpha_\infty-\alpha_k)$ ratio, where α_0 is the critical multiplier of pinned joint frame, α_∞ is the critical multiplier of rigid frame - and the non-dimensional connection stiffness K of the semi-rigid joint (fig. 4.9). In case of portal frame such linear relationship is characterized by a slope which varies with the ratio girder to column stiffness T1; adopting a log-log representation of the relationship, the obtained lines for the different values of T1 are parallel, so they can be defined by only one slope (fig. 4.10a), and according to the value of T1, it is possible to define a constant A to be used in an approximate "linear method"based on the following relation $\alpha_K - \alpha_0 = (\alpha_\infty - \alpha'_K)AK$. The error provided by this approximation is function of K; the maximum scatter between the exact and the approximate critical multipliers for portal frames is more or less about 5% (fig. 4.10b).

The same concept has been extended to the multi-story frame (fig. 4.11): a linear representation of the critical multiplier has been found and the value of the constant A to be used in the approximate linear method have been provided for the values of T2 considered in the analysis. The scatters provided by the approximation are in this case quite bigger (about 15-20%) than the ones of the portal frame case.

4.4 Inelastic behaviour

Adopting an inelastic approach, it is necessary to account for the actual inelastic relationship between moment and rotation (fig. 4.12).

As regard to this, some preliminary considerations about the shape of such curve and about the interaction between connection and girder behaviour must be made. Two main cases can be distinguished (fig. 4.13):
- the connection load carrying capacity is higher than the girder one (full-strength joint);
- the connection load carrying capacity is lower than the girder one (partial-strength joint);
The global curve accounting for the behaviour of the girder plus the connection corresponds to the minimum of both behaviours. In the former case it is obtained by considering the curve of the connection in the elastic range and the one of the girder in the plastic range, adequately joined together in the knee. In the second case, the connection plays the most important role and the global curve to adopt is practically the one of the connection.

In intermediate cases, in which there is a sort of balance between both strength of connection and member, the global curvï will always account for the worst behaviour.

Anyway, we can observe that, if the connection is of full-strength type, the interpretation of the elastic-plastic behaviour by means of a bilinear curve is quite close to the actual behaviour, while in the case of partial-strength joint, the behaviour is generically inelastic and must be interpreted by means of a continuous nonlinear M-φ law.

The results of the performed inelastic analysis are presented by means of the relationship between collapse multiplier and horizontal sway. In fig. 4.14 these curves are given for full-strength and partial-strength joints and for different values of the joint flexibility K, showing the influence on the load carrying capacity of different factors, as well as strength and rigidity of joints and amount of axial forces in columns.

4.5 Seismic behaviour

In fig.4.15 the influence of design criteria on the frame inelastic behaviour is shown: it can be observed that by increasing the column section the load bearing capacity increases, but the hinge formation path, corresponding to the final collapse mechanism, may be remarkably different.

These considerations are very important when frame structures are designed in seismic areas; in particular, some requisits of design of seismic-resistant structures are the following [4]:
a) a seismic moment connection must be detailed to withstand forces which act in either directions;
b) the problem of low cycle fatigue, which is associated with cyclic loadings at large plastic strain, becomes a factor to be considered in seismic moment connections;

c) since seismic loadings are random in nature, a probabilistic rather than a deterministic approach of analysis is necessary to assess the behaviour of a seismic moment connection;

d) the design of a framed structure must be done in such a way as to confine plastic hinges to be developed in the beams not in the columns in order to avoid local collapse mechanism [13].

As regard to hysteresis loops, a connection should possess sufficient strength and ductility and should be strong enough to allow large rotation in the beams so that plastic hinges can be formed, providing ductility to the frame.

Ductility is very important in seismic regions to absorb and dissipate energy as well as to dampen the vibrations generated by ground motions.

According to these main principles, among the hysteresis loop shape shown in fig. 4.16, the most desirable one is the first one (a). Contrary, the other ones are not desiderable because they present lack of ductility (b), lack of strength (c), lack of stiffness (d), respectively.

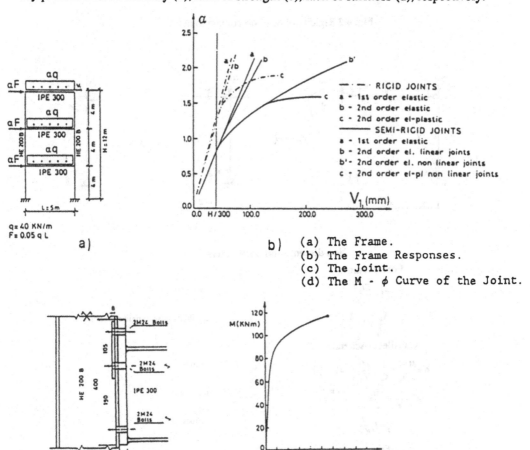

Fig. 4.1 Influence of joint flexibility on different frame analysis methods.

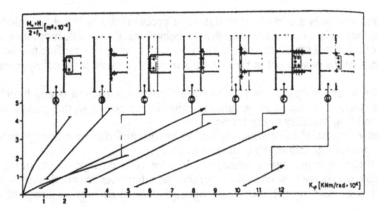

Fig. 4.2 Rigidity of beam-to-column joints.

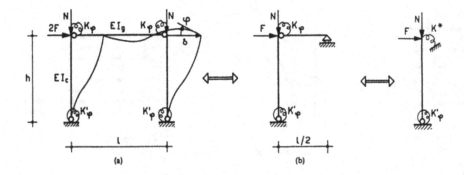

(a)   (b)

Non dimensional parameters

Portal frame:

$$T1 = \frac{EI_g/l}{EI_c/h}$$

rigidity of portal

$$K = \frac{K_\phi l}{EI_g}; K' = \frac{K'_\phi l}{EI_c}$$

flexibility of semi-rigid connections

Cantilever scheme:

$$\phi = \frac{Ml/2}{3EI_g} + \frac{M}{K_\phi} = \frac{Ml}{6EI_g}\left(1 + 6\frac{EI_g}{K_\phi l}\right)$$

rotation

$$K^* = \frac{M}{\phi} = \frac{6EI_g}{l}\frac{K}{K+6} = \frac{6EI_g^*}{l}$$

stiffness

$$I_g^* = I_g\frac{K}{K+6}$$

fictious inertia

$$T1^* = \frac{EI_g^*/l}{EI_c/h} = T1\frac{K}{K+6}$$

$$\phi^* = \phi\frac{EI_c}{Fh^2}; \delta^* = \delta\frac{EI_c}{Fh^3}$$

Fig. 4.3 Portal frame analysed

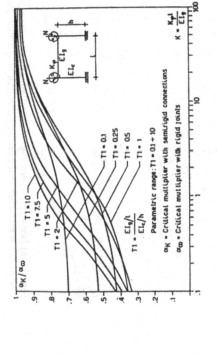

Fig. 4.4 b) Critical multiplier non-dimensionalized with respect to fixed-end case; fixed base frame.

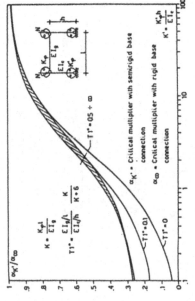

Fig. 4.5 b) Critical multiplier of semi-rigid base frame non-dimensionalized with respect to fixed base frame.

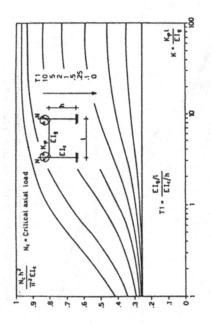

Fig. 4.4 a) Critical load variation with beam-to-column connection stiffness; fixed base frame.

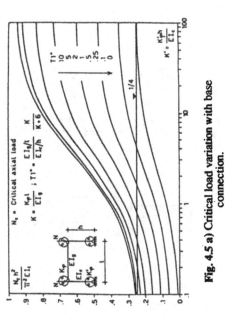

Fig. 4.5 a) Critical load variation with base connection.

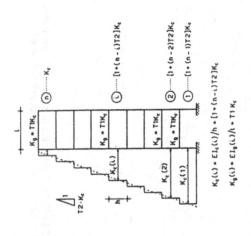

Fig. 4.7 Variation of parameters in multi-storey frames.

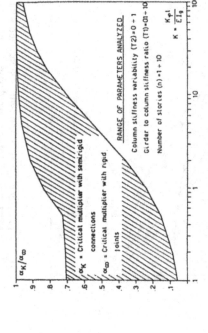

Fig. 4.8 b) Decrease of semi-rigid critical multiplier with respect to the rigid joint case in the entire parametric range analysed.

Fig. 4.6 Equivalence of multi-bay frames.

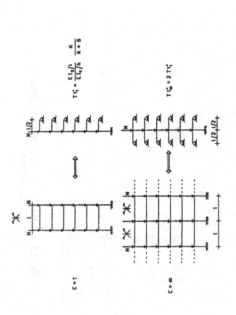

Fig. 4.8 a) Critical multiplier non-dimensionalized with respect to the rigid joint case; T1=5 and T2=1.

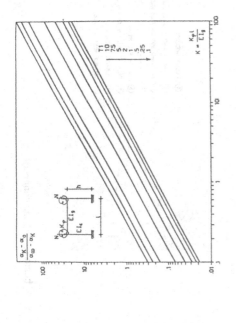

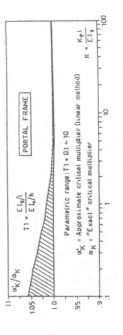

Fig. 4.9 Curves representing the critical multiplier α_z through $(\alpha_z - \alpha_0)/(\alpha_\infty - \alpha_z)$ ratio.

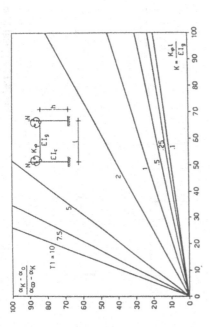

Fig. 4.10 a) Value of the constant A to be used in the approximate "linear method".

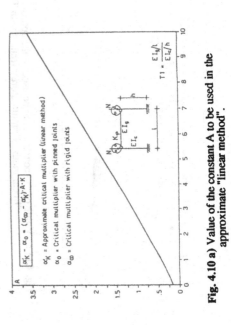

Fig. 4.10 b) Error provided by the approximate "linear method".

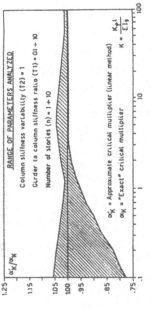

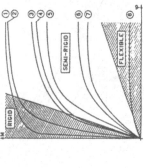

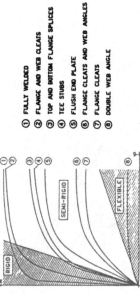

Fig. 4.11 b) Error provided by the approximate "linear method"; multi-storey frames with T2=1.

① FULLY WELDED
② FLANGE AND WEB CLEATS
③ TOP AND BOTTOM FLANGE SPLICES
④ TEE STUBS
⑤ FLUSH END PLATE
⑥ FLANGE CLEATS AND WEB ANGLES
⑦ FLANGE CLEATS
⑧ DOUBLE WEB ANGLE

Fig. 4.12 Typical M-φ curves for beam-to-column connections.

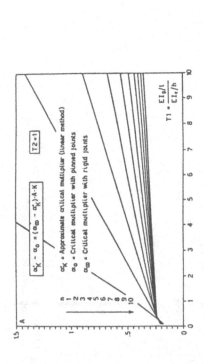

Fig. 4.11 a) Value of the constant A to be used in the approximate "linear method"; multi-storey frames with T2=1.

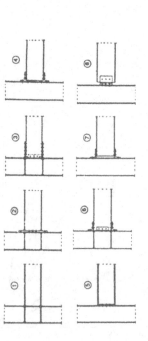

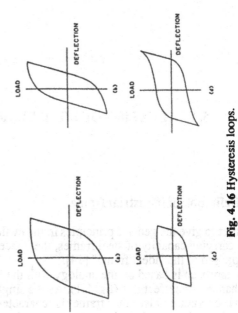

Fig. 4.14 α-δ curves.

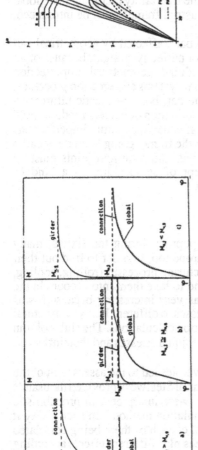

Fig. 4.13 Interaction between connection and girder behaviour.

Fig. 4.16 Hysteresis loops.

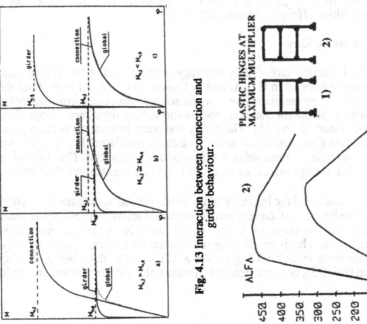

Fig. 4.15 Non linear analysis of two frames characterized by different design.

5. EVALUATION OF LOAD BEARING CAPACITY OF FRAMES

5.1 Definition of industrial frame

In order to give the general principles in the evaluation of the true safety factor against global load carrying capacity of steel frames, the concept of "industrial frame" must be introduced, as opposed to the "ideal frame" [14].

This approach is based on the analogy with the "industrial bar", affected by geometrical and mechanical imperfections (fig. 5.1). As the single bar is not perfectly straight, because of an initial curvature, also the frame is characterized by global geometrical imperfection (out-of-plumb along the columns), due to both constructional systems and erection processes. Moreover, in the analysis of the buckling behaviour of the bar, both the elastic bifurcation theory of Euler and its modification by Shanley with the introduction of tangent modulus hold for a perfect straight bar only. In order to analyze the actual behaviour, initial imperfections and plastic behaviour must be introduced. Analogously for the frame, going from the ideal to the actual behaviour, among various kinds of imperfections, the semirigid joints must be introduced. Consequently we can generalize the concept of semirigidity as a kind of mechanical imperfection which affects the frame behaviour.

5.2 Local and global imperfections

Measurements of residual stresses (fig. 5.2) were very popular during the sixties; many experimental studies were developed in U.S.A. and in Europe too, in order to find out their presence in hot rolled and built-up members due to manifacturing processes (rolling, cooling, welding...). Next step was to define mechanical models and to take them into account in the analysis of buckling behaviour of bars [5]. The result was very interesting because it was possible to associate different given models of residual stresses to different kinds of structural shapes (fig. 5.3); then these models were used in the simulation calculations. The stub-column test was used as indirect tool to emphasize the presence of residual stresses and their influence on the σ-ε law (fig. 5.4).

This approach during the sixties and the begining of the seventies led to the assessment of the stability of columns [5]. The European Convention of Structural Steelwork provided thousands of experimental buckling tests on columns, whose results were interpreted in probabilistic way and they were assumed as a basis to calibrate the simulation methods. In such a way it was possible to define the three main buckling curves a, b, c, each of them being correlated to a type of cross section (fig. 5.5), because residual stressees play different effects according to the shape of the cross section. Later these buckling curves became five and the so-called "multiple curve approach" has been introduced in the most important codes of european countries (fig. 5.6).

The same kind of approach has been more recently introduced also in the Eurocode n.3 [2], together with the concept of a generalized equivalent geometrical imperfection, which is a conventional imperfection taking into account also the effects of mechanical imperfections (fig. 5.7). This concept was firstly introduced by Ayrton and Perry more than a century ago and is now still considered in modern recommendations. This equivalent geometrical imperfection for a given buckling curve is the sum of the standard out-of-straightness $(v_0 = L/100)$ plus a second value which takes into account the effect of residual stress distribution in the cross section.

The concept of imperfection was furthermore developed and extended to all kinds of structural details. In the appendix of Eurocode n.3 different kinds of local and global imperfections are listed and for each kind the permitted deviation is given. Thus, the calculation rules included in Eurocode n.3 can be used provided that tolerances of manifacturing and erecting buildings are contained in the range given in the code.

Local imperfections are, for instance, deviations from regular geometry in the cross section of a beam and along its axis or in the connections. Global imperfections are deviations from regular geometry in the overall structure (inclination of column, scatter in overall height and distances of the structural mesh).

The code suggests to take into account these imperfections by means of equivalent forces; for istance, in the case of column out-of-plumb, horizontal forces giving the same P-Δ effect of the eccentric vertical load can be introduced. Analogously in the case of a frame, equivalent horizontal forces conventionally take into account the effect of sway imperfections by means of appropriate factors.

Eurocode n.3 is therefore based on a new design philosophy, which assumes that structures are affected by imperfections, due to manifacture and erection processes.

5.3 Analogy between bar and frame

Coming back to the analogy between bar and frame [14], it can be analyzed considering the geometrical and mechanical imperfections, and their influence the buckling behaviour (fig. 5.8).

As regard to geometical imperfections, some aspects have been already pointed out; it has been emphasized that both out-of-plumb of columns and frames are responsible to the lowering of the load bearing capacity and can be taken into account by means of fictious actions, both for column and frame. The values of such imperfections are now codified as 1/1000 of the lenght for the single column and by means of the Ψ factor for the frame.

As regard to mechanical imperfections it has been pointed out that residual stresses influence the behaviour of the industrial bar; the analogous feature in the frame is the semirigidity of the connections which also produces a reduction in the overall load bearing capacity. From the point of view of the measurement of mechanical imperfections, as residual stress presence was emphasized by means of the stub column test, which provides the actual σ-ε curve, joint tests must be performed in order to obtain the real moment-rotation curve of a semirigid connection.

This analogy can be extended to the problem of the numerical simulation of the buckling behaviour (fig. 5.9). In the analysis of the local behaviour of a bar, the cross section must be subdivided in several elements, for each of them the σ-ε relationship of the material must be defined in order to evaluate the moment versus curvature curve of the member. For a frame, the actual moment-rotation relationship must be given for each joint in order to characterize the behavioural model and, hence, to obtain the elements of the frame stiffness matrix.

In the analysis of the buckling behaviour, as the bar is subdivided in a certain number of longitudinal parts, also the frame can be interpreted by the assemblage of several members, each of them is a "finite element", provided that it has been characterized from the point of view of geometrical and mechanical imperfections. So the analysis of the bar behaviour is preliminar to the definition of the frame behaviour, for which the knowledge of the joint behaviour is also essential.

From the analysis of the buckling behaviour both of bar and frame, different curves are obtained: for bars they correspond to different types of cross section and, hence, of distribution of residual stresses, while for a given frame they correspond to different degree of semirigidity.

5.4 Analysis of the influence parameters

The industrial frame is characterized by overall deviations from the ideal scheme and by imperfections of its elements (fig. 5.10). The real distribution of geometrical imperfections of bars is unknown: so it is necessary to consider a random distribution and to find out the worst case among all the possible ones.

The main parameters necessary in the elastic analysis of sway frame behaviour are the following:
a) the ratio between the flexural stiffness of beams and the one of columns;
b) the variation of stiffness parameters along the frame height;
c) frame typology

In the plastic analysis the additional parameters are requested:
a) the ratio between ultimate strength of beam sections and the one of columns (provided that full-strength joints are adopted);
b) the non-linear M-φ curve characterizing the joint;
c) the amount and distribution of local and global imperfections.

In the following, some results of a parametrical analysis including geometrical and mechanical imperfections are shown [15]; the analysis is based on the following hypotesis:
a) an equivalent bar imperfection defined by its initial out-of-straightness v_0 equal to 1/200 of the bar lenght (this corresponds to the stability curve "c" of Eurocode n.3);
b) an initial out-of-plumb Ψ defining an initial sway equal to 1/400 of the global height H of the frame;
c) a mechanical imperfection defined by the semi-rigidity of joint, which is interpreted by an elastic-perfectly plastic M-φ relationship and therefore characterized by a stiffness K_φ and by a strength $M_{u,j}$ equal to that of the girder $M_{u,g}$ (full strength joint);
d) non-dimensional connection stiffness, given by

$$K = \frac{K_\varphi L}{EI_g}$$

The results of this analysis are presented in graphical form for a single portal frame. For each value of the connection stiffness K, varying from 0.1 to 1000, a set of dimensionless curves providing the multiplier of axial load versus sway for different non-dimensional column slenderness $\overline{\lambda}$ is given (fig. 5.11, 5.12). The same results are also presented in a different kind of diagram, which provides for each value of non-dimensional slenderness $\overline{\lambda}$ the multiplier of axial load versus sway curves for different non-dimensional stiffness K of connections (fig. 5.13). From the first type of curves, it can be observed that as the value of K decrease from 10000 (which practically corresponds to rigid joint) to 0.1 (which practically corresponds to perfectly pinned joint), the load carrying capacity decreases too and this effect is more evident for higher slenderness.

The inelastic multiplier normalized with respect to "rigid framing" case versus the stiffness K of connections for different values of non-dimensional slenderness $\overline{\lambda}$ is provided in three sets of curves, corresponding to different cross sections of the girder (fig. 5.14).

Probably the most interesting graphical representation to describe the frame behaviour is the one in which the stability curves give the non-dimensional collapse multiplier versus non-dimensional slenderness of the columns, by varying the semirigidity factor K, for different cross-sections of the girder. This kind of curves have been evaluated not only for portal frames, but also for one-bay multi-storey frames with three-storey and six-storey (fig. 5.15), in which a random distribution of geometrical imperfections in the bars has been considered. In this way the buckling behaviour of each frame, characterized by a value of the semirigidity constant K, is completely described by the corresponding buckling curve.

The effects of non-symmetry of columns, (where cross-sections vary at each storey) on stability curves is also examined and it seems to be not very important (fig. 5.16).

Comparing the frame inelastic buckling curve, the frame elastic buckling curve and the single column collapse curve of a three-storey frame, it is possible to find a similarity between frame and bar, according to the analogy introduced before (fig. 5.17).

Still according this analogy, extending the well-known concept from the bar to the frame, it is possible to define a generalized value of non-dimensional slenderness λ_F, given by the square root of the ratio between the elastic multiplier and the critical multiplier:

$$\lambda_F = \sqrt{\frac{\alpha_y}{\alpha_c}}$$

When representing the inelastic buckling multipliers of portal and multi-storey frames as functions of such "frame slenderness", it is found that these points belong to the area bounded by the single column collapse curve and the frame elastic buckling curve (fig. 5.18).

Such result implies that not important scatters exist among the behaviour of frames; so, an idea to be developed in the future, starting from these results, could be to introduce in the recommendations for frames the same approach already codified for bars. This multiple-curve approach will allow to represent the overall behaviour of each category of frames by means a single buckling curve, which takes into account all kinds of mechanical and geometrical imperfections. As said before, this is the same approach that twenty years ago was introduced for the columns, grouping them according to the shape of cross-section (what means according to the distribution of residual stresses) and that now-a-day is very well consolidated. Of course for frames this approach is quite more difficult because of the larger number of parameters involved and their larger range of variation, but going furthermore on these parametrical analysis and calibrating numerical results by means of experimental tests, a new way in the analysis of steel frames, based on the concept of "industrial frame", will be surely available.

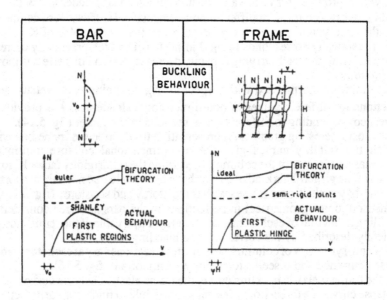

Fig. 5.1 Analogy between imperfect bar and imperfect frame buckling behaviour.

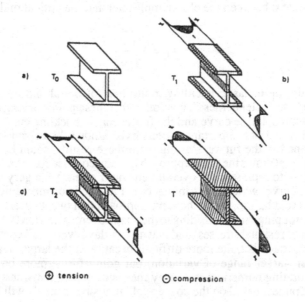

Fig. 5.2 Residual stress formation.

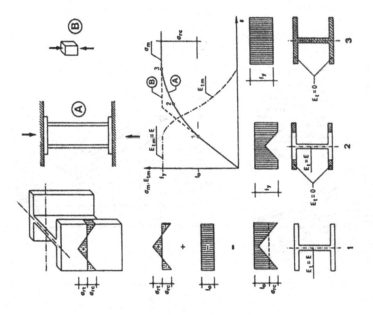

Fig. 5.4 Stub column test.

h/b	Profile	Residual stresses		t_w/i	l_w/b	l_f/h	t_f/b
		Web	Flanges				
<1.2	a			0.032 to 0.040	0.032 to 0.040	0.045 to 0.061	0.045 to 0.080
	b			0.075 to 0.100	0.078 to 0.112	0.091 to 0.162	0.093 to 0.182
>1.2	c			0.062 to 0.068	0.068 to 0.073	0.104 to 0.114	0.113 to 0.121
<1.7	d			0.031 to 0.032	0.042 to 0.048	0.048 to 0.051	0.062 to 0.080
				0.030	0.046	0.051	0.077
>1.7	e			0.018 to 0.028	0.039 to 0.056	0.025 to 0.043	0.063 to 0.085

Fig. 5.3 Residual stress distribution.

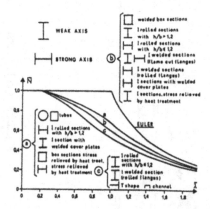

Fig. 5.5 Buckling curves.

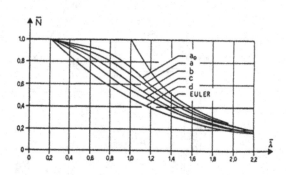

Fig. 5.6 Non-dimensional buckling curves from the
ECCS Recommendations.

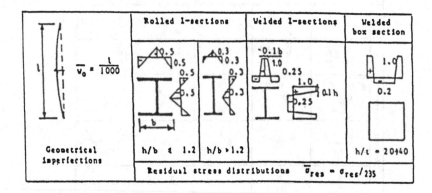

Fig. 5.7 Geometrical and mechanical imperfections

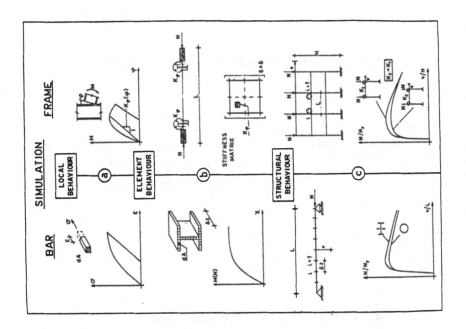

Fig. 5.9 Analogy between bar and frame: simulation.

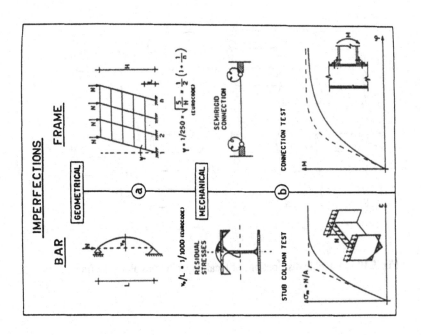

Fig. 5.8 Analogy between bar and frame: imperfections.

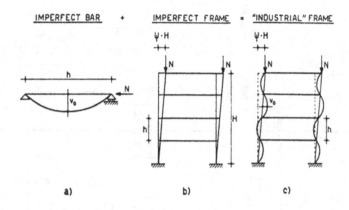

Fig. 5.10 a) Combination of local and global imperfections

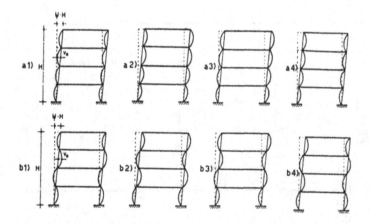

Fig. 5.10 b) Possible distributions of local imperfections along the frame.

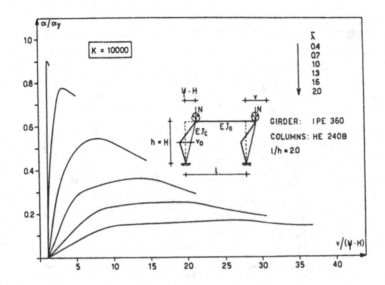

Fig. 5.11 a) Multiplier of axial load versus sway curves for different non-dimensional column slenderness;
K=10000.

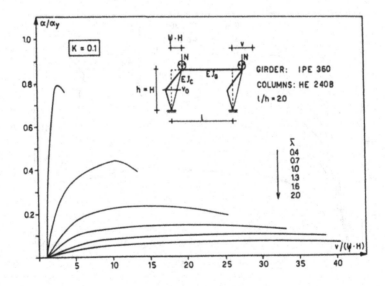

Fig. 5.11 b) Multiplier of axial load versus sway curves for different non-dimensional columns slenderness;
K=0.1.

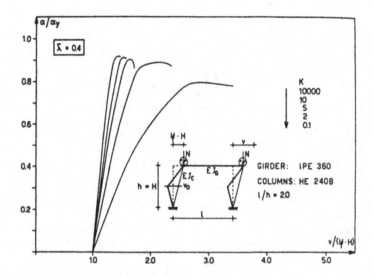

Fig. 5.12 a) Multiplier of axial load versus sway curves for different non-dimensional stiffness of connections; $\overline{\lambda} = 0.4$.

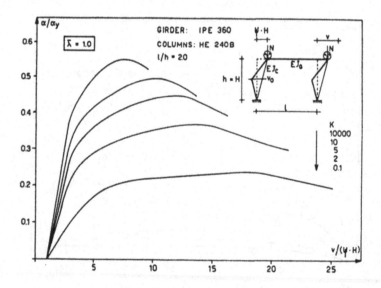

Fig. 5.12 b) Multiplier of axial load versus sway curves for different non-dimensional stiffness of connections; $\overline{\lambda} = 1.0$.

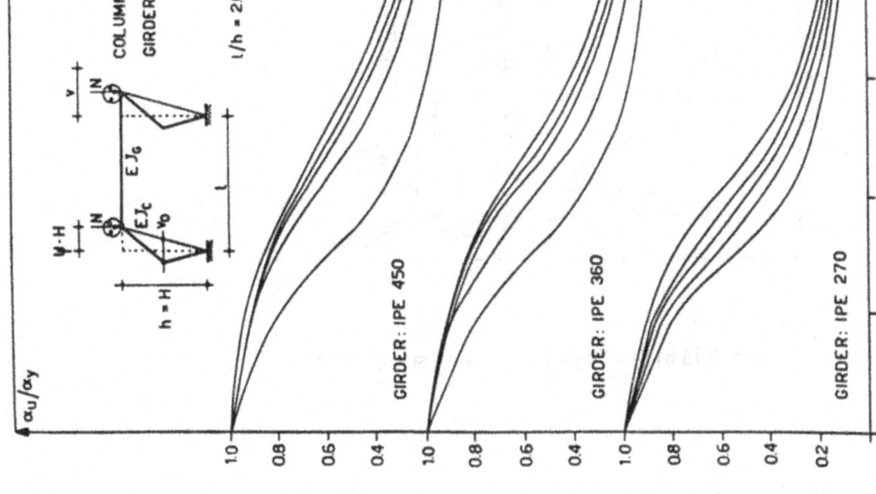

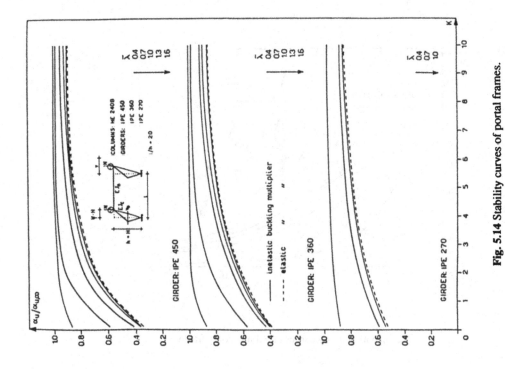

Fig. 5.14 Stability curves of portal frames.

Fig. 5.13 Inelastic buckling multiplier normalized with respect to "rigid framing" case.

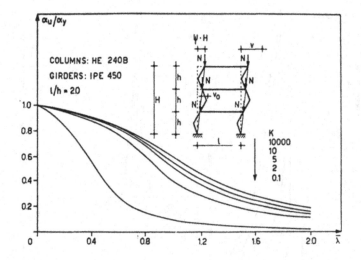

Fig. 5.15 a) Stability curves of a three-storey frame.

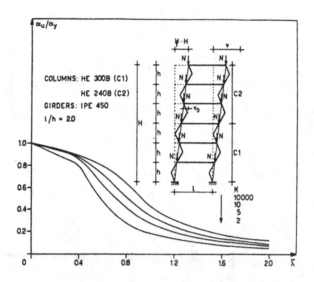

Fig. 5.15 b) Stability curves of a six-storey frame.

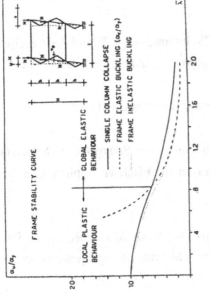

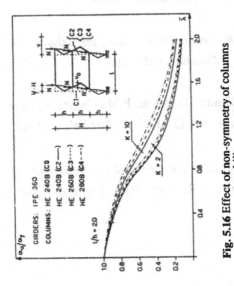

Fig. 5.16 Effect of non-symmetry of columns on stability curves.

Fig. 5.17 Comparison among stability curve, column curve and elastic critical curve of a three-storey frame.

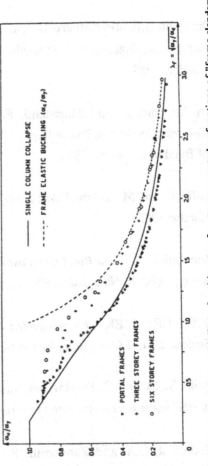

Fig. 5.18 Inelastic buckling multipliers of multi-storey frames represented as functions of "frame slenderness" λ_f.

Essential References

1. Bijlard, F.S.K., Nethercot, D.A., Stark, J.K.B., Tschemmernegg, T. and Zoetemeijer, P.: *Structural properties of semirigid joints in steel frames*, IABSE Periodica 2/1989.

2. Eurocode No. 3: *Design of Steel Structures*, Eurocode Editorial Group, 1990.

3. Davidson, J.B. and Nethercot, D.A.: *Overview of Connection Behaviour*, in: Structural Connection: Stability and Strength (Ed. R. Narayanan), Elselvier Applied Sciences, London, 1989.

4. De Martino, A. and Mazzolani, F.M.: *Inelastic Behaviour of Connection in Steel Construction Design*, Proceedings of the International Meeting on Earthquake Protection of Buildings, Ancona, 1991.

5. Ballio, G. and Mazzolani, F.M.: *Theory and Design of Steel Structures*, Chapman & Hall, London 1983.

6. Zandonini, R.: *Semi-Rigid Composite Joints*, in: Structural Connection: Stability and Strength (Ed. R. Narayanan), Elselvier Applied Sciences, London, 1989.

7. ECCS, CECM, EKS: *Recommended Testing Procedure for Assessing the Behaviour of Structural Steel Elements under Cyclic Loads*, Publication n.45, 1986.

8. Ballio, G., Calado, C., De Martino, A., Faella, C. and Mazzolani, F.M.: *Cyclic behaviour of steel beam-to-column joints: experimental research*, Costruzioni Metalliche, 2/1987.

9. Nethercot, D.A. and Zandonini, R.: *Methods of Prediction of Joint Behaviour*, in: Structural Connection: Stability and Strength (Ed. R. Narayanan), Elselvier Applied Sciences, London, 1989.

10. **Mazzolani, F.M.**: *Mathematical Model for Semi-Rigid Joints under Cyclic Loads*, in: Connections in Steel Structures: Behaviour, Strength and Design (Ed. R. Bjorhovde, J. Brozzetti, A. Colson), Elselvier Applied Sciences, London, 1988.

11. **Mazzolani, F.M.**: *Influence of Semi-Rigid Connections on the Overall Stability of Steel Frames*, in: Connections in Steel Structures: Behaviour, Strength and Design (Ed. R. Bjorhovde, J. Brozzetti, A. Colson), Elselvier Applied Sciences, London, 1988.

12. **Cosenza, E., Da Luca, A. and Faella, C.**: *Elastic Buckling of Semirigid Sway Frames*, in: Structural Connection: Stability and Strength (Ed. R. Narayanan), Elselvier Applied Sciences, London, 1989.

13. **ECCS, CECM, EKS**: *European Recommendations for Steel Structures in Seismic Zones*, Publication n. 54, 1988.

14. **Cosenza, E., Da Luca, A., Faella, C. and Mazzolani, F.M.**: *Imperfection Sensitivity of "Industrial" Steel Frames*, in: Steel Structures: Advances, Design and Construction, (Ed. R. Narayanan), Elselvier Applied Sciences, London, 1987.

15. **Cosenza, E., Da Luca, A. and Faella, C.**: *Inelastic Buckling of Semirigid Sway Frames*, in: Structural Connection: Stability and Strength (Ed. R. Narayanan), Elselvier Applied Sciences, London, 1989.

Printed in the United States
By Bookmasters